MACMILLAN DICTIONARY OF ENERGY

MALCOLM SLESSER

First edition published 1982, paperback 1985

Second edition first published 1988 by
THE MACMILLAN PRESS LTD.
London and Basingstoke

Associated companies in Auckland, Delhi, Dublin, Gaboronе, Hamburg, Harare, Johannesburg, Kuala Lumpur, Lagos, Manzini, Melbourne, Mexico City, Nairobi, New York, Singapore, Tokyo.

British Library Cataloguing in Publication Data.

Macmillan dictionary of energy — 2nd ed.
1. Energy resources
I Slesser, Malcolm II. Dictionary of energy
33.79

ISBN 0-333-45461-8
ISBN 0-333-45462-6 Pbk

Printed and bound in Great Britain by
Antony Rowe Ltd, Chippenham and Eastbourne

Transferred to digital printing 2003

Introduction

Energy pervades every aspect of modern society. Following the energy crisis of 1973, the word energy was used by almost everyone from technologists through to politicians. New words were generated, old words misused and a lot of loose thinking unleashed. As the world wrestles with a multitude of solutions to the problems of energy supply and energy costs, with antagonism to nuclear energy, with proposals for renewable energies, the range and breadth of terms used within the energy context has become wider and wider. It had become obvious that there was a need for an inter-disciplinary dictionary of energy. This is it. It covers fuel technology, science, economics, the built environment, the external environment, renewable and alternative energies, energy transformation, biology, fossil and nuclear fuel and fuel treatment.

The problem with written language is that there is a limited number of words available but there is an unlimited number of concepts and descriptions. One discipline borrows a word from another, and in the process may subtly change its meaning, extending it into an arena which renders it obscure. For example, most of us know what we mean by elasticity. It suggests stretchability with a tendancy to return, and we can imagine stretching things like catapults, corsets, car tyres and fan belts. The economists have adopted this word to suggest just such a phenomenon of how people respond to price changes. It is, if you like, the stretchability of our purses in the face of the change in price of a commodity. What better word could they pick, yet its meaning is not immediately obvious. This dictionary will explain to the non-economist in adequately simply language what is meant by the economic concept of elasticity, and go on, I am afraid, to use it frequently.

A cross-disciplinary dictionary cannot be all things to all men. It would not be useful to provide definitions of terms in nuclear engineering which could only be understood by nuclear engineers and physicists, or in economics which could only be understood by economists. There are a number of excellent definitive works to serve the experts in their own fields. This dictionary aims to provide information for intelligent people searching for concepts, ideas, definitions and explanations in areas outside that of their own expertise. Thus a definition in a technical area may not wholly satisfy an expert in that area. As editor, it has been my task to judge just how far a definition has to be precise in order to be adequately accurate, yet straightforward enough to be understood by the non-expert. There is, therefore, compromise between accuracy and simplicity. I cannot fail to be criticized. Some of it will be merited, and let us hope that it may serve to produce an even better second edition.

The dictionary is not just for experts outside their own field. I hope it will help those who claim no expertise at all but seek information in a brief, lucid and succinct form. Thus I hope it will serve the student, the businessman, the sociologist, the politician; that it will find its way into public libraries, and that the general reader will consult it to clear up points that arise in his daily work and conversation.

A fascinating aspect of editing such a dictionary is to find the extent to which some disciplines have borrowed words from others for their own needs. Thus efficiency means quite different things to the economist and the scientist. Absorption connotes quite different processes to a chemist and a nuclear engineer. Cross-section is adapted to express a very neat concept by the nuclear engineer, while the economist does the same with elasticity. Equally fascinating is the way that identical concepts are differently named in different disciplines. Calorific value to a fuel technologist is enthalpy to a chemist and combustion energy to others. It is dispiriting to find these differences between allied sciences, but it is no more than a recognition that disciplines retain their own boundaries, in spite of attempts at unification, such as the SI (Système International) system of units now used by virtually all European schoolchildren for whom the Btu or the calorie are museum pieces. It is also dispiriting to find that in spite of international solidarity among scientists, there still exist minor differences in the definition of such fundamental scientific units as a joule of energy – the ultimate unit of definition. Some conversion tables show 4.184, and others 4.187 joules per calorie. Both are correct, depending on whether you hearken to the international joule or the absolute joule. The error is too small to be measurable in the affairs of nations or firms, but could be a factor in scientific experimentation.

There are many terms in this dictionary which do not have obvious energy connotations. They appear here because they represent concepts or terms essential to understanding all the various aspects of energy. Indeed, a dictionary which included only terms including the word energy would be brief but lamentably inadequate. Consider energy as the inside doll in a set of Russian dolls, as one might, mathematically, consider a set within a set within a set and so on. Energy is the inside doll or set. It pertains to many things within the larger set, which in turn may need explanation by recourse to an even larger set of terms.

There are omissions, some deliberate, some accidental. Words that could be found in a standard English language dictionary are omitted, unless they have a special energy connotation. For example, one can find the term energy in every such dictionary, where it will be described along the lines 'power of doing work'. That is a pretty good succinct definition, but not enough to understand all that needs to be understood about energy and work, and begs the question of what is meant by power. In the ever-expanding set of Russian dolls that comprise an inter-disciplinary dictionary one must call a halt somewhere. It is arbitrary. I admit that. Nevertheless I believe this is the most comprehensive energy dictionary ever produced. It is, however, a dictionary and not an encyclopedia. The first part of each definition is intended to be elementary. Often a second part goes into more detail. Sometimes a diagram is inserted to assist understanding. There are some long entries: in particular, that of nuclear reactors. To have described each reactor individually and unambiguously would have required enormous repetition and constant referral to other terms in the dictionary. The manner of presentation allows the reader to go from the simple to more complex, gathering his or her impression of this or that type of reactor in increasing depth, with each reactor type given its own unique space.

Units. A considerable problem for anyone entering another discipline is to have a feel for the units used. If somebody tells you the cost of something in Japanese yen, and you've never been to Japan and don't know the rate of exchange, the information is meaningless. The dictionary has decided to reduce all terms to those proposed under the Système International d'Unités – the International System of Units (SI). A barrel – a measure used in the oil industry – is related to litres. Temperature is expressed in Kelvin, which relates to the degree centigrade. The standard unit of length is the metre. Nevertheless all the old favourites are listed, and a conversion table provides a means of arriving at a numerical answer if that is wished for. Similarly, but without the benefit of an international convention, the US dollar has been chosen as the unit of money throughout.

Finally, an inter-disciplinary dictionary should be international. Though foreign terms are not listed, many institutions in many countries are included.

MALCOLM SLESSER
Dunblane, February 1982

Acknowledgements

The dictionary is the work of five advisory editors, and two specialists who have selected the entries and formed the definitions. Their names are listed on the title page. However, special thanks must go to those who typed the entries, read them to ascertain their comprehensibility, and helped with the unrewarding task of filing, noting, sorting and annotating. In particular, I should like to thank Mrs Evelyn Smith and Mrs Ann Lawson of the Energy Studies Unit, University of Strathclyde, and my wife, Janet Slesser.

Accreditation

Accreditation is noted at the end of each entry by initials, indicating the writer. Thus:

MS Malcolm Slesser
TM Thomas Maver
JWT John Twidell
DB Donald Bennet
CL Christopher Lewis
PH Patricia Howell
WG William Gibb
MC Mick Common

SUBJECT EDITORS

Nuclear Energy:
Dr D.J. Bennet–University of Strathclyde

Mechanical & Electrical Engineering
The Built Environment:
Professor T. Maver–University of Strathclyde

Renewable Energy & Physics:
Dr J. Twidell–University of Strathclyde

Chemistry, Fuels and Fuel Technology:
Dr W. Gibb– University of Strathclyde

Economics:
M. Common–University of Stirling

SPECIAL ASSISTANTS:

Environmental Matters:
Dr P. Howell–Independent Consultant, Swansea

Bioenergies:
Dr C. Lewis–University of Strathclyde

Conversion Tables

Table 1: energy, heat, work
To convert horizontal row into vertical column multiply by

Units	electron-volt	joule	calorie	Btu	kWh	thermie	*TCE	*TOE	quad
electronvolt	1	6.24×10^{8}	—	—	—	—	—	—	
joule	—	1	4.184	1055	3.6×10^{6}	4.184×10^{6}	30×10^{9}	44×10^{9}	1.055×10^{18}
calorie	—	0.239	1	252.5	—	10^{6}	7.17×10^{9}	10.52×10^{9}	2.525×10^{17}
Btu	—	9.48×10^{-4}	3.96×10^{-3}	1	3412	3.96×10^{3}	2.85×10^{7}	4.17×10^{7}	10^{15}
kWh	—	2.78×10^{-7}	6.64×10^{-8}	2.93×10^{-4}	1	1.163	8.34×10^{3}	1.22×10^{4}	2.93×10^{12}
thermie	—	2.39×10^{-7}	1×10^{-6}	2.52×10^{-5}		1	1.25×10^{5}	1.84×10^{5}	2.525×10^{11}
*TCE	—	3.33×10^{-11}	7.96×10^{-12}	3.51×10^{-8}	1.198×10^{-4}	7.96×10^{-6}	1	1.467	3.51×10^{7}
*TOE	—	2.27×10^{-11}	5.43×10^{-12}	2.39×10^{-8}	8.174×10^{-5}	5.43×10^{-6}	0.68	1	2.39×10^{7}
quad	—	9.479×10^{-19}	3.96×10^{-18}	10^{-15}	3.41×10^{-13}	3.96×10^{-12}	2.85×10^{-8}	4.184×10^{-10}	1

*Assume TCE = 30×10^{9} J or 30 GJ/tonne
TOE = 44×10^{9} J or 44 GJ/tonne

Table 2: power
To convert horizontal row into vertical column multiply by

Units	watt	calories per sec.	Btu/hour	kilowatts	horsepower	terajoule/y	*TCE/y
watt	1	4.184	.293	1000	746.3	3.175	952.3
calories/sec	.239	1	4.20	2.23×10^{2}	178.1	758.7	227.8
Btu/hour	3.413	.238	1	3412	2545	1.082×10^{5}	3246×7
kilowatts	10^{-3}	4.48×10^{-3}	2.93×10^{-4}	1	.746	3.17×10^{-2}	.952
horsepower	1.34×10^{-3}	5.615×10^{-3}	3.93×10^{-4}	1.34	1	42.5	1.27
terajoule/year	3.156×10^{-5}	1.318×10^{-3}	9.24×10^{-6}	3.156×10^{-2}	2.35×10^{-2}	1	.03
*TCE/y	1.05×10^{-3}	4.39×10^{-3}	3.08×10^{-4}	1.05	0.784	33.3	1

*Assume 1 TCE = 30×10^{9} joules

How to use the dictionary

1. The dictionary is not in sections but in alphabetical order.
2. Acronyms and abbreviations are not listed in the dictionary but in the list of acronyms on pages x–xii. This will lead to the correct entry location. Thus Btu stands for British thermal unit, which is listed in the dictionary under British.
3. Institutions are listed within their appropriate place in the dictionary. Those with abbreviations are listed on pages x–xii.
4. Scientific notation is not used – see note below on combination units.
5. A brief list of symbols is given below.
6. Conversion tables for energy and power units are given opposite.
7. Asterisk before a word indicates it is defined elsewhere in the dictionary.

Units and symbols

°C	degrees centigrade
cal	calorie
d	day
eV	electron-volt
g	gram, gramme
h	hour
ha	hectare
J	joule
K	kelvin
kW	kilowatt
m	metre
N	newton
s	second
W	watt

Multiples

k	kilo	10^3
M	mega	10^6
G	giga	10^9
T	tera	10^{12}
P	peta	10^{15}

Submultiples

m	milli	10^{-3}
μ	micro	10^{-6}
n	nano	10^{-9}
p	pico	10^{-12}

Combination units

Scientific notation is *not* employed. Thus joules of heat passing a square metre of surface per second (energy flux) is shown as $J/m^2.s$ not, as in scientific notation $J.m^{-2}.s^{-1}$.

List of Acronyms

ACRS	Advisory Committee on Reactor Safeguards
AEE	Atomic Energy Establishment, Winfrith
AERE	Atomic Energy Research Establishment, Harwell
AGR	advanced gas-cooled reactor
ANS	American Nuclear Society
API	American Petroleum Institute
ARAMCO	Arabian American Oil Company
ASHRAE	American Society of Heating, Refrigerating and Air Conditioning Engineers
ASTM	Americann Society for Testing Materials
Avgas	aviation gasoline
AVR	West German pebble bed reactor
Avtag	aviation turbine gasoline
Avtur	aviation turbine kerosene
BEIR	Biological Effects of Ionizing Radiation (Committee)
BIS	Bank for International Settlements
BNES	British Nuclear Energy Society
BNF	British Nuclear Forum
BNFL	British Nuclear Fuels Limited
BOD	biological oxygen demand
BSI	British Standards Institution
BTU/Btu	British thermal unit
BWEA	British Wind Energy Association
BWR	boiling water reactor
CANDU	Canadian heavy water reactor
CEA	Commissariat a l'Energie Atomique
CEGB	Central Electricity Generating Board
CHU	centigrade heat unit
CIBS	Chartered Institution of Building Services
cif	cost insurance freight
CND	Campaign for Nuclear Disarmament
CNEN	Comitato Nazionale per l'Energía Nucleare
COMECON	Council for Mutual Economic Assistance
DAF	dry, ash-free
DCE	domestic credit expansion
DERV	diesel-engined road vehicle
dmmf	dry, mineral-matter free
DNA	deoxyribonucleic acid
DOE (UK)	Department of Energy (UK)
DOE (US)	Department of Energy (US)
EA	energy analysis
ECCS	emergency core cooling system
EDRP	European Demonstration Reprocessing Plant
EEC	European Economic Community

EFTA	European Free Trade Association
emf	electromotive force
EMS	European Monetary System
EPA	Environmental Protection Agency
ERE	energy requirement for energy
ERSU	Energy Research Support Unit of Scientific and Engineering Research Council
ETSU	Energy Technology Support Unit of UK Department of Energy
Euratom	Agency of the European Economic Community dealing with nuclear research and inspection
eV	electronvolt
FAO	Food and Agriculture Organization
FBP	final boiling point
fob	free on board
FOE	Friends of the Earth
GATT	General Agreement on Tariffs and Trade
GCGR	gas-cooled graphite-moderated reactor
GCV	gross calorific value
GDP	gross domestic product
GER	gross energy requirement
GJ	giga–joule
GNP	gross national product
GW	gigawatt
hex	uranium hexafluoride
HLW	high-level waste
HP	horsepower
HTGR	high-temperature gas-cooled reactor
IAEA	International Atomic Energy Authority
IBP	initial boiling point
IBRD	International Bank for Reconstruction and Development
ICRP	International Commission on Radiological Protection
IEA	International Energy Agency; Institute for Energy Analysis
IEJE	Institut Economique et Juridique de l'Energie
IFIAS	International Federation of Institutes of Advanced Study
IIASA	International Institute for Applied Systems Analysis
IMF	International Monetary Fund
IMCO	International Governmental Maritime Consultative Organization
INFCE	International Nuclear Fuel Cycle Evaluation
I.Nuc.E	Institute of Nuclear Engineers
ISES	International Solar Energy Society
ITDG	Intermediate Technology Development Group
JET	Joint European Torus
kg	kilogram
kW	kilowatt
kWh	kilowatt-hour

LDF	light distillate feedstock
LMFBR	liquid metal-cooled fast breeder reactor
LNG	liquefied natural gas
LOCA	loss of coolant accident
LPG	liquefied petroleum gas
MJ	megajoule
MPBB	maximum permissible body burden
MPC	maximum permissible concentration
MPD	maximum permissible dose
MPL	maximum permissible level
MUF	materials unaccounted for
NCV	net calorific value
NEA	Nuclear Energy Agency
NII	Nuclear Installations Inspectorate
NNPT	Nuclear Non-proliferation Treaty
NPT	Nuclear Non-Proliferation Treaty
NRC	Nuclear Regulatory Commission
NRPB	National Radiological Protection Board
NSHB	North of Scotland Hydro-electric Board
NTP	normal temperature and pressure
OAPEC	Organization of Arab Petroleum Exporting Countries
OECD	Organization for Economic Cooperation and Development
OPEC	Organization of Petroleum Exporting Countries
ORNL	Oak Ridge National Laboratory
OTEC	Ocean Thermal Energy Conversion
PAN	peroxyacetyl nitrate
PER	process energy requirement
PFR	Prototype Fast Reactor
PSALI	permanent supplementary artificial lighting of interiors
PWR	pressurized water reactor
RfF	Resources for the Future
ROV	remote-operated vehicle
SCF	standard cubic feet
SDR	special drawing rights
SGHWR	steam-generating heavy-water-moderated reactor
SI	Système International
SNG	synthetic natural gas
SSEB	South of Scotland Electricity Board
TVA	Tennessee Valley Authority
TCE	tonnes of coal eqivalent
TMI	Three Mile Island
TOE	tonnes of oil equivalent
TW	terawatt
UKAEA	United Kingdom Atomic Energy Authority
UNCTAD	United Nations Conference on Trade and Development
UNSCEAR	United Nations Scientific Committee on the Effects of Atomic Radiation
VAT	Value-Added Tax
WAES	Workshop of Alternative Energy Strategies
WECS	Wind Energy Conversion System

A

absolute price. The price of a commodity expressed in terms of the rate at which it exchanges for a unit of *money, as opposed to the *relative price which is the quantity of another commodity which exchanges for a unit of the commodity in question. Thus, following the action on oil of OPEC in 1973, the absolute price of oil rose much more than the relative price of oil since the absolute prices of all other commodities rose together with, though typically less than, that of oil.

MC

absolute temperature scale. A scale of temperature that (*a*) has zero at the lowest attainable temperature when all atoms of matter have effectively zero kinetic energy, and (*b*) is defined without reference to any physical properties of materials. Absolute temperature scales have particular importance for studies in *thermodynamics and *radiation. Historically the concept of absolute zero temperature was obtained by extrapolating changes in volume with temperature of an ideal perfect gas to the temperature where the volume would be zero.

The Celsius (also called *Centigrade) and *Fahrenheit temperature scales are now defined against the corresponding absolute scales of Kelvin (symbol K) degrees and Rankine (symbol R) degrees. Thus degree intervals in the Celsius and Kelvin scales are equal, and similarly with intervals in the Fahrenheit and Rankine scales. The freezing point of pure water (strictly the triple point when ice, vapour and liquid water are all in equilibrium) is at 273.16 K (0°C) and 491.69 R (32°F). Termperatures of p°C and q°F are related to absolute temperatures °K and °R by

$$°K = (p + 273.2)\ °C$$

$$°R = (q - 32 + 491.69)\ °F$$

One degree C is 1.8 degrees F.

JWT

absorbed dose. *See* radiation dose.

absorber (solar). A device that absorbs solar energy as *heat. It is usually a system of black surfaces or fluids heated by the sunshine and having pipes fixed to take away the heat for use. *See also* collector.

JWT

absorptance. The fraction of incident radiation absorbed by a material over a specified or implied range of wavelengths. It is loosely synonymous with *absorption coefficient and *absorptivity. Strictly defined, absorptance varies with the angle of incidence and the wavelength of the radiation (*see* monochromatic). However, the term is often defined, rather imprecisely, such that if a radiation beam of intensity I has an amount ΔI absorbed then the absorptance equals $\Delta I/I$. Note that materials with one value of absorptance in visible light may have a quite different value in infrared radiation. For example, glass is transparent (has low absorptance) to solar radiation in the visible and near visible region but is opaque (has high absorptance) to the infrared radiation from hot surfaces (*see* greenhouse effect). *See also* reflectance; transmittance.

JWT

absorption. (1) (chemical) The removal of one or more components of a gas mixture by causing them to dissolve in a liquid or solution (*see* amines).

(2) (of a neutron) The interaction of *neutron with an atomic *nucleus and its absorption into it. The compound nucleus thus formed is in an excited state and may de-excite by emitting *gamma radiation, in which case the process of absorption is known as a capture, radiative capture or (n,γ) reaction. For example,

$$^{23}_{11}\text{Na} + ^{1}_{0}\text{n} \rightarrow ^{24}_{11}\text{Na (excited)}$$

$$\rightarrow ^{24}_{11}\text{Na (ground state)} + \gamma$$

Alternatively, the reaction may produce an *alpha particle in which case it is called an (n,α) reaction. For example,

$$^{10}_{5}\text{B} + ^{1}_{0}\text{n} \rightarrow ^{11}_{5}\text{B (excited)}$$

$$\rightarrow ^{4}_{2}\text{He} + ^{7}_{3}\text{Li}$$

In the case of the heaviest elements, the absorption of a neutron may cause the *fission of the compound nucleus. The rate at which the absorption process takes place in any isotope is characterized by its absorption *cross-section.

The importance of neutron absorption is that it can produce, from naturally occurring isotopes, a large number of radioactive isotopes, including the *transuranium elements. The importance of fission is that it is a source of energy. The operation of a *nuclear reactor depends on the correct balance between fission and neutron capture, and the reactor must be designed accordingly.

(3) (of gamma radiation) The interation of radiation with an atom, thereby destroying the energy of radiation. Gamma radiation absorption refers to any process in which all the energy of a gamma *photon is absorbed in the interaction between it and an atom. As a result the gamma photon ceases to exist. The rate of this type of reaction in any material is characterized by the absorption coefficient, μ, of the material according to the following equation:

$$\text{rate of absorption of gamma radiation per unit volume per second} = \text{flux of gamma radiation} \times \text{number of atoms per unit volume of material} \times \mu$$

An important example of this process in nuclear engineering is the absorption of gamma radiation in the biological shield of a *nuclear reactor, required to contain biologically damaging radiation.

WG, DB, DB

absorption chiller. A refrigeration plant in which a liquid, known as the refrigerant (commonly water), is continuously evaporated from an aqueous solution under pressure, condensed, allowed to evaporate (so absorbing heat) and then reabsorbed into the aqueous solution (commonly lithium bromide or ammonia). The principal advantage of the absorption chiller over *vapour-compression chillers is that it can operate on low-grade heat energy and has few moving parts.

TM

absorption coefficient. The fraction absorbed by a material. A term that can refer to both *radiation (*see* absorptivity) and mass (*see* absorption).

JWT

absorptivity (Symbol: α). That property of the surface of a body which determines, in relation to its temperature, the quantity of heat radiation it absorbs. Absorptivity is measured as the fraction of incident radiation of a given wavelength that is absorbed by the body.

TM

accelerator (particle). The general name for machines which, by strong electric and magnetic fields, accelerate charged particles such as *protons, *electrons and *alpha particles to the high speeds and energies at which they can cause nuclear reactions.

DB

acclimatization. The process whereby a living organism gradually readjusts following a change in its environment. This adjustment or *adaptation may take several generations before completion: such a time lag must be taken into consideration when selecting fuel-crop species for rapid growth under unfamiliar conditions within an *energy farm.

CL

accumulator. A rechargeable electric *battery for storing electricity. The most common form of electric accumulator has lead-based plates in sulphuric acid. Each section (cell) of this accumulator has a voltage of two volts when charged.

JWT

acetylene. The *alkyne C_2H_2, the first in the alkyne series. It has a triple bond $HC{\equiv}CH$. At normal pressures it is a highly flammable gas which may be burned to provide a luminous flame. Since the flame is intensely hot the gas is used in welding and metal cutting. Acetylene may explode when under pressure and when in contact with copper or silver. Its gross *calorific value is 50.02 MJ/kg.

WG

acetylenes. *See* alkynes.

acid dew point. As hot gases containing water vapour are cooled, a temperature is reached at which the water vapour starts to condense as a liquid. This is known as the *dew point temperature. When chimney gases from a combustion process containing water vapour and small amounts of sulphur trioxide are cooled, it is found that condensation takes place at a higher temperature than predicted. This is known as the acid dew point temperature, due to the fact that the sulphur trioxide reacts with the condensed water to form sulphuric acid. Since this acid will cause corrosion of chimney and heat-recovery appliances, the chimney gases may be cooled only to a limited extent and the heat which may be recovered is reduced.

WG

acid mine drainage. Drainage produced when water, passing through surface or underground mines, interacts with pyrites, an impurity in coal, to form sulphuric acid via bacterial action. This acidic water drains into local streams and may destroy aquatic *ecosystems.

PH

acid rain. Rainfall occurring when atmospheric water vapour combines with oxides of sulphur and nitrogen released through the combustion of fossil fuels. The sulphur dioxide combines with water vapour to form sulphuric acid. The very tall chimneys which have been built at power stations in recent years to ensure optimal dispersion of emissions have meant that pollutants are discharged high into the atmosphere and transported great distances by prevailing winds, often across national boundaries, before being washed out by rainfall.

Acid rain is almost entirely a man-made problem, one that has become progressively more serious over the last decade. Rainfall with high levels of acidity has been recorded in many part of the world.

PH

acid tar. *See* acid treatment.

acid treatment. The refining of petroleum distillation fractions and lubricating oils by contact with concentrated sulphuric acid. Reactive molecules such as unsaturated hydrocarbons dissolve in and react with the acid to form sulphonic acids. The used sulphuric acid, containing the sulphonic acids, when separated from the refined oil is dark-coloured and known as acid sludge or acid tar.

WG

actinides. The collective name for the elements of *atomic numbers 89 to 103, so-called after the first of the series, actinium. The first four of the series (actinium, *thorium, protactinium and *uranium) occur naturally. The others, sometimes called the *transuranium elements, can be produced by nuclear reactions. All the *isotopes of the actinides are radioactive.

DB

activated carbon. Amorphous carbon with a porous structure which is capable of the *adsorption of large amounts of matter from a gas mixture or from solution. It has a considerable surface area per unit weight, typically 1000 m^2/g. The carbon may be prepared by the *pyrolysis of a variety of substances such as wood, peat, coal, animal bones and petroleum. The activation is achieved by treating the carbon with steam or carbon dioxide at about 900 °C.

WG

activation. The process whereby non-radioactive *isotopes are converted to radioactive ones as a result of neutron capture (*see* absorption, neutron). For example, in the reaction

$$^{23}_{11}\mathrm{Na} + ^{1}_{0}\mathrm{n} \rightarrow ^{24}_{11}\mathrm{Na}$$

stable sodium-23 is converted into radioactive sodium-24.

Activation analysis is the procedure whereby traces of elements can be analysed by activating the element itself and measuring the *radioactivity of the isotopes so formed.

DB

active solar energy system. (1) A system in which a purpose-built device absorbs solar radiation and the resulting heat or work is passed for use at a distance from the device. In contrast a *passive solar energy system uses the energy directly.

(2) In the USA especially, a system that uses energy forms such as wind or *biofuels, that arise from solar radiation. One example is a wind turbine. *See also* renewable energy.

JWT

activity. *See* radioactivity.

acute radiation syndrome. *See* biological effects of radiation.

acute toxicity. The capacity of an agent or material to cause injury to an organism as a result of a single *exposure. Acute toxicity tests usually involve the measurement of the lethal dose to kill 50% of the exposed population.

PH

adaptation. Some alteration(s) in the morphological, physiological, or behavioural aspects of an organism, resulting in greater compatibility between itself and its *environment. Thus a fuel crop may have to change in some way when confronted by unfamiliar growth conditions. An example would be the case of a plant forced to grow in an arid zone where water conservation is a priority. Such a plant would have to adapt by keeping its stomata (leaf pores) closed during the hot day and open them only at night when water loss through evaporation is less likely.

CL

additive. A substance added to a product to modify its properties. Additives may be used with petroleum fuels to improve their combustion characteristics or in lubricating oils to improve their ability to prevent metal wear, etc. Additives used in motor spirit include antiknock agents, antioxidants, anti-icing agents, upper-cylinder lubricants and dyes.

WG

adiabatic lapse rate. The rate at which the temperature of air decreases with the height above the ground. A meteorological term.

PH

adiabatic process. A process in which in heat is supplied to or rejected from a fluid during its *change of state.

TM

admittance. (1) (electrical, symbol: *Y*). The inverse of *impedance in an electric circuit. Admittance is measured in *siemens.

(2) (thermal, symbol: *Y*). The property of building materials and the spaces they enclose which determines their response to sinusoidal variation in external temperature. Admittance values (in units of W/m^2.k) have been estimated for a range of wall and roof types.

TM

adsorption. The removal of one or more components of a gas mixture, or the removal of a dissolved substance from an aqueous solution, by contact with a solid surface. The adsorbed components are held on the surface of the solid by physicochemical forces and thus porous substances with a large surface area per unit weight are most effective. *See also* activated carbon.

WG

ad valorem tax. A tax for which the rate is specified as a percentage of the price, before tax, of a commodity, as distinct from a specific tax for which the rate is specified per unit of the commodity. Examples of ad valorem taxes are local sales taxes in the USA and *value-added tax in Europe. Taxes on *natural resources, such as *severance taxes on oil extraction, may be either ad valorem or specific. As the price of the taxed commodity changes, an ad valorem tax yields a constant percentage of the net of tax price, whereas a specific tax yields a changing percentage of the net of tax price.

MC

advanced gas-cooled reactor (AGR). *See* nuclear reactor.

Advisory Committee on Reactor Safeguards (ACRS). A statutory group of 15 scientists and engineers which advises the NRC on safety matters referred to it and on the adequacy of proposed reactor safety standards.

Address: 1717 H St. N.W., Washington, D.C. 20555, USA.

PH

aerobic. Surviving in or requiring air and oxygen. It is used of bacteria that develop on material open to oxygen. *See also* anaerobic.

JWT

aerofoil. *See* airfoil.

aerogenerator [wind generator; *wind turbine; *windmill]. A machine that generates electricity from the wind. All such machines require a tower or support for the *turbine which will have a number of blades. The axis of the turbine may be horizontal or vertical (or very occasionally at an intermediate slope). An electric *generator is connected to the turbine, often through a gear box or hydraulic connection. The fewer the number of blades (optimally one), the higher the frequency of rotation in a given wind speed and the easier the generation of electricity. However, fewer blades produce increased structural problems for the turbine so that aerogenerators commonly have two or three blades. An aerogenerator with blades 2m long will produce about 15 kW of electricity in a strong wind of 20 m/s. Such machines are called small aerogenerators. Large machines with blades about 30m long, may produce about 3MW maximum power. *See also* wind power.

afterdamp. The non-flammable mixture of gases found in the tunnels of a coal pit after a gas explosion. The major components are carbon monoxide, carbon dioxide and nitrogen, thus the mixture is likely to be toxic.

WG

agriculture. The practice of cultivating the land to produce crops for the use of the human population or domestic animals. More precisely, it is the art and science of using plants to capture energy by *photosynthesis. This energy capture can be enhanced by the application to the crop roots of fertilizers, natural or synthetic. Synthetic fertilizers are energy-intensive products.

CL

airchange rate. The actual or anticipated rate at which the volume of air within a built space is replaced, either by means of mechanical air-handling plant or by natural infiltration. Control of airchange rate is a significant factor in the energy-conscious design of buildings; and where *radon consitutes a health hazard.

TM

air conditioning. The means by which an atmospheric environment is controlled, either for the comfort of human beings or animals or for the proper performance of some industrial or scientific process. Full air conditioning implies that the purity, movement, temperature and *relative humidity of the air is to be controlled within the limits imposed in the design specification.

Air-conditioning systems in buildings are typically of seven types: *perimeter induction, *fan coil, *chilled ceiling, *single duct, *dual duct, *variable air volume, or *terminal heat-recovery systems.

TM

air-dried coal. Finely divided coal prepared for analysis by being exposed to the air of the laboratory or test room until it reaches a constant weight. The remaining *moisture content of the sample is then in equilibrium with the humidity of the air in the room.

WG

airfoil [aerofoil]. A specially shaped surface, such as the blade of an *aerogenerator or propeller, that is usually designed to produce maximum *lift force and minimum *drag force as the air passes.

JT

air fuel ratio. The ratio (normally in terms of mass) of air to fuel that is fed to a combustion unit operating under steady-state conditions. With gas-fired furnaces the ratio may be expressed by volume.

WG

air mass. A term used in the analysis of the absorption of *solar energy passing through the earth's atmosphere. In the figure, direct solar radiation is passing at

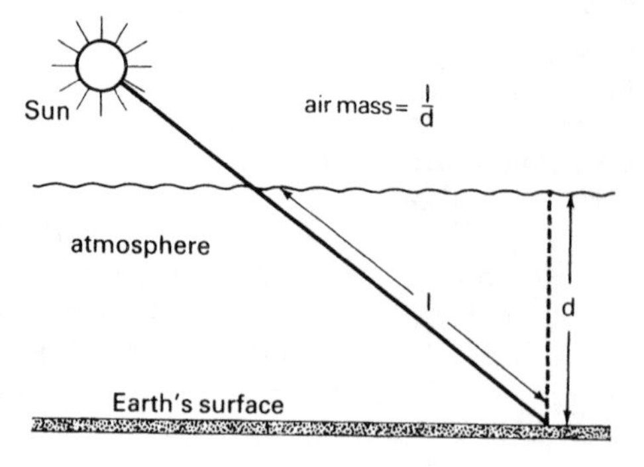

an angle through the atmosphere. The air mass is the ratio of this path length, *l*, through the atmosphere to the path length, *d*, when the sun is directly above. The use of the word mass is misleading and the quantity is better called air mass ratio.

JWT

air temperature. *See* temperature.

albedo. A term synonymous with *reflectance but usually only referring to *radiation of visible or near-visible wavelengths. The albedo of the earth is the proportion of *solar energy reflected by clouds, snow, ice and the natural environment. This amounts to about 30% of the incoming solar radiation.

JWT

alcohols. A generic name for organic compounds containing one or more hydrocarbon groups (C_nH_m-) and one or more hydroxyl groups ($-OH$). Ethyl alcohol (ethanol), C_2H_5OH, is a monohydric alcohol; ethylene glycol, $CH_2OH.CH_2OH$ is a dihydric alcohol.

WG

alder. The common name for the tree genus *Alnus*, which includes several excellent fast-growing fuelwood species (*see* firewood). Many can carry out biological *nitrogen fixation, thereby reducing the need for energy-intensive *fertilizer applications. The alder is also a useful tree for soil reclamation and reforestation. Most species grow well in tropical highland regions; for example, *Alnus nepalensis* grows in the hills of Burma, northern India, Nepal and parts of China. *Alnus rubra* (the red alder) is one of the most productive trees of North America and is used for domestic heating in western Canada and the western United States including Alaska. This tree is a candidate for fuelling wood-burning power stations. Its wood is moderately dense (specific gravity, 0.39), has a high heat content of about 19.25 GJ/t and produces good quality charcoal. For short coppice rotations (*see* coppicing), total stem yields are 17 – 21 m^3/ha.y, giving a gross energy yield of 130 – 160 GJ/ha.y.

CL

algae. A diverse group of plants which possess *chlorophyll and other pigments enabling them to perform oxygen-evolving *photosynthesis. Some are unicellular while others are filamentous, colonial, or multicellular. The are dependent upon light as an energy source and carbon dioxide as their source of carbon. Certain *blue–green algae are capable of atmospheric *nitrogen fixation and so can be used as *fertilizers in agriculture to substitute for energy-intensive synthetic fertilizers. Other algae, in combination with aerobic bacteria, can purify waste waters at a significant energy saving over conventional methods of treatment. Yet other algae can be anaerobically digested to *biogas, whereby up to 60% of the algal energy content can be converted to *methane gas.

CL

alicyclic hydrocarbons. Hydrocarbons in which the skeleton of carbon atoms is arranged in one or more closed rings. Such compounds behave as *aliphatic hydrocarbons rather than *aromatic hydrocarbons.

WG

aliphatic. An organic chemical compound in which the carbon atoms in the molecule are linked in open or branched chains. *Paraffins and *olefins are aliphatic hydrocarbons.

WG

aliphatic hydrocarbons. Hydrocarbons such as an *alkane, in which the carbon atoms in the molecule are arranged as straight or branched chains.

WG

alkanes. The systematic name for straight-chain saturated hydrocarbons with the general formula C_nH_{2n+2}. *See also* paraffin.

WG

alkenes. The systematic name for straight-chain unsaturated hydrocarbons with the general formula C_nH_{2n}. *See* ethylene; olefin.

WG

alkylation. The addition of an alkyl group, $-C_nH_{2n+1}$, to an organic compound, generally a hydrocarbon. This is carried out industrially at high temperature and pressure. If a catalyst is used, e.g. aluminium chloride, hydrofluoric acid or sulphuric acid, more moderate process conditions are possible. For example, high *octane gasoline may be produced by reacting *ethylene with isobutane.

WG

alkynes. The systematic name for straight-chain unsaturated hydrocarbons with the general formula C_nH_{2n-2} in which at least one pair of carbon atoms in the molecule are joined by a triple valency bond. With from two to 12 atoms in the molecule, alkynes are gases or volatile liquids. They are very reactive and readily oxidized. *Acetylene is the first in the series.

WG

allocation problem. The problem of finding an arrangement of the activities of production and consumption in an economy such that valuable resources are not wasted. In *neoclassical economics the objective is *efficiency in allocation, also known as Pareto optimality. An arrangement is an efficient one if it is impossible to alter it so as to make any individuals worse off. Given the production possibilities open to the economy and the preferences of individuals with respect to those possibilities, there exist for that economy a large number of arrangements of activities which satisfy the objective of efficiency in allocation – the objective does not identify a unique arrangement. Under certain stringent conditions a *market equilibrium produces an efficient solution to the allocation problem, but the particular efficient solution arising depends on the distribution of endowments across individuals and would not necessarily be one that could widely be regarded as fair between those individuals. The achievement of an efficient allocation does not ensure the solution of the *distribution problem.

Where the necessary conditions for a market equilibrium to coincide with an efficient allocation are not satisfied, *market failure is said to exist and the realization of efficiency in allocation calls for government intervention in economic activity. The conditions are such that market failure must be presumed as operative in actual economies. It arises in connection with the extraction and use of energy resources, especially where they are, or their use involves the use of, *common property resources. Even if market failure were not present, there can be no presumption that markets would, for example, result in the depletion of oil stocks over time in a way that could be regarded as fair between successive generations. *See also* compensation principle; welfare.

MC

allochthonous coal. *Coal formed from a large variety of plant remains brought down rivers to a lake or an estuary and deposited there.

WG

alpha emitter. *See* alpha radiation.

alpha particle (Symbol: ^4_2He, α). A nucleus of helium-4, consisting of two *protons and two *neutrons, that is emitted by certain radioactive substances of high mass number. A typical example of an alpha decay process is that of radium-226 to form radon-222, which is part of the natural *radioactive decay chain of uranium-238:

$$^{226}_{88}Ra \rightarrow {}^{222}_{86}Rn + {}^4_2He$$

Alpha particles may also be produced by a variety of nuclear reactions, such as

$$^{10}_{5}B + {}^1_0n \rightarrow {}^4_2He + {}^7_3Li$$

Alpha particles produced by radioactive decay or nuclear reactions usually have an energy of a few MeV. They are positively charged particles which cause intense *ionization and have a very short range in matter.

DB

alpha radiation. A form of *ionizing radiation that is comprised of *alpha particles and is emitted by a certain class of radioactive substances called alpha emitters. Alpha radiation can penetrate a few centimetres of air but is stopped by a sheet of paper or a very thin layer of aluminium foil. Consequently alpha emitters are not hazardous to humans unless they are taken into the body by ingestion, inhalation or through a wound in the skin.

PH

alternating current (ac). An electric current which reverses its direction of flow at a regular rate. Most electricity for public supply is in the form of alternating current at 50 hertz (cycles per second) in Europe or 60 hertz in the USA. *See also* direct current.

TM

alternative energy. Energy sources and systems that could supplant or substitute for traditional fossil fuel supplies, such as coal, oil and natural gas. Alternative energies may be divided into *renewable energy from the natural environment and *nuclear energy. The renewable supplies may be classified by such terms as solar, wind, wave, hydro, tidal, biomass, photoelectric, ocean thermal energies and photosynthesis. *Geothermal energy is an alternative supply that is not strictly renewable.

JWT

alternator. *See* generator.

aluminium. A light element that is the most abundant metal in the earth's crust. It is obtained from its ores by the electrolysis of alumina in the Hall–Heroult process. Aluminium has a high thermal conductivity and so finds many applications in solar-energy development, heat transfer and for light-weight construction. Its principal use in nuclear power is as the *cladding material for the fuel in low-power research reactors.

Symbol: Al; atomic number: 13; atomic weight: 26.98; density: 2700 kg/m^3.

DB

ambient temperature. The temperature of the immediate surroundings. The ambient temperature to a person in a room is the air temperature of that room.

MS

American Institute of Mining, Metallurgical and Petroleum Engineering. A publishing society.

Address: 345 East 47th Street, New York, NY 10017, USA.

WG

American Nuclear Society (ANS). An international scientific, technical and educational society devoted to the advancement of nuclear science and technology.

Address: 244 East Ogden Avenue, Hinsdale, Illinois 60521, USA.

DB

American Society for Testing and Materials (ASTM). A society which sets standards and publishes them annually for a wide variety of materials and their testing within the energy field.

Address: 1916 Rose Street, Philadelphia, Pennsylvania, USA.

WG

American Society of Heating, Refrigerating and Airconditioning Engineers (ASHRAE). A professional body within the USA with members engaged in the design, maintenance and operation of plant for environmental control.

Address: 345 East 47th Street, New York, NY 10017, USA.

TM

amines. Primary amines are organic compounds with the general formula RNH_2, where R– is an alkyl or aryl radical linked to the nitrogen-containing amino group. Secondary amines contain two radicals and an imino group (=NH), and tertiary amines have three radicals linked to a nitrogen atom. These substances are non-acidic (basic) in character and when the molecular weight is low they resemble *ammonia in behaviour. Solubility in water and volatility decrease with increase in size of the attached radicals (R–). Aniline, $C_6H_5NH_2$, is perhaps the best-known aromatic amine. Low molecular weight aliphatic amines are used in gas absorption solutions to remove hydrogen sulphide and carbon dioxide from fuel gases.

WG

ammonia. A colourless gas, a compound of hydrogen and nitrogen (NH_3), which is intensely irritating to the eyes and lungs. It is combustible and toxic. The gas is prepared on an industrial scale by the reaction over a catalyst of a mixture of *hydrogen and nitrogen at high pressure and elevated temperature and by the *Haber process. Small amounts of ammonia (about 0.3% of the weight of coal) may be recovered as a byproduct during the *carbonization of coals. The gas is readily soluble in water forming a strongly alkaline solution. When reacted with mineral acids, e.g. hydrochloric acid (HCl), a neutral salt is formed, e.g. ammonium chloride (NH_4Cl).

WG

Amoco Cadiz. The supertanker which was wrecked on rocks near Portsall on the NW coast of Brittany on 16 March 1978. The quantity of crude oil spilled on this occasion was far greater than in any previous accident; it was spilled very close to shore and the entire cargo of 220,000 tonnes was released over a very short period during which time the winds were mainly onshore from the west. As a result more than 200 km of the Breton coastline were affected, including areas of outstanding ecological and biogeographical interest, as well as important shellfish and kelp industries. About 30% of the fauna and 5% of the flora was destroyed locally. However the overall damage to the environment was moderate in comparison with the **Torrey Canyon* accident, probably because there was only a limited use made of toxic *dispersants, and vigorous measures were taken to physically remove the oil from the coastline.

PH

amorphous. Denoting a solid in which the atoms have no regular order or which does not appear to have a crystalline structure. *Soot is considered to be an amorphous substance.

WG

amortization. The repayment of a debt by a series of usually equal instalments. Each instalment consists of two components. The first is the interest on the debt. The second is a contribution to a fund which will, with *compounding, accumulate to a sum equal to the original loan by the end of the loan period. The capital cost element in the cost of electricity from a given generating plant is, for example, usually calculated by finding the series of equal annual payments necessary to amortize the initial cost of the plant over the lifetime of the plant.

MC

ampere (amp; symbol: A). The SI base unit of electric *current, defined as that constant current which, if maintained in each of two infinitely long straight parallel wires of negligible cross-section placed one metre apart, in a vacuum, will produce between the wires a force of 2×10^{-7} newtons per metre length. A current of one amp will pass along a conductor of *resistance of one ohm when a *potential difference of one volt is applied to it.

TM

amp-hour. A term related to the energy stored in a *battery or *accumulator. It is equal to the product of the electric current in amperes multiplied by the time in hours during which the battery produces a current at a fixed voltage. The energy produced in *joules is the number of amp-hours times the voltage times 3600.

JWT

Anabaena. A *blue–green alga capable of *nitrogen fixation. Since this property reduces the need for energy-demanding fertilizer, it has been used in mixed algal culture on farmland in India and surrounding countries to raise crop yields. *Anabaena azolla* and the water fern, *Azolla*, grown together in *symbiosis are utilized for the same purpose in China and the Far East. Another species, *Anabaena cylindrica* can produce hydrogen gas directly in a process called *biophotolysis, the biological splitting of water molecules in the presence of light.

CL

anaerobic. Surviving in or requiring the absence of oxygen. It is used of bacteria that develop in enclosed or enveloped conditions where air or oxygen become excluded. A product of the digestion of organic material by anaerobic bacteria is *methane, CH_4. *See also* aerobic.

JWT

anemometer. An instrument for measuring wind speed. A common form used at meteorological stations is the cup anemometer, mounted at a standard height (usually 10 m). Some anemometers, e.g. the hot wire anemometer, operate with no moving parts.

JWT

anhydrous compound. *See* hydrate.

aniline point. The lowest temperature at which a petroleum fraction is completely miscible with an equal volume of the aromatic *amine aniline. It is used as a measure of the proportion of aromatic hydrocarbons in the fraction and is particularly important in assessing its suitability as a *diesel fuel.

WG

annihilation (nuclear). A process whereby matter is destroyed and energy released. An important example occurs when a *positron (produced by the radioactive decay of certain isotopes) reacts with an *electron and both are annihilated. Energy is released as *gamma radiation, which is then known as annihilation radiation. The process can be represented as

$$-{}^{0}_{1}e + {}^{0}_{1}e \rightarrow 2\gamma \text{ (0.51 MeV each)}$$

DB

annual limits of intake. *See* radiological protection standards.

anthracene. A crystalline solid aromatic hydrocarbon, $C_{14}H_{10}$, in which the carbon atoms in the molecule are arranged in three fused rings. It occurs in tars from high-temperature coal *carbonization processes and is a raw material for certain dyestuffs.

WG

anthracite. The highest grade or *rank of coal. Lustrous and hard, it does not soil the hands. It can have a carbon content of over 93% and a volatile-matter content of less than 9%. When carbonized it does not soften nor form a coke. Since it burns virtually without smoke it is convenient fuel for use in domestic stoves and central-heating installations. It also finds application in steam raising and in malting and hop drying. A typical value of gross *calorific value on a dry ash-free basis would be 36.0 MJ/kg, a value much higher than most coals.

WG

anthraxylon. A constituent of coal which has a bright appearance and is formed from woody tissue, i.e. from tree trunks and branches. This term was used in a US system of coal classification and is approximately equivalent to the terms vitrain and clarain used in the UK.

WG

anticline. A folded rock structure within the earth in the form of an arch. When the rocks forming it comprise an impermeable layer above a permeable layer of sedimentary rock, the structure is likely to be an underground trap for hydrocarbon oils and gases. Considerable amounts of oil and gas have been found in anticlines.

WG

antiknock. A chemical compound added in small concentration to motor spirit to reduce the tendency to ignite spontaneously (to *knock) when the vapour mixed with air is subjected to heat and pressure in the cylinder of an *internal combustion engine. A popular antiknock is *tetraethyl lead (TEL). Fears about the health hazard resulting from exhaust gases containing lead are leading to the reduction or elimination of such lead compounds from motor spirit. Other compounds have been tried but as yet none are as effective as TEL and all are toxic or produce toxic compounds in the exhaust gases.

The elimination of lead from motor spirit is a question of economics. Petroleum refineries can produce motor spirit with little tendency to knock, i.e. high *octane rating, but at subtantially increased cost. Alternatively, engines can be designed with lower *compression ratios and consequently lower thermal efficiencies.

WG

antinuclear. A term applied to the popular movement against the expansion of the civilian nuclear industry which originated in the US and Europe in the 1960s. More recently in Europe the term is being used to describe the Campaign for Nuclear Disarmament (CND).

PH

antioxidant. A substance added in low concentration to an oil or other oxidizable substance in order to inhibit oxidation, i.e. reactions with *oxygen. An antioxidant must be soluble in the oil, have a low volatility, be easy and safe to handle and be low in cost. It acts through its ability to react with and thus tie up one of the intermediates in the oxidation reaction chain. Antioxidants for petroleum hydrocarbons are generally complex *phenols.

WG

API gravity (American Petroleum Institute gravity). An arbitrary scale based on the formula

$$\text{degrees API} = \frac{141.5}{\text{specific gravity}} - 131.5$$

where the *relative-density measurement is made at 60°F. Its application enables a linear scale to be used on the stem of a density-measuring device like a *hydrometer.

WG

appreciation. A rise in the value of an *asset, usually reflected in its market price. This comes about because of an increase in the demand for, or a decrease in the supply of, such assets and the commodities that their utilization provides. Thus, an increase in energy prices would lead to an appreciation of energy-related assets such as oil wells, coal stocks, hydroelectric sites, etc. Where the *exchange rate is determined under a regime of floating rates, a rise in the price of the home *currency in terms of foreign currency is known as appreciation of the home currency. *See also* depreciation.

MC

appropriate technology. Technology chosen to be particularly beneficial to the local population who are involved in the manufacture, growth or consumption of the products. The phrase 'appropriate technology' became commonly accepted in the 1970s with attempts to improve the development of communities in non-industrialized Third World countries, since it was assumed that technology applied from the industrialized countries did not fully benefit the local people. An 'appropriate technology' considers the skills, culture and resources of a particular community within its own local environment and seeks to improve the quality of life of that community. This form of technology is considered to contrast with those technologies encouraged for the sake of the financial or material benefit of outside agencies.

In practice it is never possible to decide explicitly whether a particular technology is entirely appropriate or inappropriate. The term 'appropriate' is perhaps best used to demonstrate an ethical standpoint that seeks to improve the human and environmental aspects of a technology. In general the soft technologies using the *renewable forms of energy (*solar, *wind, *biomass, etc.) encourage local participation in dispersed communities and are considered 'appropriate'. Examples are the generation of methane fuel from animal waste (*biogas) and the harnessing of *hydropower even on a small scale.

The success of the concept of appropriate technology, in less-developed countries, has led to the term being used also in the industrialized countries. Of importance here are technologies that encourage worthwhile employment to the satisfaction of employees and employers alike. The scale of the manufacturing units or the technological application need not be specified, however in general, small or intermediate scale is favoured (*see* intermediate technology).

JWT

aquaculture. In its broadest sense, the cultivation of *algae, water plants, fish or any other water-borne organism. Algae and water plants can be grown for their fuel value. After harvesting they can undergo *anaerobic digestion to produce *biogas. However, the main value of aquaculture lies in its potential for energy conservation. By producing algae as sources of protein, fertilizers, drugs and other materials, conventional energy-intensive production routes are replaced, thus giving rise to energy savings. The same concept applies to algae as components of advanced waste water treatment and recycling processes for *waste.

CL

aqueous reactor. *See* homogeneous reactor.

Aramco. Arabian American Oil Company.

arbitrage. The buying and selling of commodities, currencies, bills, etc., in different markets in order to take advantage of a temporary difference in the rates. The markets can be in different parts of the world (space arbitrage) or at different times (time arbitrage). Arbitrage can also be distinguished by the number of currencies involved (e.g. triangular arbitrage), by whether a trader is involved or not (trader or non-trader arbitrage), or whether it refers to assets (e.g. interest arbitrage). In its pure form arbitrage involves no *risk; where risk is involved there are elements of *hedging or *speculation. A person operating in such a market is called an arbitrageur.

MC

arenes. The systematic name for unsaturated cyclic hydrocarbons. *See* aromatic hydrocarbons.

WG

argon. An inert gas occurring naturally as 0.94% of air, from which it is separated by fractionation. Its many uses as an inert gas include the filling of certain types of *radiation detectors and the gas cover over the free surface of sodium or sodium–potassium in reactors cooled by *liquid metal. In air-cooled reactors traces of argon in the coolant can cause a hazard due to the formulation of radioactive argon-41 by neutron capture in the naturally occurring argon-40.

Symbol: Ar; atomic number: 18; atomic weight: 39.948.

DB

Argonne National Laboratory. A unit of the US Department of Energy operated by the University of Chicago. It undertakes research and development in all aspects of science and engineering related to both nuclear and non-nuclear energy.

Address: 9700 South Cass Avenue, Argonne, Illinois 60439, USA.

DB

armature. The part of an electric machine in which, in the case of a *generator, the electromotive force is produced or, in the case of an *electric motor, the *torque is produced. It includes the winding through which the main current passes and the portion of the magnetic circuit upon which the winding is mounted.

TM

aromatic. An organic chemcial compound, the molecules of which are unsaturated but which undergo chemcial substitution reactions, e.g. the sustitution of a chlorine atom for hydrogen atom, rather than an addition reaction, e.g. the addition of two atoms of chlorine to the molecule by the breaking of a double bond. Typical aromatic hydrocarbons contain a basic *benzene six-membered ring structure and react in a manner indicating that the single and double bonds in the ring are arranged alternately. Such hydrocarbons have good solvent properties and those in the *gasoline boiling range have good *antiknock characteristics.

WG

aromatic hydrocarbons. *Unsaturated hydrocarbons in which the carbon atoms in the molecule are arranged in rings of six with the bonds between the carbon atoms alternately single and double. *Benzene, C_6H_6, has one such six-membered ring, naphthalene, $C_{10}H_8$, has two and anthracene, $C_{14}H_{10}$, has three. Such hydrocarbons are obtained from the distillation of *coal tar. Larger tonnages are available from the *catalytic reforming of petroleum fractions.

WG

ash. The incombustible residue remaining after the complete combustion of solid or liquid fuel. It is determined by burning a small sample of the fuel in a laboratory furnace under standard conditions, and is derived from the adventitious *mineral matter associated with the fuel and inorganic elements chemically combined with the organic matter. The ash content of coals may range from about 1% in a so-called 'clean' coal to 30% in poor-quality coal burned as *pulverized fuel in large steam generators. Heavy fuel oils may contain as much as 0.2% ash. The 'ash' from a solid fuel combustion unit is rarely the result of complete combustion and thus will still contain some combustible matter.

WG

ash content. The weight of *ash expressed as a percentage of the weight of fuel sample burned under standard conditions in a laboratory furnace.

WG

ash-free basis. A basis for calculating and reporting the analyses of coals and other solid fuels after the ash content has been subtracted from the total.

WG

ash fusion temperature. The ash from the combustion of solid and liquid fuels is a mixture of metallic oxides and inorganic salts. When heated to a high temperature it does not have a single melting point but softens and melts over a temperature range, the ash fusion temperature. In a furnace molten and sticky ash may cause the formation of deposits which may affect the entry of air, and in a boiler produce deposits on the boiler tubes which interfere with the transfer of heat. The behaviour of the ash can be predicted by a standard test in which a moulded shape of ash particles is heated to a high temperature in a laboratory furnace and its change of shape observed. Ash fusion temperatures normally lie in the range 1200–1600°C. *See* *also* clinker.

WG

asphalt. A term that tends to be used interchangeably with *bitumen, but should be reserved for mixtures, either natural or synthetic, of bitumen with mineral matter.

WG

asphaltenes. Complex hydrocarbons having a number of linked aromatic rings and long side chains. They are found in the high boiling fractions of most crude oils but predominate in *asphaltic crude petroleum, giving it a black and viscous character. Asphaltenes may be precipitated from petroleum fractions by light petroleum spirit. In a standard test, *heptane is employed.

WG

asphaltic crude. A crude petroleum which contains dark-coloured complex cyclic hydrocarbons called *asphaltenes. When the crude is distilled they remain as a residue of *bitumen. The other hydrocarbons in the crude are generally of a cyclic nature and thus provide good petrol, poor kerosene, poor diesel oil, no lubricants and good heavy fuel oil.

WG

Asse. An abandoned salt mine in the Federal Republic of Germany which is now used as a disposal site for *radioactive wastes.

PH

assessment of technology. *See* technology assessment.

asset. Something which is ownable and tradeable, and which has therefore a price in some market. An important distinction is between real assets and financial assets. Real assets are things directly useful in production or consumption: *commodities, machines, buildings, *natural resources. Financial assets are titles to receive income or services: money, *securities, *bills. Individuals may prefer to hold their wealth, at least in part, in the form of financial assets, which typically have more *liquidity (are more easily converted into money) than real assets. Oil in the ground is a real asset: a security representing part ownership of an oil company is a financial asset.

MC

asynchronous. Not at the same frequency or phase as in some forms of electric generators. It is possible however for certain generators to rotate asynchronously with respect to a mains *grid supply yet produce electricity at the same frequency and phases as the grid.

JWT

atom. The smallest unit of any matter that still retains a specific chemical and physical characteristic. Thus each element is comprised of atoms. An atom itself is composed of a nucleus of *neutrons and *protons, surrounded by a shell of *electrons.

There are approximately 6.02×10^{23} atoms in a gram atom, i.e. the atomic mass expressed in units of grams mass.

MS

atmosphere. The gaseous envelope surrounding the earth. Dry air is composed of 78% nitrogen, 21% oxygen, 0.3% carbon dioxide, together with trace amounts of neon, helium, methane, krypton, nitrous oxide, hydrogen and xenon. It can also contain up to 3% water vapour according to local conditions.

PH

atmospheric dispersion. The manner by which the gaseous *effluent from an industrial plant is dispersed to or by the *atmosphere. The way in which an effluent plume disperses depends on many factors, including its composition, temperature, buoyancy and initial momentum, the chimney height and prevailing meteorological conditions. The presence of *temperature inversions greatly influences atmospheric dispersion because the intermixing of air layers lying above and below an inversion is virtually excluded. If an inversion forms above a chimney the effluent plume will usually have insufficient buoyancy to penetrate the inversion and disperse upwards so that the plume is trapped close to the ground.

PH

atmospheric pressure. The pressure exerted by the mass of air surrounding the planet Earth. It varies from place to place and time to time according to the meteorological conditions, humidity and height above sea-level.

MS

atomic energy. A popular term now little used. *See* nuclear energy.

MS

Atomic Energy Establishment (AEE), Winfrith. An establishment of the United Kingdom Atomic Energy Authority which is primarily concerned with research in reactor physics, shielding, heat transfer, fluid dynamics and reactor control and safety. Winfrith is also the site of the steam-generating heavy-water-moderated reactor (SGHWR) and the OECD Dragon high-temperature reactor, now closed down.

Address: Winfrith, Dorset, England.

DB

Atomic Energy Research Establishment (AERE). The largest research laboratory of the United Kingdom Atomic Energy Authority and one of the largest in the world. About half the work is devoted to nuclear-reactor development and nuclear power in general, with particular reference to materials research. In addition work of a more varied nature related to energy and technology is carried out on behalf of the government and industry.

Address: Harwell, Didcot, Oxfordshire, England.

DB

atomic mass. Of an isotope, the mass of one atom of that isotope expressed in *atomic mass units. For example, the atomic mass of hydrogen, the lightest element, is 1.00797 while that of uranium is 238.0508.

DB

atomic mass unit. The unified (internationally accepted) atomic mass unit (symbol: u), defined as one-twelfth of the mass of a neutral carbon-12 atom:

$$1\,\text{u} = 1.6605 \times 10^{-27}\,\text{kg}$$

DB

atomic number. Of an element, the number of protons in the *nucleus of the element.

DB

atomic weight. Of an element, the sum of the *atomic mass of each isotope of the element multiplied by its fractional abundance in the naturally occurring element.

DB

atomization. The dispersion of a liquid fuel as a spray of fine droplets so that it may be mixed with air and burned rapidly and efficiently. Atomization may be achieved by high-pressure injection through a suitably designed orifice, by impingement on a spinning disc, or by impact with a high-velocity jet of air or steam.

WG

attritus. A dull constituent of coal composed of a variety of plant debris. This term, used in a US system of coal classification, is approximately equivalent to the UK term *durain.

WG

authorized discharge limit. The maximum amount of radioactivity which may be released into the environment by a particular nuclear facility in one year. For example, the nuclear-fuel reprocessing plant at Sellafield (Windscale) England is authorized to discharge up to 300,000 *curies of total beta radioactivity and up to 6000 curies total alpha activity, in any one year. Discharge limits are based on ICRP dose-limit recommendations.

PH

autochthonous coal. Coal formed *in situ* from trees and other plant materials previously growing in that location.

WG

autoignition temperature. When a mixture of a flammable gas and air is heated, the temperature ultimately reached at which the rate of oxidation of the gas becomes so rapid that the mixture ignites spontaneously. For hydrocarbons it lies in the region 250–500°C. Paraffin hydrocarbons have lower autoignition temperatures than aromatics and are thus preferred in a diesel fuel, which is required to ignite spontaneously when injected into the hot cylinder of the *diesel engine. WG

automatic stabilization. The operation of relationships in the economy which work to reduce the amplitude of fluctuations in economic activity, independently of discretionary changes in *fiscal and *monetary policy. The principal example is the effect of direct taxation. In the expansionary phase of the business cycle, rising personal incomes mean that with constant allowances and rates the income-tax yield increases so cutting back the rate at which *national expenditure grows, assuming constant government expenditure. With automatic stabilization there is an element of *negative feedback built into the operation of the economy. MC

autotrophic organisms. Organisms that require carbon dioxide as their main source of carbon. They can be divided into two groups. Photoautotrophic organisms are dependent on light as their energy source and utilize carbon dioxide as their principal carbon source. This category includes higher plants, algae and some photosynthetic bacteria. Chemoautotrophs are also dependent on carbon dioxide as their main source of carbon but are incapable of *photosynthesis. They use instead a supply of *chemical energy such as ferrous iron, hydrogen, reduced nitrogen compounds (NH_3, NO_2), or reduced sulphur compounds (H_2S, S, $S_2O_3^{2-}$). CL

available energy. That portion of energy actually able to do work. It is an engineering term, very close numerically to *negentropy. *See also* free energy. MS

availability. A *thermodynamic concept used in engineering calculations to determine the work that may be obtained from a unit of *heat.

availability factor. Primarily of wind generators, the proportion of time that the generator supplies power at or above a specified level. The availability factor depends on the distribution of wind speeds over an extended time period. JWT

average cost pricing. A situation where a producer sets the price to be charged per unit of output as the total cost of production divided by the number of units produced. Unlike *marginal cost pricing, setting price equal to *marginal cost, it is not consistent with *efficiency in allocation.

average cold spell. A period during which the average outdoor temperature over a defined number of days has been below some defined limit. The occurrence of such an 'average cold spell' may occasion the supplementary payments for home heating to people receiving government benefit. TM

Avgas. Abbreviation for aviation gasoline.

aviation spirit. A volatile liquid fuel burned in the spark-initiated *internal combustion engine (piston engine) of an aeroplane. A number of grades are available at different *octane ratings, some of which are in excess of 100. WG

Avtag. Abbreviation for aviation turbine gasoline.

Avtur. Abbreviation for aviation turbine kerosene.

azimuth. The angle, usually in a clockwise direction from north, made on the horizontal (horizon) plane by the projection of the sun–earth direction. It is given by the angle γ in the figure.

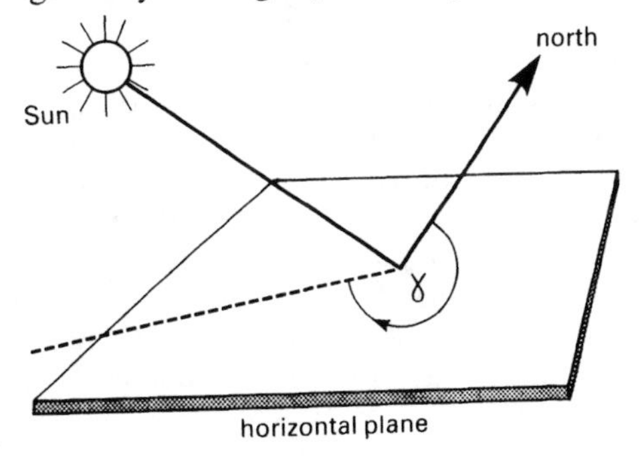

B

background radiation. Radiation originating from natural as opposed to man-made sources. The sources of background radiation are *(a)* cosmic, and *(b)* very long-lived radioactive nuclides in the earth's crust. Cosmic radiation, originating in outer space, consists of high-energy radiation and particles which are to some extent absorbed in the earth's atmosphere. The intensity of cosmic radiation consequently increases at high altitude, and aircrew in high-flying aircraft receive a significantly greater dose of this type of radiation than do people at ground level.

Very long-lived naturally occurring nuclides in the earth's crust include uranium-238, uranium-235, thorium-232, their daughter products (among which radium-226 and the inert gas radon-222 are particularly important) and potassium-40. Another naturally occurring source of radiation is carbon-14 (half-life = 5730 years) which is formed in the atmosphere by the interaction of cosmic radiation and nitrogen.

Radiation levels due to background radiation vary from one part of the earth to another, the average dose equivalent being about 0.002 *sievert/year. Abnormally high background radiation levels occur in granite (uranium bearing) regions or in cities with granite buildings (e.g. Aberdeen), in the Kerala district of India where the thorium-bearing Monazite sands give dose rates thirty times the average and in high-altitude areas (e.g. Mexico City, 2240 m, whose inhabitants receive over twice the average sea-level cosmic radiation dose).

DB

back pressure. Resistance to the flow of fluids, such as that produced by a constriction in a pipe, by surface roughness or by passage through a filter. It has the effect of reducing the pressure head of the flowing fluid and thus its rate of flow.

WG

bacteriorhodopsin. A purple pigment which is present in the salt-tolerant bacterium *Halobacterium halobium* and allows this organism to carry out *photosynthesis without *chlorophyll. The pigment is more stable than chlorophyll and, if integrated within a *photosynthetic artificial membrane, can set up an electric current through the transfer of electrons. Development of this system has possibilities for a novel form of electricity production sometime in the future.

CL

bagasse. The cellulosic residue of the sugar-cane plant after the sugar has been extracted. It constitutes about 50% of the cane stalk by weight and has a calorific value of 6.4–8.6 GJ/t. Bagasse may be combusted to raise steam for both the extraction and purification plant in a sugar mill and for ethanol concentration in an alcohol distillery, thereby improving the energy balance of the overall process. Excess bagasse is often utilized in particle board manufacture. World production is approximately 40 – 50 million tonnes per annum, mostly in Australia, Thailand, Indonesia, Brazil, the southern USA and the Caribbean countries.

CL

balanced budget. A situation where revenues match expenditures, used especially to refer to government revenues and expenditures. According to *Keynesian economists, a balanced budget is desirable only when the economy is operating at full employment. Otherwise the government should run a surplus or a deficit to restrain or stimulate economic activity via the *multiplier. According to *monetarist economists, a balanced budget is more generally desirable: deficits should always be avoided since they must either be financed by government borrowing, giving rise to the *crowding-out effect, or by increasing the *money supply, giving rise to *inflation.

MC

balance of payments. A record of a country's debits and credits with the outside world, arising from the flow of goods, services and capital between the country and the rest of the world. By definition total credits equal total debits, so that the sum of entries in the record is zero. Deficits and surpluses on the balance of payments refer to subsections of the record, and ambiguity arises if subsections are not clearly defined. The table shows, with hypothetical numbers, how the accounts are put together in most countries. The *visible trade balance is the difference between the *export and *import of goods; the *invisible trade balance is the difference between the export and import of services. The *current account balance is

Balance of payments for a country

Visible exports	1000	
Visible imports	1100	
Visible balance		– 100
Invisible credits	500	
Invisible debits	250	
Invisible balance		+ 250
CURRENT BALANCE		+ 150
Net capital flows		– 300
Balancing item		+ 50
BALANCE FOR OFFICIAL FINANCING		– 100
OFFICIAL FINANCING		+ 100

the visible plus the invisible balance, which is what is usually meant by a reference to the balance of payments. In the table the country has a surplus on its balance of payments.

Net capital flows relate to financial transactions which are not related to current movements of goods and services. For example, if a firm in the country buys an overseas coal mine this gives rise to an outward capital movement, while an inward movement would arise if an overseas firm bought an asset from a resident of the country. In the table, the net capital flow is outward from the country. The balancing item covers items not allocated elsewhere in the accounts, arising from over- or underrecording or from lack of information. The current balance plus net capital flows plus the balancing item gives the balance for official financing which is the change in the country's official foreign exchange reserves made necessary by the outcome of private transactions between inhabitants of the country and the rest of the world. Following the rise in energy prices in the 1970s, the balance of payments position of many countries was greatly affected by their trade in energy, especially in the form of oil. Countries which were dependent on imported oil had to meet the increased visible imports item in respect of oil and of visible and invisible export earnings, or by net capital inflows, or by running down their reserves of foreign exchange.

MC

ballast water (from oil tankers). After an oil tanker has discharged its cargo some oil remains adhering to internal tank, pump and line surfaces. Some of the dirty tanks have to be ballasted before the vessel can safely leave port, resulting in dirty ballast water. Other tanks have to be washed to remove accumulated sludge. In the past the dirty ballast water and tank washings were discharged to the sea causing pollution, but now most tankers operate the load-on-top (LOT) system. In this procedure all tank washings are collected in slop tanks; dirty ballast separates into aqueous and oily layers and the aqueous layer is pumped to sea while the upper oily layer is added to the slop tanks. When the contents of the slop tank have separated, the water layer is discharged to sea and the oily layer remains as slops oil. The next cargo of crude oil is then loaded on top of the slops oil, which is later discharged at the oil terminal along with the cargo.

PH

bank. An institution engaging in various financial activities based on the acceptance of deposits of money with it. Commercial, or (in UK usage) clearing, banks accept *current account deposits on which cheques can be drawn and make advances in the form of loans and overdrafts. There are other types of bank and financial institution which carry out some, but not all, of these functions. In the UK the distinctions between the various types of financial institution have become less marked in recent years. Thus, for example, whereas formerly building societies did not offer deposits on which cheques could be drawn, they now do. However, it is still true that building societies do not offer overdraft facilities on these accounts, and that they lend only for the purpose of house purchase. Again, whereas this type of lending was formerly exclusive to building societies, it is now the case that commercial banks and some other financial institutions will lend money long term for house purchase.

MC

banking. A technique of piling coal on a grate, whatever the size, to provide a supply of fuel during a period of slow burning when there is a low demand for heat or when labour is not available to tend the fires.

WG

bar. A unit of pressure equivalent to 1.0×10^5 *newtons per square metre.

WG

barn. The unit of neutron *cross-section. 1 barn is equal to 10^{-24} cm^2, i.e. 10^{-28} m^2.

DB

Barnwell nuclear fuel reprocessing plant. A *reprocessing plant whose construction, by Allied General Nuclear Services at Barnwell, S. Carolina, was started in 1971. It is designed to handle 1500 tonnes of spent nuclear fuel annually. It has not yet started operating, partly as a result of the Carter administration's decision in 1977 to defer indefinitely the reprocessing of spent nuclear fuel.

PH

barrel. A unit of volume popular in the oil industry, usually taken as 42 US gallons or 35 imperial gallons, i.e. 159 litres.

MS

barrier. A physical impediment to the movement of mass, such as water dammed for the purpose of *hydroelectrical generation, or to the transfer to radiation, as in the biological shield surrounding a *nuclear reactor.

MS

barter. A form of exchange in which goods and services are exchanged directly for other goods and services, without the intermediation of *money. It is found in primitive societies, sometimes in international trade (especially with the Communist bloc), and sometimes during periods of very high *inflation in advanced economies. To take place, barter requires the 'double coincidence of wants'—each party wanting the particular item offered by the other. Since the use of money avoids this requirement it facilitates exchange and trade and leads to its increase.

MC

base load. *See* load.

basin. The water catchment area behind a *dam or *barrier, especially for *tidal range power.

JWT

Battelle Memorial Institute. A research laboratory engaged in a wide range of nuclear and energy studies under US Government and industrial sponsorship. *Address:* 505 King Street, Columbus, Ohio 43201, USA.

DB

battery. A system in which stored chemical energy is converted directly into electrical energy. Batteries and *fuel cells are similar in operation, the major difference being that a battery contains a fixed quantity of fuel or chemical energy whereas a fuel cell operates with a continuous supply of fuel. A battery typically consists of one or more cells, each containing two electrodes – the anode and the cathode – separated by an electrolytic solution or matrix. Chemical reaction between the electrodes and the electrolyte causes the anode to build up a positive charge in relation to the cathode; as a consequence, when the electrodes are connected to the two ends of a circuit, an electric current flows.

Ultimately, the chemical reaction slows up and the battery becomes discharged. With certain types of battery, recharging is possible; an electric current is passed through the battery in the reverse direction thus effecting a reversal in the chemical reaction. *See also* accumulator.

TM

beam radiation. Solar radiation directly incident from the disc of the sun. It is direct solar radiation as contrasted with *diffuse radiation from the sky, clouds and nearby objects.

JWT

bear. A stock exchange expression for a participant who sells *securities in the belief that the price will fall, and then buys them back at the lower price. If the bear does not have the securities he sells, he is said to be 'selling short'; if he does, he is said to be a 'covered bear'. The very act of selling could cause the price to fall. *See also* bull; speculation.

MC

Beaufort scale. A wind-speed scale relating to visual observations of the effect of wind on sea (*see* table) but also used for the effect on vegetation, e.g. scale 7, moderate gale, wave height 4–6 m with streaks of foam.

		Wind speed at 6 metre height			
Beaufort scale	Description	(km/h)	(m/s)	(mph)	(knots)
0	calm	0–1	0–0.3	0–1	0–1
1	light air	1–5	0.3–1.5	1–3	1–3
2	light breeze	6–11	1.6–3.1	4–7	4–6
3	gentle breeze	12–19	3.2–5.4	8–12	7–10
4	moderate breeze	20–28	5.5–7.9	13–18	11–16
5	fresh breeze	29–38	8.0–10	19–24	17–21
6	strong breeze	39–49	11–13	25–31	22–27
7	moderate gale	50–61	14–17	32–38	28–33
8	fresh gale	62–74	18–20	39–46	34–40
9	strong gale	75–88	21–24	47–54	41–47
10	whole gale	89–102	25–28	55–63	48–55
11	storm	103–117	29–32	64–75	56–65
12	hurricane	118–	33–	76–	66–

becquerel (Symbol: Bq). The SI unit of *radioactivity:

$$1\ \mathrm{Bq} = 1\ \text{disintegration/second}$$

In terms of the previously used unit of radioactivity, the *curie,

$$3.7 \times 10^{10}\ \mathrm{Bq} = 1\ \text{curie}$$

DB

BEIR Committee. *See* Biological Effects of Ionizing Radiation Committee.

benefit. The total of the values (i.e. price times quantity) of the outputs, intended and incidental, of some economic activity. It will vary with the level of the activity. It should be compared with the *cost of the activity in order to determine the desirable level of the activity which gives the greatest excess of benefit over cost. Where there is no *market failure, benefit is the same as receipts from sales, and private and social benefit are the same.

In *project appraisal the benefit stream extends over time and has to be reduced to the present value of benefit using the *discount rate. If the project arises in the *public sector, or if there is market failure affecting its outputs so that they are not correctly priced by markets, a social project appraisal, or *cost benefit analysis, is called for. The construction of a dam for hydroelectric generation may create recreational opportunities (fishing, sailing) as well as electricity, in which case the value of these should be included in benefit along with the value of the electrical output.

MC

benzene. The basic aromatic hydrocarbon, C_6H_6, with a molecule consisting of six carbon atoms arranged in a hexagon and with one hydrogen atom attached to each carbon atom – the so-called benzene 'ring'. Such a structure contains three double bonds between the carbon atoms arranged alternately around the ring and it thus has a high chemical reactivity. Benzene is a volatile liquid at normal temperatures. The vapour is highly flammable.

Benzene may be obtained by the distillation of coal tar or by catalytic reforming reactions carried out in a petroleum refinery. It may be blended into gasoline since it has a high *octane rating. However its ability to dissolve rubber leads to problems with gaskets in pumps and fuel lines. Its gross *calorific value is 42.33 MJ/kg.

WG

benzine [petroleum spirit]. One of a number of highly volatile petroleum hydrocarbon distillation fractions with a short boiling range, e.g. 40–60°C, 60–80°C, 80–100°C, 100–120°C. It is highly flammable and finds use as a solvent.

WG

benzole. The first fraction (boiling range, 99–165°C) obtained in the distillation of *coal tar. It is a mixture of the aromatic hydrocarbons: benzene, toluene and the xylenes.

WG

benzopyrene [benzpyrene]. An *aromatic hydrocarbon present in coal tar, cigarette smoke and in the atmosphere as a result of incomplete combustion. It is a potent *carcinogen.

PH

Bergius process. *See* coal liquefaction.

beryllium. A light metallic element with certain properties that are useful in nuclear reactors. It has a low atomic weight and a very low neutron capture *cross-section, which make it apparently attractive for use in nuclear reactors as the *moderator, *reflector or fuel *cladding. However other characteristics, including high cost and toxicity, have prevented its use on a large scale. The oxide beryllia, BeO, has been used as the moderator and reflector in one or two reactors.

Symbol: Be; atomic number: 4; atomic weight: 9.01; density: 1850 kg/m³.

DB

beta particle. A high-energy *electron emitted in the decay of certain radioactive *isotopes, known as beta emitters, an example of which is cobalt 60:

$$^{60}_{27}\mathrm{Co} \rightarrow {}^{60}_{28}\mathrm{Ni} + {}^{0}_{-1}\mathrm{e}$$

Beta particles so emitted have a spectrum of energies rather than a single energy, and consequently they do not have a well-defined range in matter. The stream of beta particles, known as beta radiation, from beta emitters is a form of *ionizing radiation. Typically it can penetrate 3–4 metres of air, several millimetres of aluminium, or around a centimetre of biological tissue. Beta emitters are therefore hazardous to both external superficial tissues and to internal organs if they are ingested or inhaled.

PH, DB

betatron. A particle *accelerator designed to produce high-energy *electrons.

DB

Betz theory. The classical theory for explaining the efficiency of a *wind turbine using basic dynamics. Modern theories include the more complicated processes of vortex movement in the air passing across the blades, but the criteria for maximum efficiency as determined by Betz is still used as a standard. By Betz theory, the maximum power that can be extracted from wind is 59% of the power passing in the wind across the swept area of the turbine. *See also* cube factor.

JWT

bilge water. Dirty water that collects in a vessel's bilge. The bilge water from oil tankers usually contains an accumulation of oily waste which should be passed through an oil separator and disposed of on-shore. Sometimes this dirty bilge water is discharged at sea.

PH

bill [short-term paper]. A short-term, usually up to six months, debt instrument or *security. It orders the drawee (the debtor) to pay the drawer (the creditor) a stated sum on a particular date, or sometimes to pay on demand. Bills are issued by private companies and by governments.

MC

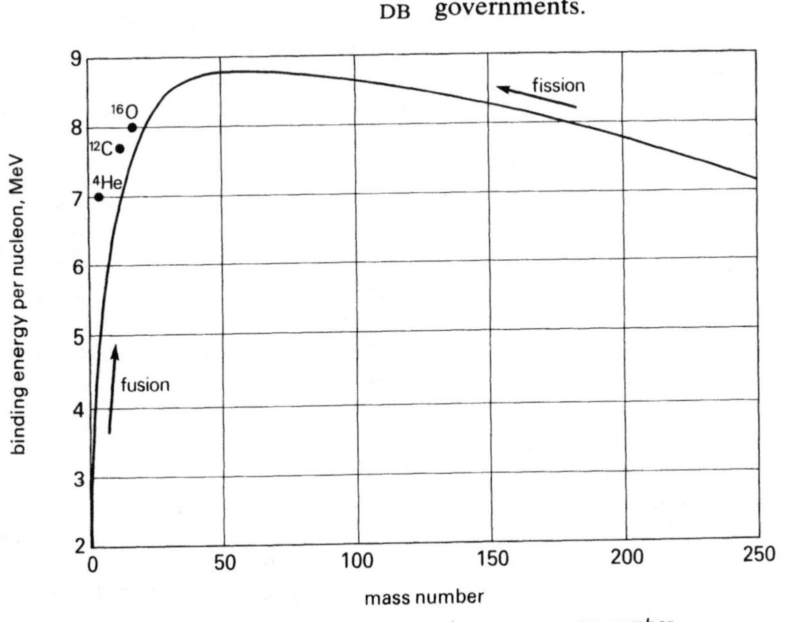

binding energy *Curve of binding energy per nucleon versus mass number*

binding energy (nuclear). The energy equivalent of the difference in mass between an atom or an atomic *nucleus and its constituent particles. For example, the mass of an atomic nucleus is less than the sum of the masses of its constituent protons and neutrons. This difference is called the mass defect of the nucleus and is given by the equation:

$$\text{mass defect} = Z \times \text{proton mass} + (A - Z) \times \text{neutron mass} - \text{mass of nucleus}$$

where Z is the *atomic number and A is the *mass number.

According to *Einstein's equation, $E = mc^2$, the binding energy is equal to the mass defect times c^2, where c is the velocity of light. The binding energy of an atomic nucleus can be regarded as the energy that would be required to split it into its constituent particles, or alternatively as the energy released when the nucleus is formed from its constituent particles. The binding energy per *nucleon of any nucleus is equal to the binding energy divided by the mass number.

A plot of binding energy per nucleon against mass number for all known nuclides shows a curve of well-defined shape (*see* figure). A study of this curve reveals that the *fusion of very light nuclides and the *fission of very heavy nuclides are (provided these reactions are possible) *exothermic. This is because the binding energy per nucleon of the products of these reactions is greater than the binding energy per nucleon of the original nuclides.

DB

bing. A large man-made deposit of solid fuel or mine refuse arranged on the ground as a mound or hillock. After many years it can weather into the shape of a small mountain and may support vegetation.

WG

bioconversion. *See* biological energy conversion.

biodegradation. The breakdown of materials by means of biological processes. In natural decay processes dead plant material is broken down by soil micro-organisms to form carbon dioxide, ammonia, water and humus. Controlled decay under *anaerobic conditions is the basis for *biogas production, in which organic wastes such as animal manure are degraded by micro-organisms in a digester to produce *methane gas. Biodegradation plays an important role in the removal of petroleum from the marine environment after oil spills: certain species of marine bacteria, fungi and yeasts break down petroleum hydrocarbons by utilizing them as a food source. It is also significant in the corrosion of oil-exploration and production platforms in the marine environment.

PH

bioenergy. (1) Solar energy captured by plants during *photosynthesis and used in the creation of carbon compounds, usually *carbohydrates.

(2) The energy content of a *biofuel, e.g. ethanol, which has been produced from the *fermentation of a substrate such as *sugar-cane in which the solar energy was initially captured.

CL

biofertilizers. Substances used to enrich soil but derived from a biological rather than a synthetic process. They may be byproducts of the *anaerobic digestion of organic matter, nitrogen-fixing *algae, or *legumes. Compounds of nitrogen, phosphorus and potassium are the main constituents, with most being especially rich in nitrogen.

CL

biofuels. Fuels derived from organic plant matter, i.e. *biomass. They include primary sources such as wood as well as derivative fuels like *ethanol, *methanol and *biogas. The derivative fuels themselves originate from primary production after undergoing a *biological energy conversion such as fermentation or anaerobic digestion.

CL

biogas. The fuel produced following the microbial decomposition of organic matter in the absence of oxygen. It consists of a gaseous mixture of methane and carbon dioxide in an approximate volumetric ratio of 2:1. In this state the biogas has a *calorific value of around 20–25 MJ/m^3 but this can be upgraded by removing the carbon dioxide. Sewage and human and animal manure are the most common organic wastes used for biogas production, though waterweeds, algae and agricultural residues such as straw can also be inputs. Biogas is a clean-combusting fuel that can be used for cooking, space and water heating, lighting and providing mechanical power in agricultural machinery and water pumps and as a generator of electricity.

Two major classes of bacteria are involved in biogas production: non-methanogenic organisms, which hydrolyse and ferment the raw-material inputs to (chiefly) organic acids such as acetic acid, and *methanogenic bacteria, which complete the conversion to biogas as follows:

$$CH_3COOH \rightarrow CH_4 + CO_2$$

Biogas offers an alternative fuel source to wood, and so can ease the fuelwood scarcity in many developing countries, lessening the need for firewood energy, which in turn reduces the possibility of deforestation and soil erosion.

CL

biogas generator. *See* Gobar gas plant.

biological concentration. The means by which certain materials are concentrated by biological process as they move from stage to stage through a *food chain in the aquatic or terrestial environment. For example, radioactive ruthenium-106, which is released into the Irish Sea from the nuclear-fuel reprocessing plant at Sellafield, is taken up from the sea water and concentrated by the edible seaweed *Porphyra*. It is accumulated to a level some 1500 times as great as the concentration prevailing in the sea water, i.e. with a concentration factor of 1500. A second radionuclide released from Sellafield, caesium-137, is taken up from the sea water by phytoplankton and zooplankton; these are consumed by certain species of fish, resulting in a concentration factor of about 40.

PH

Biological Effects of Ionizing Radiation (BEIR) Committee. The Committee established by the Division of Medical Sciences of the US National Research Council. It has prepared three reports (latest 1980) which have reviewed the scientific basis for the establishment of radiation protection standards. *See* biological protection.

PH

biological effects of radiation. Effects that have been studied for many years in radiation-accident victims, survivors of the atomic bombs at Hiroshima and Nagasaki, patients undergoing radiotherapy, persons exposed to radiation in their work, and in animal experiments. These biological effects can be broadly classified as somatic, occurring within the exposed individual, and genetic (hereditary), affecting the descendants of the exposed individual. Exposures can be classified as acute (brief) and chronic (continuous).

An acute exposure of the whole body to a radiation dose of thousands of *greys produces almost immediate effects, known as the acute radiation syndrome. These usually include nausea, vomiting, loss of hair, severe blood changes with haemorrhage and increased susceptibility to infection. Acute exposure to a small dose of radiation or continuous exposure to low levels may result in delayed (late) somatic effects.

The most apparent late somatic effects are an increase in the incidence of *leukaemia and other types of cancer, especially tumours of the lung, thyroid, breast and bone. The time between the exposure to a dose of radiation and the appearance of the cancer is called the latent period. The average latent period is about 15 years, but may range from a year or two for leukaemia to more than 30 years for bone cancer. The UNSCEAR estimate for the *risk factor for leukaemia is about 1 in 500 per *sievert. The foetus is especially sensitive to radiation: exposure *in utero* during the early weeks of pregnancy may result in developmental defects, while exposure later in pregnancy increases the risk of early childhood malignancies, especially leukaemia.

The genetic effects of ionizing radiation arise from the *mutations induced in the *DNA of germ cells (eggs and sperm). These changes to the genetic material are usually considered to be detrimental and will be transmitted to any offspring, in whom their effects may range in severity from inconspicuous to lethal and may not be revealed for many generations. The UNSCEAR estimate for the risk factor for severe genetic damage to humans is about 1 in 50 per sievert when all generations subsequent to the irradiation are taken into account. About half the damage would be expressed in the first two generations.

PH

biological energy conversion [bioconversion]. The overall process in which organic plant matter is produced, collected and either converted to or used as a fuel. In the strict sense the term refers only to conversions mediated by microorganisms, such as *fermentation to alcohol and *anaerobic digestion to methane. Generally the thermochemical conversion routes of gasificaticn and *pyrolysis are also included within this definition, as is the combustion of biomass as a means of electricity generation.

CL

biological oxygen demand (BOD). A measure of the amount of oxygen required for the *biodegradation of a sample of water.

CL

biological shield. The radiation barrier surrounding the core of a *nuclear reactor. It is designed to prevent the escape of *neutrons and *gamma radiation from the core into the surrounding space and to reduce their intensity so that they are no danger to human health. In large power stations the biological shield is concrete a few metres thick. In ships and submarines it is a more compact assembly of layers of steel, lead and water.

DB

biomass. A mass of naturally grown biological material. The term is usually used in the context of useful energy or material sources. *See* biofuels.

JWT

biome. *See* ecosystem.

biophotolysis. The biological splitting of water molecules into hydrogen and oxygen in the presence of light. This experimental process has been accomplished by living organisms such as the blue–green alga **Anabaena cylindrica* under conditions of nitrogen starvation: in the absence of molecular nitrogen the enzyme *nitrogenase liberates hydrogen from water instead of beginning the usual reactions of *photosynthesis. Hydrogen has also been evolved by adding extracted *hydrogenase enzyme to isolated plant chloroplasts as well as to *photosynthetic artificial membranes. The latter are more stable than living systems and generate a potential difference through the initial transference of electrons, thereby developing an electric current. As yet, yields from such *photobiological energy-production systems have been low but are increasing as advances in the science are made. Thus, in principle, there is a route to producing hydrogen gas as a fuel directly from solar energy.

CL

biosphere. That part of the earth where biological life exists and continues. It extends about 10 m down into the soil, about 100 m into the sea or fresh water, and into the *atmosphere about 300 m above land or water surfaces. *See also* ecosystem.

JWT

biota. The sum total of living organisms inhabiting a specific region.

CL

bitumen. A heavy dark-coloured viscous or semi-solid hydrocarbon residue from the distillation of certain crude oils. It is also found as a natural resource in Colorado and Utah and in the form of a pitch lake in Trinidad. The natural material may contain up to about 30% finely divided mineral matter. It is used for waterproofing and for surfacing roads, where the mineral matter imparts skid resistance.

WG

bituminous coal. Black banded coals covering a wide range of characteristics and thus requiring further definition as they may be either *caking or non-caking in character. The carbon content ranges from 75–90% associated with volatile matter from 45–20%. Over this range there is a marked variation in caking power. The bituminous coals are the general-purpose coals of industry and of the household. The gross *calorific value on a dry ash-free basis will generally lie within the range 31–35 MJ/kg.

WG

black body. A notional surface or structure that totally absorbs all light and other radiation falling on it. Therefore there is no reflection and the surface will appear black at visible and all other frequencies of the incident radiation. By *Kirchoff's law a black-body surface does not reflect any radiation but it does emit its own radiation: it has, by definition, an *emissivity of unity. It follows that no other type of surface can emit more strongly at the same temperature. Thus a good absorber of radiation also becomes a good emitter of radiation. For instance a (near) black-body solar-collector panel will lose energy by infrared radiation emitted from the surface. *See also* selective surface.

JWT

blackdamp. A non-flammable mixture of gases, predominantly carbon dioxide and nitrogen, found in old coal mine workings. The oxygen of the air reacts over a long period with oxidizable material in the coal forming the above mixture of inert gases. In conditions of poor ventilation these gases may form a layer close to the floor of the tunnel so that it is safe to walk through but not to lie down.

WG

blackout. The loss of electric light, etc., resulting from the cut-off of power from an electrical supply.

MS

blade. The part of a turbine or propeller that interacts with the moving fluid, air or water. *See* airfoil.

JWT

blanket (nuclear). *Fertile nuclear material surrounding the core of a *breeder reactor. The blanket is composed of uranium-238 or thorium-232. Some of the neutrons produced by *fission in the core escape into the blanket, where they are captured in the fertile material to produce new fissile material. For example, uranium-238 becomes transformed into the fissile element plutonium-239. *See also* breeding.

DB

blast furnace gas. Waste combustible gas generated in a blast furnace when iron ore is being reduced with *coke to metallic iron. The gas contains high proportions of nitrogen and carbon monoxide and smaller amounts of carbon dioxide and hydrogen. It has a low *calorific value, typically 3.7 MJ/m^3, and is utilized as a fuel within the steel works.

WG

blowout preventer. Equipment installed at an oil well-head for the purpose of controlling pressures in the annular space between the casing and drill pipe, or in an open hole during drilling operations.

PH

blue–green algae. Organisms which are dependent on light as their energy source and on carbon dioxide as the main source of carbon. They may be unicellular, helicoidal or filamentous. Some are capable of *nitrogen fixation and can thus act as *biofertilizers. *Anabaena cylindrica* has been used to produce hydrogen in *biophotolysis. However, their energy-related role at present is more as a conserver than as a supplier by virtue of their use as biofertilizers, substituting for energy-intensive chemical fertilizers in agriculture. Important genera are *Anabaena, Nostoc, Oscillatoria* and the high-protein *Spirulina*.

CL

blue water gas. Gas generated by blowing steam through red-hot coke, forming a mixture of carbon monoxide and hydrogen. Since, however, reaction is not complete, carbon dioxide and water vapour are also present. It burns with a blue flame. Its manufacture was previously associated with high-temperature coal carbonization for producing town gas since it provided a convenient way of using unsaleable coke. However, its low *calorific value, 11.2 MJ/m^3, makes it unsuitable for blending with town gas without further enrichment. *See also* carburetted water gas.

WG

boghead coal. A type of coal which is similar in appearance to *cannel coal and may contain as much as 10% hydrogen. Microscopic analysis shows that it contains spores and numerous round flattened yellow bodies which are considered to be the remains of *algae.

WG

boiler. Any of a range of devices in which heat is transferred from a primary source (typically by the combustion of fossil fuel) to a medium (typically water), which is then distributed to the point of end use or to the point of transfer to a secondary medium. The term was originally specific to a vessel in which water was boiled in order to raise steam for use in *central heating systems.

TM

boiling point. That temperature at which a liquid becomes vapour, i.e. when its vapour pressure is equal to that of the surrounding *atmospheric pressure.

MS

boiling water reactor (BWR). *See* nuclear reactor.

Boltzmann constant. *See* Stefan–Boltzmann constant.

Boltzmann law. *See* Stefan–Boltzmann law.

bomb calorimeter. A laboratory apparatus which measures accurately the *calorific value of a fuel. A small measured weight of a solid or liquid fuel is placed within a closed stainless-steel pressure vessel which is then charged with oxygen gas to a pressure of at least 25 atmospheres. The vessel is immersed in water in an insulated tank and the fuel ignited electrically. The resultant rise in temperature of the surrounding water is measured. From a knowledge of the weights and specific heats of the various parts of the system, the gross calorific value of the fuel may be calculated.

WG

bond. A *security or debt instrument. As used originally the term denoted a fixed-interest security with payment of such made regularly until the maturity date (in the case of a redeemable bond) or in perpetuity (on a non-redeemable bond), which could be traded in the *capital market, especially on the *stock exchange. Nowadays the term is sometimes used for securities which do not carry fixed interest and which cannot be traded.

MC

bore. The size, usually the diameter, of a cylinder tube or circular hole.

JWT

boron. An important element in nuclear engineering due to the very high neutron capture *cross-section of its naturally occurring *isotope boron-10. Boron, enriched in this isotope, is used as a reactivity control material in many *nuclear reactors. In some designs the boron is alloyed in steel, in others the compound B_4C is compacted in steel tubes. In the form of boric acid it can be used as a *burnable poison for the long-term control of *reactivity in pressurized water reactors, the boric acid being dissolved in very dilute solution in the reactor water.

Another important use of boron is in neutron-detecting instruments. Neutrons cannot cause *ionization directly but they produce *alpha particles by the reaction

$$^{10}_{5}B + ^{1}_{0}n \rightarrow ^{7}_{3}Li + ^{4}_{2}He$$

The alpha particles can be detected and measured by the ionization which they cause in *gas-filled radiation detectors. The boron in neutron detectors is enriched in boron-10 and is either coated on the electrodes of the detector or used as the gas boron trifluoride to fill the detector.

Symbol: B; atomic number: 5; atomic weight: 10.81; density: 2300 kg/m^3.

DB

bottled gas. Hydrocarbon gas which can be liquefied under pressure. It is normally sold in light cylindrical pressure vessels (bottles) of steel or aluminium. The gas may be propane or butane.

WG

boundary condition. The laminar or turbulent characteristics of fluid flow in the fluid layer closest to the solid surface over which the flow is taking place. These characteristics affect the *heat transfer by *convection between the fluid and the solid.

TM

Boyle's law. A relation between pressure (P), volume (V) and temperature (T) of a gas: such that

$$PV = nRT$$

where n refers to the number of *mols of the gas and R is the *universal gas constant. In fact the relation applies only to 'ideal' gases, that is to say gas at comparatively low pressure. *See also* Charles' law.

MS

branched-chain hydrocarbons. Hydrocarbons in which at least one of the carbon atoms in the molecules is linked to three other carbon atoms. Iso-octane is a branched chain molecule with the formula

$$CH_3C(CH_3)_2\,CH_2\,CH(CH_3)_2$$

WG

Brayton cycle. *See* heat engine.

breeder reactor. A type of *nuclear reactor designed to produce more fissile fuel than it consumes, and thus have a *breeding ratio greater than one. In general, the breeding ratio of any nuclear reactor tends to increase as the energy of the neutrons causing *fission increases; consequently fast reactors are more effective as breeders than thermal reactors (*see* nuclear reactor). Fast reactors using uranium-233, uranium-235 and plutonium-239 as the fissile fuel can all be designed as breeder reactors, but the only type of thermal reactor that can act as a breeder is one fuelled with uranium-233. Breeder reactors, by converting the non-fissile but *fertile isotopes uranium-238 and thorium-232 to fissile plutonium-239 and uranium-233, considerably extend the potential of nuclear fuels as an energy resource.

DB

breeding. The process of converting non-fissile material to *fissile nuclear fuel. The non-fissile but *fertile isotopes uranium-238 and thorium-232 are converted to the fissile isotopes plutonium-239 and uranium-233. These processes are initiated by neutron capture (*see* absorption) in the fertile isotopes, and take place in nuclear reactors. The sequences of reactions are as follows:

$$^{238}_{92}U + ^{1}_{0}n \rightarrow ^{239}_{92}U$$

$$^{239}_{92}U \rightarrow ^{239}_{93}Np + ^{0}_{-1}e$$

$$^{239}_{93}Np \rightarrow ^{239}_{94}Pu + ^{0}_{-1}e$$

and

$$^{232}_{90}Th + ^{1}_{0}n \rightarrow ^{233}_{90}Th$$

$$^{233}_{90}Th \rightarrow ^{233}_{91}Pa + ^{0}_{-1}e$$

$$^{233}_{91}Pa \rightarrow ^{233}_{92}U + ^{0}_{-1}e$$

The plutonium-239 and uranium-233 produced as the result of these reactions are not only fissile but are also radioactive, with very long *half-lives. *See also* breeder reactor; breeding ratio.

DB

breeding ratio. A parameter describing the effectiveness of a *nuclear reactor for producing new *fissile material. The breeding ratio of a nuclear reactor is defined as the number of atoms of new fissile fuel produced per atom of fissile fuel used in the reactor. When this is greater than one, then the reactor is producing more fissile fuel than it is consuming. The breeding ratio depends on the energy of the neutrons causing fission, and is in general higher for *fast neutrons than for *thermal neutrons. *See also* breeder reactor; breeding; conversion ratio.

DB

breeze. *See* coke breeze.

bremsstrahlung (from the German: braking radiation). Electromagnetic radiation produced as the result of the interaction between high-energy *electrons and atomic nuclei and electrons.

DB

Bretton Woods Agreement. An agreement, made at Bretton Woods, New Hampshire, USA in 1944, which set out the system by which deficits and surpluses on the balance of payments were to be financed and the type of world *liquidity that would be available. It was based on a system of fixed exchange rates, except for variations of 1% either side. The rate could be altered when a country was in 'fundamental disequilibrium' (never defined by the IMF). The major institutions of the system were the

*International Monetary Fund and the *International Bank for Reconstruction and Development. The system initially broke down in 1971 and was salvaged by the *Smithsonian Agreement, but finally broke down in 1973 when European countries initiated a joint floating exchange-rate system.

MC

bright coal. A coal which contains bands of shiny minerals such as *vitrain or *clarain.

WG

briquettes. Conveniently sized shapes of solid fuel prepared from small-sized coal or coke particles bound with bitumen or pitch and moulded in a heated press.

WG

British Coal. The present (1987) name of the nationalized UK coal industry, formerly known as the Coal Board.

Address: Hobart House, Grosvenor Place, London, SW1X 7AE.

MS

British Nuclear Energy Society (BNES). A society established to provide an independent forum for the discussion and presentation of papers and lectures on nuclear energy.

Address: 1–7 Great George Street, London, SW1P 3AA.

DB

British Nuclear Forum (BNF). An industrial association which exists to coordinate the activities and promote the interests of the nuclear-energy industry in the UK.

Address: 8 Leicester Street, London, WC2H 7BN.

DB

British Nuclear Fuels plc (BNFL). An organization which operates factories at Capenhurst, Cheshire for uranium enrichment; at Springfields, Lancashire for the production of uranium fuel elements and uranium hexafluoride, and at Sellafield, Cumbria for the reprocessing of irradiated nuclear fuel and the production of irradiated nuclear fuel and the production of plutonium for the UK and other countries. It also operates the plutonium and power-producing reactors at Calder Hall, Cumbria and Chapelcross, Dumfries.

Address: Risley, Warrington, Cheshire, England.

British Standards Institution (BSI). An organization which lists as its objects 'To coordinate the efforts of producers and users for the improvement, standardization and simplification of engineering and industrial materials, . . . to set up standards of quality and to promote the general adoption of British Standards'. The Institution publishes over 5000 British Standards.

Address: British Standards House, 2 Park Street, London, W1A 2BS.

WG

British thermal unit (BTU, Btu). The amount of heat energy necessary to raise the temperature of one pound weight of water from 39.2 to 40.2°F. The preferred SI unit of energy is the *joule (J); one BTU equals 1.05506×10^3J.

TM

British Wind Energy Association (BWEA). An organization primarily of engineers and scientists who are professionally involved in wind energy research, development and commerce. Conferences, technical workshops and publications are arranged.

Address: Central Electricity Generating Board, Laud House, 20 Newgate Street, London EC1A 7AX.

JWT

Brookhaven National Laboratory. A laboratory which is operated by Associated Universities Inc. under contract to the US Department of Energy and carries out basic and applied research in nuclear energy and other fields.

Address: Upton, Long Island, New York, NY 11973, USA.

DB

brown coal. A brown form of the poor-quality coal, *lignite. It may be further subdivided into fibrous brown coal having the structure of wood, earthy and friable brown coal, and dark brown coal with a slight woody structure and slaty cleavage. These are characterized by an 'as mined' moisture content of 45–55%, and a carbon content of under 75% with oxygen over 20% on a dry ash-free basis. The calorific value on a dry ash-free basis can vary from 20–26 MJ/kg. The need to remove most of the contained moisture before burning reduces the attraction of brown coal as a fuel. It may however be hydrogenated (*see* hydrogenation) to obtain liquid fuels. WG

brown-out. A US term indicating a lowered voltage of electrical supply during times when *peak load exceeds peak capacity of the generating system.

MS

buffer stock. An *inventory of a *commodity held by an intervention agency, usually under a *commodity agreement, so that sales from and purchases for the buffer stock can be used to even out price fluctuations. The size of the stock held depends upon the costs of storage and finance in relation to anticipated price movements. A buffer stock is useful only for short-term price stabilization; it is not effective against long-term trends in a commodity price. Agreements to hold stocks of oil are intended to meet supply interruptions rather than to stabilize oil prices, and cannot affect the long-term trend in oil prices. MS

buildings, energy usage. One of the main functions of buildings is to provide shelter from the *climate. In modifying climatic extremes to provide an acceptable internal environment, buildings currently consume, in Europe for example, approximately 51% of the total primary energy.

TM

bulk supply tariff. The price structure where large supplies are sold at one tariff for distribution at another. For example, in the electricity supply industry in England and Wales, the price structure for sales of electricity from the *Central Electricity Generating Board to the 12 area boards which sell electricity to industrial and domestic consumers according to their own retail tariffs. MC

bull. A *stock-exchange expression for a participant who buys *securities now in the belief that the price will soon rise, when the securities can be sold and a profit made. Buying on a large scale may cause the price to rise. *See also* bear; speculation.

MC

bunker. A storage container for solid fuels or a compartment on board ships for storing fuels.

WG

Bunker B, Bunker C. *See* heavy fuel oil.

bunker oils. Very heavy fuel oil burnt in the steam-generating plants of large ships and in some land-based installations.

WG

burnable poison [burnable absorber]. A means of controlling the *reactivity and neutron *flux distribution in *nuclear reactors. One method, applicable to pressurized water reactors, uses *boron in the form of boric acid (H_3BO_3) in very dilute solution in the reactor water to capture *neutrons and control excess reactivity in a newly fuelled reactor. As operation of the reactor proceeds the boron is eliminated, or burnt up, by capturing neutrons and its 'negative reactivity' effect decreases, thus compensating for the reactivity changes due to the burn-up of the fuel. In another method, applicable to pressurized water reactors, boiling water reactors and gas-cooled reactors, the poison in the form of boron carbide (B_4C) or gadolinia (Gd_2O_3) is introduced into the reactor either incorporated in the fuel elements or in separate tubes to control reactivity.

DB

burner. A device designed to burn a fuel under controlled conditions; a jet from which a flame issues. It is used primarily for fluid fuels and comprises an orifice through which the fuel is injected and a means for the efficient mixing of the fuel stream with the air necessary for combustion.

The various types of gas burner are differentiated by the extent to which air is mixed with the gas before combustion occurs. In the Neat gas burner little or no mixing occurs: air diffuses through the flame front to react with the gas. In the aerated burner about two-thirds of the air required for combustion is induced to mix with the gas before combustion takes place, and the remainder diffuses through the flame front, as in the Bunsen burner. In the premixed burner all the air required is mixed with the gas before combustion occurs.

Oil burners are characterized by the means used to disperse the oil as droplets (atomization). This may be achieved by vaporizing in which the preheated light oil rises up a wick and burns as a vapour above it, or by means of a pressure jet, in which the oil is squirted under high pressure from a specially designed orifice, or by mechanical means, whereby the oil flows into whirling blades which disperse it as a cloud of droplets. A further alternative is steam or air atomizing, in which a jet of the oil under relatively low pressure is mixed with a steam or air jet as it leaves the orifice.

WG

burner reactor. A *nuclear reactor which, not being designed for *breeding, uses more fissile material than it produces during the course of its operation. All existing thermal reactors, particularly those fuelled with enriched *uranium, come into this category.

DB

burning oil. An alternative name for *kerosene (paraffin), the light petroleum distillate burnt in domestic heaters and in many central heating systems. It also finds use for start-up and flame stabilization in large coal-fired steam-generating stations.

WG

burnup. The total energy released by nuclear fuel during its time of residence in a *nuclear reactor. The burnup is usually measured in units of MW d/t. Typical values are:

Magnox reactors	3500 MWd/t
advanced gas cooled reactors	18,000 MWd/t
fast breeder reactors	50,000 MWd/t

DB

busbar. A main electrical connection point in a power station or subsidiary network.

JWT

business cycle [trade cycle]. The more or less regular fluctuations about the long-term trend in the level of economic activity. The usual economic indicator considered is the level of *national income, and a typical representation is shown in the diagram. The amplitude of the cycle is the difference between national income at the upper turning point B and the lower turning point D. The period of the cycle is the time elapsed in a complete cycle, i.e. from A to E. The phase from B to D is the contraction or recession, and that from D to F is the expansionary phase. Some business cycles have longer contractionary than expansionary phases, and vice versa. The pattern of business cycles varies across countries and over time in any one country: the variation is so great that not all economists agree that they are a useful concept.

MC

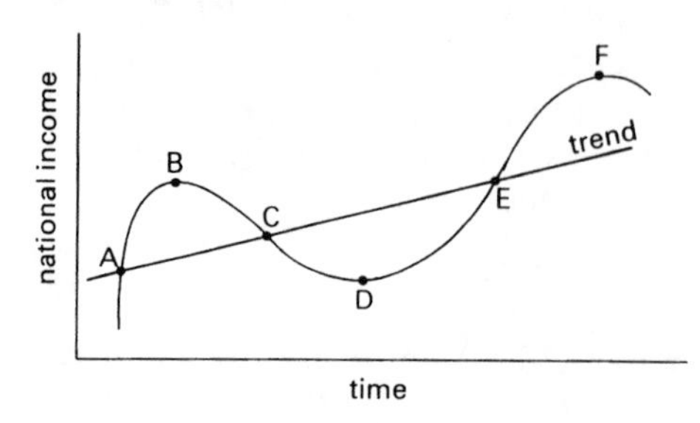

butane. A flammable gaseous hydrocarbon, C_4H_{10}, of the *alkane series which occurs in some crude oils and from which it may be separated by distillation under pressure. At ambient temperature it can be liquefied by pressure. It is sold as bottled gas for domestic and industrial use. Since the boiling point of the liquid is 0°C, gas cannot be obtained from a cylinder below that temperature. Butane cylinders provide a very convenient portable supply of a gaseous fuel, and are used extensively with caravan and boat heaters.

Butane exists as two different molecular structures (i.e. *isomers): n-butane, in which the four carbon atoms are arranged in a line, and iso-butane, which has a *branched chain. The gross *calorific value of n-butane is 51.38 MJ/kg while that of iso-butane is 51.23 MJ/kg.

WG

butene [butylene). A flammable gas which is an unsaturated hydrocarbon, C_4H_8, of the *alkene series with one double bond in the molecule. It is frequently a constituent of *refinery gases. It has a gross *calorific value of 50.32 MJ/kg.

WG

C

Cadarache. A French research centre devoted to industrial nuclear research mainly in the areas of electricity generation and propulsion. It also carries out fundamental research in the fields of radiological safety, radioecology and radioagronomy.

Address: B.P.I, 13115 Saint-Paul-les-Durance, Bouches-du-Rhône, France.

PH

cadmium. An element which is of importance in nuclear engineering because of its very high neutron capture *cross-section, which makes it suitable as a material for control rods (*see* control systems) in *nuclear reactors.

Symbol: Cd; atomic number: 48; atomic weight: 112.4; density: 8650 kg/m^3.

DB

caesium [US: cesium]. A metallic element which is of importance in nuclear energy because its radioactive form, caesium-137, is a *fission product in *nuclear reactors. Its *half-life of 30 years makes it one of the most hazardous of the fission products from the point of view of the storage and disposal of *radioactive waste.

Symbol: Cs; atomic number: 55; atomic weight: 132.9; density: 1900 kg/m^3.

DB

caking character. The behaviour of a coal when carbonized and the properties of the coke when formed. There would appear to be no internationally agreed test which quantifies caking character. Tests have been devised to measure the expansion or shrinkage during the transformation of coal to coke, the mass, density and hardness of the coke formed and its resistance to abrasion. Generally, for routine testing of coals, standard samples of coke are prepared and qualitative comparisons made. *See also* Gray–King assay.

WG

caking test. A laboratory test to determine the ability of a sample of powdered coal to form an adherent mass or 'cake' of coke when carbonized by rapid heating in a covered crucible. The test may be used to compare coals rather than to predict their behaviour in a coke oven where the rate of heating is much lower.

WG

calandria. A device for facilitating exchange of heat between two fluid streams. In chemical engineering it is part of an evaporator consisting of a bundle of tubes through which the heating fluid flows. In certain types of *nuclear reactor, namely CANDU and the steam-generating heavy-water-moderated reactor (SGHWR), the *core vessel is called a calandria. It is a closed cylindrical vessel of stainless steel with a large number of *Zircaloy calandria tubes passing through it parallel to its axis. Inside the calandria tubes, and thermally insulated from them by annular gas gaps, are the Zircaloy pressure tubes which contain the *fuel elements. The *moderator (heavy water) is contained in the calandria at low pressure and temperature, and the *coolant (heavy water in the case of CANDU, light water in the case of the SGHWR) passes through the pressure tubes at high pressure and temperature.

DB

Calder Hall. The site adjacent to *Sellafield on which the first large UK nuclear reactors were built in 1956 for the production of electricity as well as of plutonium. The power station, consisting of four reactors, was operated by the UK Atomic Energy Authority, and more recently British Nuclear Fuels Limited. At the time of its commissioning in 1956 it was the first large nuclear power station (160 MW) built in the world which supplied electricity for national use, although in 1954 a much smaller Russian power reactor was commissioned.

The name Calder Hall has become generic for the gas-cooled, graphite-moderated reactors with Magnox clad uranium fuel elements which have been developed from the original Calder Hall reactors. In the decade from 1960 to 1970 nine electricity power stations were built in the UK with reactors based on this design with a total generating capacity of 4.8 GW. *See also* nuclear reactor.

DB

calorie (cal). A unit of heat energy equal to the amount of heat required to raise the temperature of one gram of water through 1 °C. The preferred SI unit is the *joule (J); one calorie equals 4.184 J.

TM

calorific value. The amount of heat evolved when unit mass of a fuel is burned completely under standard conditions.

Two values are recognized: the gross (or higher) calorific value, GCV, is the amount of heat evolved in a *bomb calorimeter when the products of combustion are cooled to ambient conditions (25°C) and water vapour has condensed, evolving its latent heat; the net (or lower) calorific value is calculated by deducting the latent heat of condensing water from the gross value. The net calorific value is considered to reflect more accurately the heat evolved under operating conditions. However it will be only an estimate unless the concentration of water vapour in the chimney gases is known accurately. The calorific value of a gas may be stated per unit mass of the gas, or per unit volume at standard temperature and pressure.

WG

calorifier. a type of *heat exchanger used in the space heating and hot-water supply systems of buildings. Typically, high-pressure hot water or steam from the *boiler plant passes through a tube bundle around which water for space heating or hot-water supply circulates. In the latter case the calorifier may be sized to provide a reservoir of hot water to meet the peak demands made by building users.

TM

Calvin–Benson cycle. The reaction sequence, occurring in most plant chloroplasts, which converts light energy into chemical energy during *photosynthesis. It occurs in four phases, involving mainly 3-carbon compounds, and results in the conversion of carbon dioxide in the air to sugars in the plant. The first or carboxylation phase consists of the addition of carbon dioxide to the 5-carbon sugar, ribulose diphosphate, forming two molecules of phosphoglyceric acid. A reduction phase then reduces the carbon dioxide to the energetic level of a 3-carbon sugar, triose phosphate, thus accomplishing the energy-conserving portion of photosynthesis. The third or regenerative phase regenerates ribulose diphosphate for further carbon dioxide fixation reactions, and finally comes the product synthesis phase for the production of carbohydrates, fats, fatty acids, amino acids, etc. This overall cycle forms the basis for primary production (*see* primary productivity) in the predominant *3-carbon plants, and hence is vital to most global *bioenergy supplies.

CL

Cambell–Stokes recorder. An instrument for recording periods of direct bright sunshine. A glass ball is mounted so as to focus the sunlight onto a calibrated strip of paper, on which a charred track is made when there is direct radiation from unclouded sunshine. JWT

candela (Symbol: cd). The SI unit of luminous intensity, of the same order as that provided by an ordinary candle. It is defined as the light flux, measured in *lumens, over the solid angle of flux emission.

TM

candu. *See* nuclear reactor.

cannel coal. A hard coal with a matt surface which breaks with a concoidal fracture. Having a high hydrogen content, suitably shaped pieces may be induced to burn like a candle. It is composed of unusually high concentrations of spores and other plant remains which have suffered relatively little structural change. Its hydrogen content, in excess of 6%, distinguishes it from *bituminous coal of similar carbon content. On carbonization, cannel coal gives extremely high yields of liquid hydrocarbons. It has been considered in the past to be a valuable raw material for a low-temperature carbonization process in which fuel oil and a solid smokeless fuel could be produced simultaneously. A typical gross *calorific value on a dry ash-free basis is 35 MJ/kg. WG

canopy cover. A numerical description of the extent to which the branches and leaves of plants cover a sunlit area. In the case of land plants this amounts to a maximum of 80% of the solar radiation usable in *photosynthesis which falls on the land area upon which the plants grow.

CL

capacitance (Symbol: *C*). The property of a system of conductors and insulators which allows them to store an electric charge when potential difference exists between the conductors. Capacitance is measured in *farads (F). TM

capacitor. Any system or component possessing appreciable *capacitance.

TM

capacity factor. The average power produced by an electrical generator divided by the maximum power that could be produced. JWT

capital. A stock existing at a point in time, the result of the accumulation over time of flows per period of time. Three different usages exist in economics.

(1) As a *factor of production, capital is a stock of *durable goods used in production, e.g. machines and buildings. It is distinguished from other factors of production by being itself produced by economic activity for the sole purpose of providing input to further economic activity. It is difficult to measure, since this requires the use of, for example, the prices of machinery and buildings to add together the *value of machinery and buildings, so as to produce a single figure for capital. The rate at which the capital stock grows is the rate of *investment. Capital and investment can be considered *net or *gross. Capital accumulation is important in *economic growth, with more capital raising the *productivity of labour. An important, but unresolved, question is the *elasticity of *substitution between capital and energy in producing goods and services. If this is

large, so that capital can easily replace energy as the price of energy rises, then energy availability need not be a constraint on economic growth.

(2) An individual's capital is his personal *wealth, equal to his accumulated savings plus any inherited accumulated savings, net of *consumption out of such. This personal capital is held in the form of *assets of varying degrees of *liquidity such as *money, *securities, or *durable goods.

(3) In financial terms, companies are said to have various forms of capital, in which usage the term refers to the financial instruments corresponding to the capital, as a factor of production, employed by the company. The most basic distinction is between *share (or equity) and debenture capital. A company must pay out a fixed *interest rate on its debenture capital. On share capital it pays out at a rate which varies with the company's profitability. The company's capital structure is the ratio in which it has these two forms of capital financing.

MC

capital cost. The initial cost of acquisition of equipment, plant, buildings, etc., in distinction from the *recurring costs, which are associated with their maintenance and energy consumption. Accountancy techniques allow capital and recurring costs to be combined into *costs-in-use, thus facilitating *cost–benefit analyses.

TM

capital energy requirement. Energy expended in producing capital goods. Where capital (money) is spent on goods and services, rather than land or the purchase of technology, those goods and services have to be manufactured and delivered. Thus they have their own *energy requirement. In *energy analysis, the energy requirement of capital is generally amortized over the life of the capital, and so apportioned to the product(s) produced as a result of that capital investment.

MS

capital gain. The difference between the purchase price of an *asset and the price realizable from sale at a later date. Thus an increase in the price of oil leads to capital gains for the owners of oil wells. If the percentage capital gain in money terms is less than the rate of *inflation, there is a capital loss in real terms.

MC

capital market. A group of closely related markets where *securities are traded. The trade involves both new issues of securities, corresponding to companies seeking increases in their *capital, and existing securities changing hands. To the extent that *investment is financed in the capital market, rather than from the retained profits of companies, the market is supposed to ensure that investment occurs where it is most worthwhile. Thus, for example, with high prices for energy, companies in the business of selling energy should be able to offer participants in the capital market sufficiently attractive terms to ensure that there is no shortage of investment in energy supply. The major institutions in the capital market are *banks, merchant banks, new-issue houses, and the *stock exchange.

MC

caprock. An impermeable layer of rock below ground which caps oil-bearing rock. It stops the migration of oil and thus promotes the formation of an oil and gas reservoir.

WG

capture. *See* absorption (of a neutron).

carbohydrates. A class of naturally occurring compounds which contain the elements carbon, hydrogen and oxygen only and have the general molecular formula $(C_6H_{10}O_5)_n$. The simpler carbohydrate molecules, such as sugars and starches, are easily fermented to *ethanol by yeasts. The more complex celluloses found in wood and straw, for example, are better utilized as *biofuels by means of *combustion, gasification or *pyrolysis. Carbohydrates have *calorific values around 20 GJ/t in the dry state. Worldwide annual production of carbohydrates via *photosynthesis is over 100 billion tonnes.

CL

carbon. (1) (chemical) A non-metallic element which occurs in a number of forms, including two crystalline forms, diamond and graphite, and several amorphous forms, e.g. soot and charcoal. Although chemically inactive at normal temperatures, when heated above about 500 °C carbon reacts readily with oxygen to form *carbon dioxide. This oxidation reaction is the basis of all *combustion processes in which carbon and its organic compounds are involved.

Carbon/(graphite) also finds use in electrodes and as a filler in rubber and other structural materials.

Carbon has the ability to form both inorganic compounds, such as metal carbides and carbonates, and organic compounds, in which the carbon atoms are linked to each other in long chains and rings. It is an essential part of the process of *photosynthesis in which the action of light aids in the transformation of carbon dioxide and water vapour into the living matter of plants. It is thus the basis of life itself.

(2)(nuclear) Because of its fairly low atomic weight and its very low neutron capture cross-section, carbon is extensively used in nuclear engineering as the *moderator in certain types of thermal *nuclear reactor. For this purpose it is used in the form of graphite (density about 1600 kg/m^3), which must be of very high purity; in particular there should be no traces of *boron which also has a very high neutron capture cross-section. A very important property of graphite is its good tensile strength, which increases with increasing temperature to a maximum at 2500°C, well above the temperature in any existing reactor. Graphite is therefore an ideal structural material and moderator for high temperature reactors. *See also* carbon-14.

Symbol: C; atomic number: 6; atomic weight: 12.011; valency: 4; GCV: 34 MJ/kg.

WG, DB

carbon-14. A radioactive form of carbon which has a *half-life of 5730 years and is produced in nature as a result of the bombardment of nitrogen atoms by cosmic rays in the upper atmosphere. Carbon-14 is found in the atmosphere as carbon dioxide and is therefore taken up by plants during *photosynthesis and eventually, via various food chains, incorporated into human tissues. Carbon-14 is also produced during the *fission process and is released in the routine gaseous effluent from *nuclear reactors and nuclear processing plants. Since it is long-lived it eventually becomes globally dispersed by air movements. Carbon-14 concentration and distribution is used to assess the long-term impact of the nuclear *fuel cycle on the *biosphere.

PH

carbonaceous coal. A coal whose analysis is such that in a system of classification it would lie between *anthracite and *bituminous coal. It has a volatile matter content on a dry ash-free basis in the range 10–20% and has poor coking characteristics. Such coals are very valuable as fuels for steam raising in industrial and marine boilers.

WG

carbonates. Chemical compounds formed by the reaction of carbonic acid (H_2CO_3) with a metal or base to form a salt, or with an *alcohol to form an ester. Inorganic carbonates evolve *carbon dioxide when heated or when treated with strong acids. Calcium carbonate and to a lesser extent magnesium carbonate are the major components of limestone. Carbonate esters find application in polymers.

WG

carbon black. Finely divided *amorphous carbon, prepared by the *pyrolysis of hydrocarbon oils and gases. It finds application as a component of rubber, car tyres, plastics, paper and inks. The manufacturing process is carefully controlled to obtain the particle size and surface area required for a particular application.

WG

carbon dioxide (CO_2). (1) (chemical) A colourless, odourless, dense and nontoxic gas at normal temperature, produced by the combustion or bacterial degradation of organic substances. It has a low chemical reactivity and will not itself support combustion. At high concentrations the gas acts as an asphyxiant. Carbonic acid, H_2CO_3, a weak acid, is formed when the gas is passed into water.

Carbon dioxide is present in air at concentrations of 0.03–0.04% by volume. This atmospheric concentration is the result of the balance between processes producing carbon dioxide and those removing it, e.g. *photosynthesis by green vegetation. An increase in the atmospheric content will cause greater absorption by the atmosphere of heat radiated from the earth and thus an increase in ambient temperature.

Carbon dioxide may be generated under controlled conditions by the action of heat or strong acids on inorganic carbonates, e.g. limestone. The gas is also used in the production of carbonate beverages. Carbon dioxide may be cooled to form a white solid (dry ice), which finds use as a refrigerant.

(2)(nuclear) Because of its very low neutron capture *cross-section and its low chemical reactivity, it is used as the *coolant in gas-cooled graphite-moderated *nuclear reactors. However, a reaction between carbon dioxide and graphite sets in at temperatures greater than about 400°C and imposes an operating limitation on this type of reactor unless the reaction is inhibited by the addition of small amounts of methane to the carbon dioxide.

WG, DB

Carboniferous period. A period of about 80 million years during the Palaeozoic era of geological time (270–350 million years ago) in which the majority of the *bituminous coal measures were laid down and a considerable proportion of the oil reserves originated.

WG

carbonization. The process of thermal degradation which occurs when coal or other organic matter is heated in the absence of air. On the industrial scale, coal is heated in silica-lined *retorts or ovens to produce a range of products whose composition depends on both the rate of heating and the final temperature. They include gases (*coal gas and *coke oven gas), liquids (*ammonia liquor and *coal tar) and a solid porous residue (*coke). Carbonization may be carried out with the coal at rest in an oven or moving slowly through a retort. The coking characteristics of the coal and the process conditions determine the properties of the products. With a suitable coal, carbonization to 600 – 700°C (*low-temperature carbonization) produces a readily combustible *smokeless fuel; heating to 1200–1300°C may produce a hard and less reactive coke suitable for metallurgical purposes. High temperatures also favour the production of gas whose quantity may be further augmented by blowing steam through the retort. Processes have also been developed for the carbonization of domestic waste for the production of combustible gases and liquids.

WG

carbon monoxide (CO). A colourless, odourless, toxic gas at normal temperature. It is produced in combustion processes under fuel-rich conditions which result in incomplete combustion. This may be due to insufficient secondary air or to poor mixing of air and fuel gases in the combustion chamber. Carbon monoxide may be generated by the action of air (*see* producer gas), steam (*see* water gas) or carbon dioxide on red-hot coke or carbon. It is a constituent of blast-furnace gas, and to a smaller extent of coke-oven gas. It is highly flammable and burns with a characteristic blue flame.

The poisonous nature of the gas arises out of its vigorous reactivity with the haemoglobin in blood, stifling haemoglobin's ability to absorb oxygen. The effects produced depended both on the concentration of carbon monoxide and the length of period over which it has been inhaled.

Carbon monoxide is used as a chemical reducing agent in the chemical and metallurgical industries, its toxicity making it unsuitable as a fuel gas for small-scale application. When mixed with hydrogen gas it may be passed over catalysts at elevated temperatures to form a variety of organic compounds. *See also* synthesis gas.

WG

carbon residue. The coke-like residue formed when oil is carbonized under standard test conditions. The test enables the analyst to judge the proportion of residue fuel mixed with distillate fuel in the sample under test. It is employed in determining the suitability of a diesel fuel for a particular engine.

WG

carbolic acid. *See* phenol.

carburetted water gas. A fuel gas prepared by treating mid-range hydrocarbons (gas-oil) to high temperature, i.e. *cracking. The low molecular weight hydrocarbons so produced serve to enrich other low calorific value gases, such as *water gas, so that their calorific value becomes compatible with that of town gas and blending becomes possible. The calorific value of the mixture may be varied at will by the proportion of gas-oil and the cracking temperature. A typical value would be 18.6 MJ/m^3.

WG

carcinogen. A substance or agent which causes cancer.

PH

Carnot cycle. A theoretical analysis by the French physicist Carnot of the performance of an ideal heat engine. Thermodynamic theory proves that no heat engine can be more efficient than an ideal engine working with a Carnot cycle. If an engine accepts heat at *absolute temperature T_{-1} and rejects heat at a lower temperature T_{-2}, the Carnot ideal efficiency equals $(T_{-1}-T_{-2})/T_{-1}$; this is the maximum fraction of heat energy input obtainable as work from the engine. *See also* second law of thermodynamics.

JWT

cartel. An association of producers who agree between themselves to regulate prices, output or some market conditions. The restriction in competition is for the purpose of securing greater joint profits. Since the most efficient can probably make greater profits in isolation, there is always the possibility of the cartel breaking up. Cartels can be national or international. OPEC is an international cartel of oil-producing nations.

MC

casing head gas. Low molecular weight hydrocarbon gases which separate at the well head of an oil-producing well. They may be burned in a flare or to provide power at the rig.

WG

casing head gasoline. Volatile hydrocarbons which separate as liquids from natural gas as it rises from the reservoir rock and the pressure on it falls. The liquid is separated from the natural gas at the well head.

WG

cassava [manioc]. A starchy edible tuberous crop, *Manihot esculenta*, that can be converted into ethanol. It can yield over 70 t/ha.y in the wet tropics, though 40–50 t/ha.y is considered to be a good yield since the world average is below 10 t/ha.y. The roots contain 25–40% starch, which can be fermented to ethanol following mild hydrolysis treatment. However, unlike sugar-cane *bagasse, there is little cellulose elsewhere in the plant for burning as a fuel in the conversion process. Should non-renewable energy sources be required to drive the process, then the overall *net energy will be reduced. Cassava *energy farms are being cultivated in Brazil as part of that country's national alcohol programme and so it is important that as far as possible the external energy inputs be renewable. Cassava's principal use today is, however, as an animal foodstuff.

CL

Casuarina equisetifolia. One of over 30 *Casuarina* tree species providing good quality fuelwood in many parts of the developing world. It has a calorific value of 21 GJ/t and a specific gravity of 0.8–1.2. The wood burns fiercely and is used domestically and industrially. Yields per hectare vary from 75–200 tonnes on a 7–10 year rotation on good soils, so that *wood energy yields of over 400 GJ/ha.y can be attained. Cultivation under increased organization within a plantation could increase this to a high 700 GJ/ha.y. *Casuarina* furthermore makes excellent charcoal and is capable of *nitrogen fixation, thus minimizing chemical fertilizer requirements. The trees are rapid growing, salt tolerant, wind resistant and adaptable to moderately poor soils and to climates as varied as coastal sand dunes, high mountain slopes, the hot humid tropics and semi-arid regions.

CL

catalyst. A substance which facilitates a chemical reaction without itself appearing to undergo any permanent change.

WG

catalytic converter. A device which can be fitted to the exhaust system of a motor vehicle to reduce the levels of hydrocarbons, carbon monoxide and nitric oxide in the exhaust gas. In a dual catalytic system the exhaust gas is sent first through a chamber where nitrogen oxides are reduced to form nitrogen; air is then added to the exhaust gas and it passes through a second chamber where hydrocarbons and carbon monoxide are oxidized to carbon dioxide and water. Various metals and alloys can be used as catalysts, e.g. the noble metals platinum and palladium and the base metals such as copper, nickel and chromium. *Tetraethyl lead which is added to petrol as an antiknock compound is a very efficient catalyst poison so that motor vehicles fitted with catalytic converters must use lead-free petrol.

PH

catalytic cracking. A petroleum-refinery process in which a heavy petroleum distillate is brought in contact with a solid catalyst at suitable temperature, pressure and contact time in order to produce high-octane gasoline and smaller proportions of other lower molecular weight hydrocarbons. The process is superior to *thermal cracking in that considerably less gas and coke are formed. The catalysts are aluminium silicates of controlled composition. In the original process the feedstock was cycled through a fixed-bed catalyst. This has been replaced by either a moving-bed or a fluidized-bed process.

In the fluidized-bed process the heated feedstock is injected into a stream of regenerated catalyst entering a fluidized bed of the catalyst. There conditions are controlled to obtain the desired conversion. The products leave the bed as vapour, are separated from dust, and fed to a fractionating column. A gasoline fraction is recovered and higher boiling material recycled to the catalytic cracking unit or reactor. The catalyst particles acquire a coating of carbon and lose their activity. Regeneration is achieved by continuous removal of catalyst from the cracking unit to a second fluidized bed where the carbon is burned off in hot air. The regenerated catalyst flows back to the reactor.

In the moving-bed process the reactor is arranged above the regenerator or kiln and the catalyst particles fall by gravity and are returned to the top of the reactor by a lift.

WG

catalytic reforming. Restructuring of one type of hydrocarbon molecule into another more valuable form. It is a petroleum-refinery process developed about 1940 to convert straight-run *naphtha into high-octane petrol (gasoline). The feedstock is pretreated with hydrogen to remove sulphur and other non-hydrocarbon constituents, and then passed with hydrogen over a platinum *catalyst in a series of reactor vessels at a temperature of about 500°C and a pressure in the range of 15–40 atmospheres. The product mixture is separated into liquid and gaseous components and the latter recycled to the reactors. The liquid is distilled to give a stable product suitable for blending into petrol. By this means it is possible to produce a petrol with an *octane rating in excess of 100. *Moving-bed and *fluidized-bed versions of the process have also been developed. *See also* hydrotreating; reforming; thermal reforming.

WG

cellulose. The structural material of all vegetable matter. It has a fibrous nature, is the basis of cotton and rayon textiles, and may also be obtained from wood pulp. Cellulose can be hydrolysed to sugars, which in turn ferment to alcohol. Thus a conversion from solid fuel to liquid fuel is achieved. Cellulose has the formula $(C_6H_{10}O_5)_n$ where *n* refers to a chain of molecules of this structure.

WG

Celsius scale. *See* centigrade scale.

centigrade heat unit (CHU). The amount of heat required to raise one pound mass of water through 1 degree C. Since 1 °C is 1.8 °F, a CHU is equal to 1.8 British thermal units. The unit is now little used, but was at one time popular in British and American engineering calculations.

MS

centigrade scale [Celsius scale]. A scale for measuring temperature in which 100 indicates the boiling point of water at standard *atmospheric pressure, and 0 the *freezing point of water. Thus the scale between these two points can be divided into 100 equal divisions, or degrees centigrade (°0C). 0°C is 273.16 Kelvin on the *absolute or thermodynamic scale of temperature.

PH

central bank. A *bank which is responsible for controlling and implementing the country's monetary policy. Although usually public institutions, central banks have varying degrees of independence. The main functions include being banker to the commercial banks and to the government, acting as lender of last resort, managing the *exchange rate and *foreign exchange *reserves and implementing the government's *monetary policy.

MC

Central Electricity Generating Board (CEGB). A government-created organization which owns and operates the power stations and main transmission lines in England and Wales, and is responsible for the supply of electricity to 12 area electricity boards of England and Wales. The Electricity Council however is responsible for forecasting of load estimates from which new capital programmes for generation and distribution are decided. It is the largest electricity-supply system under unified control in the western world: in 1978 the Board's system comprised 137 power stations with a net capability of 56,326 MW.

Address: Sudbury House, 15 Newgate Street, London, EC1.

PH

central heating. The system of space heating within a building which raises heat in some centrally sited location and distributes it, through the medium of steam, high-pressure hot water, low-pressure hot water, air or the like, to other locations throughout the building. Central heating may be thought of as a part of a full *air-conditioning system.

TM

ceramic. Refractory material of very high melting point. It is used as fuel and *moderator in high-temperature *nuclear reactors. For example, the high-temperature gas-cooled reactor has fuel in the form of very small spherical particles of uranium oxide or carbide coated in layers of carbon and silicon carbide. These particles are dispersed in *graphite to form a homogeneous mixture of fuel and moderator capable of withstanding very high temperatures.

DB

cesium. *See* caesium.

cetane. A straight-chain liquid hydrocarbon, $C_{16}H_{34}$, of the *alkane series. It is used as a reference fuel for *diesel engines since it is found to have excellent ignition characteristics. On the *cetane number scale it is given the value 100. *See also* diesel fuel.

WG

cetane number. A measure of the ignition quality of a fuel for *diesel engines when tested in a standard engine under specified conditions. The fuel under test is compared with a series of reference fuels. The cetane number is the volume percentage of cetane ($C_{16}H_{34}$) in a mixture with α-methylnaphthalene ($C_{10}H_7CH_3$) in the reference fuel which has the same ignition quality as the fuel under test. Values obtained range from 35 for a poor fuel to 65 for a high-quality fuel.

WG

C_4 plants. *See* 4-carbon plants.

chain reaction. A reaction, chemical or nuclear, in which the product of one stage becomes the reactant of the next. Such a reaction is established in a *nuclear reactor when, once *fission has been initiated, the neutrons produced by this process are sufficient to cause further fission, thus establishing a self-sustaining process. Fission, however, is not the only process which can occur since neutrons may also be captured in the materials of the reactor and leak out of it, as is illustrated.

The fraction of fission-produced neutrons which is captured in the reactor (*C*) depends on its composition. The fraction

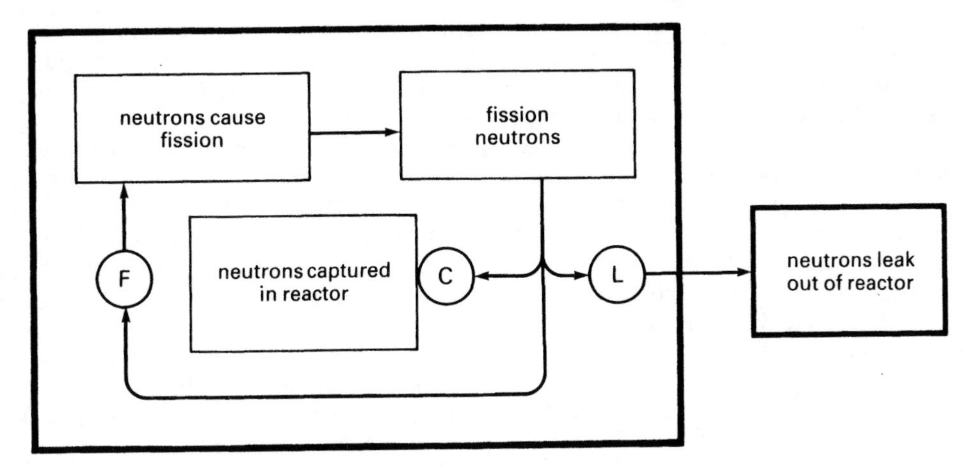

of fission-produced neutrons which leaks out of the reactor (*L*) depends on its composition and size. The fraction of fission-produced neutrons which cause further fission (*F*) depends on the fissile isotopes in the reactor.

An important factor for any fissile isotope is the average number of neutrons produced per fission, symbol ν. Its value is usually between 2 and 3. The condition for a steady chain reaction is that the value of νF should be exactly 1. If νF is less than 1 a chain reaction is not possible; if νF is greater than 1 a diverging chain reaction is established. The values of *C, L* and *F* are determined by the design of the reactor, which specifies its composition and size. *See also* criticality; multiplication factor.

DB

change of state. Matter can exist in three states — solid, liquid and gas — and can change from one state to another according to its temperature, pressure and composition.

MS

char. Any solid carbonaceous matter deposited on the internal surfaces of a retort in which a fuel gasification process is being carried out.

WG

charcoal. An amorphous form of carbon obtained by the *carbonization of wood. It can be used in any process calling for *carbon, such as smelting of ores, but has generally been replaced by *coke. Its major use today is as a cooking fuel, mainly in developing countries, for it may be efficiently combusted and produces little in the way of fumes if it has been well made. Unfortunately for the world *firewood shortage, charcoal is often produced in an inefficient manner.

MS

charge. (1) (Symbol: *Q*) The quantity of unbalanced electricity in a body, i.e. an excess or deficiency of electrons giving the body negative or positive electrification, respectively. The unit of electric charge is the *coulomb (C).

(2) An amount of fuel fed from time to time to a *furnace.

TM

charge–discharge machine. A device for facilitating the fuelling and refuelling of a *nuclear reactor without closing it down to do so. Many reactors are designed for on-load refuelling, as well as for fuel being loaded into them at the start of their operating life and unloaded at the end. In all the British gas-cooled graphite-moderated reactors (except Hunterston A) the charge–dis-

charge machine is located above the core and during fuelling and refuelling operations it is connected to standpipes passing through the upper *biological shield and the *pressure vessel. Fuel elements are lowered and raised through these standpipes. The Canadian CANDU reactor is also designed for on-load refuelling, but the design of the charge–discharge machine is different due to the different layout of the reactor core.
DB

Charles' law. The law stating that a gas at constant pressure will change volume in inverse proportion to its absolute temperature. *See also* Boyle's law.
MS

Chartered Institution of Building Services (CIBS). A professional body within the UK with members engaged in the design, maintenance and operation of the mechanical and electrical services within buildings.

Address: Delta House, 222 Balham High Street, London, SW12 9BS.
TM

char test. A burning test to evaluate the suitability of *kerosene (paraffin) as an illuminant and as a fuel for space heaters, etc. The kerosene is burned in a standard lamp for 24 hours and at the end of the period the consumption of kerosene and the amount of *char formed on the wick are measured. *See also* char value.
WG

char value. A standard test to determine the quality of *kerosene (paraffin). The weight of charred fuel on the wick of a standard lamp is measured after fuel has been burned in the lamp at a standard rate for 24 hours. It should be less than 10 mg per kg of kerosene of premium grade.
WG

Chelyabinsk. The site of a very serious nuclear accident which is thought to have occurred around 1957 in the Southern Urals, USSR. As a result of this accident hundreds of square kilometres of land were contaminated with radioactive materials. The cause of the accident is still not known, at least in the western world, but is generally believed to have been connected with the Soviet industry for production of plutonium for military purposes. Possible events which may have caused the accident include an accidental bomb explosion, a fire at a nuclear reactor or an explosion at a radioactive waste-storage facility.
PH

chemical bond. The linking of one chemical species to another. Bonds can be strong, and hence not easily broken by physical or chemical action (e.g. water — H_2O) or weak (e.g. hydrogen peroxide — H_2O_2). Very often the breaking of a bond requires energy while the formation of one releases energy. Thus when, say, *methane (CH_4) is burnt, the bond between carbon and hydrogen is ruptured, absorbing energy, but is replaced by the greater energy released as the carbon and hydrogen unite with oxygen in the air, thereby causing a net release of energy to the surroundings.
MS

chemical energy. Energy obtained from the reaction of chemicals. Since the reaction is usually controllable and may be initiated when desired, chemical energy systems are of great importance for *energy storage systems.
JWT

chemoautotrophs. *See* autotrophic organisms.

Chernobyl. A nuclear power station about 100 km north of Kiev in the USSR with 4 *RBMK type reactors. In April 1986 a serious accident occurred in one of the reactors resulting from prompt criticality during low-power testing. The resulting rapid and uncontrolled power rise melted the fuel cladding and caused a steam explosion in the core which blew off the top *biological shield. The graphite core of the reactor went on fire, fuel meltdown followed, and there was a massive and uncontrolled release of radioactive *fission products from the molten fuel through the shattered containment to the atmosphere. The fire burnt for almost a week, and the reactor was totally destroyed. It has since been encased in a concrete structure to prevent further release of radioactivity from the core into the environment.

In the months following the accident about 200 deaths due to acute radiation sickness resulted, most of these deaths being among the firemen and plant operators working at the power station at the time of and in the days following the accident. In addition, many thousand people living close to Chernobyl were subjected to radiation levels high enough to cause delayed radiation-related diseases such as cancer. These effects may cause a few hundred early deaths among the exposed population in the next few decades.

About 135,000 people living within a 30 km distance of the reactor were evacuated within a week of the accident. Considerable radioactive contamination affected this area, and some areas up to a distance of 60 km also showed significant contamination levels. Many years will elapse before these areas can be used again for agriculture unless extensive and successful decontamination can be carried out.

Very large amounts of radioactive material were released into the atmosphere at the time of the accident and during the ensuing fire. This was carried by winds over Russia and Europe and deposited by rain in many countries several hundred kilometres from Chernobyl. Higher than normal levels of radioactivity and contamination of vegetation and soil were recorded in many parts of Europe and restrictions were placed on the consumption of vegetables and meat in some areas. *See also* nuclear reactor.

DB

chilled ceiling system. A system of *air conditioning a building in which the ceiling or ceiling space of a room is cooled by a supply of chilled water or chilled air supplied from a centrally sited plant room.

chlorophyll. A light-absorbing pigment in plant cells. It is present in bodies called *chloroplasts in which light energy is converted into chemical energy. It therefore plays a dual role in the process of photosynthesis. Chlorophyll, of which there are at last seven varieties, absorbs light intensely in two spectral regions: the violet around 400 nm and the red or near infrared at 600–800 nm. Photosynthetic energy conversion is initiated when *photons of light are absorbed by a chlorophyll molecule, which is thereby oxidized by the ejection of an electron. This electron then passes along a photosynthetic electron transport system so as to generate biologically useful forms of chemical-bond energy via the *Calvin-Benson cycle. Chemically, chlorophyll has a molecular weight of 892 and consists of a porphyrin ring chelated with magnesium and an attached lengthy phytol alcohol) chain.

CL

chloroplast. A body within plant cells which contains the light-absorbing pigment *chlorophyll, and converts light energy to chemical energy through *photosynthesis. There may be from one to over 100 chloroplasts per cell, depending on species and growth conditions. The chlorophyll is embedded in double-layer membranes consisting of roughly 50% fats and 50% protein. *4-carbon plants contain two types of chloroplast: mesophyll, which are structurally similar to the chloroplasts of *3-carbon plants, and the larger bundle sheath type.

CL

chocolate mousse (in oil spills). The stable water-in-oil emulsions formed when oil is mixed vigorously with sea water; it may contain up to 80% water. It is highly viscous, floats on water and clings to solid surfaces. Although it is markedly less toxic than fresh crude oil it requires larger doses of *dispersant to disperse it.

PH

chord. The distance from a point on the leading edge of an airfoil blade to the trailing edge, in the direction of movement.

JWT

Christmas tree. The assembly at the top of a gas or oil well comprising many different types of valves.

MS

chromosomes. *See* DNA.

cladding. The material which is used to enclose the *fuel in nuclear reactors. Its purpose is to prevent the escape of *fission products, provide structural support for the fuel and (in the case of gas-cooled reactors) provide extended surfaces in the form of fins to enhance heat-transfer rates from the fuel to the *coolant. The three most common alloys used as cladding are *Magnox, *stainless steel and *Zircaloy.

DB

clarain. A recognizable constituent of coal, seen as irregular bright black bands. It is less bright and uniform than *vitrain, being composed of woody tissue mixed with other plant remains, which may be seen when a thin section of the coal is viewed under the microscope. Lumps of coal composed largely of clarain tend to break irregularly.

WG

clean coal. Coal with the minimum possible ash content, generally less than 1% by weight. Particles of mineral matter and of higher ash-content coal have been removed by a washing process in water where their greater density causes them to settle first to the bottom of the container. Clean coal is used to prepare electrodes for the aluminium industry.

WG

climate. The integrated experience of variations in temperature, humidity and air movement occurring at any particular locality on the earth's surface. The degree of seasonal change in climatic conditions at any locality depends on latitude and results from the fact that, due to the earth's tilt, the amount of solar energy received at that locality alters throughout the year. The geography of a locality also influences its climate, determining how much solar energy is absorbed by the earth, how much is stored and how readily it is released to the atmosphere.

The demographic distribution of animal, vegetable and mineral resources throughout the world, and of the consequent utilization of energy, is largely determined by climate.

In the context of building design and performance, climate is often referred to as macroclimate in distinction from microclimate, the local variation of climate affected by and affecting, respectively, an individual building and a group of buildings.

TM

Clinch River breeder reactor. A demonstration fast breeder *nuclear reactor under assembly near Oak Ridge, Tennessee, USA. Work on the project began in 1972. $1.8 billion has already (1981) been spent on the project and the final cost is expected to be $3.0 billion.

PH

clinker. Irregular rock-like lumps which form on a *grate burning solid fuel when fused ash cools and solidifies. Pieces of partly burned fuel may be incorporated in the mass. The formation of clinker will tend to obstruct the flow of primary air through the fuel bed and thus cause irregular patterns of burning. The fused masses of ash will also adhere to the furnace brickwork around the grate and their forcible removal will damage the brick surfaces. Clinker formation may be avoided by selecting solid fuel with a high *ash fusion temperature, by avoiding thick fuel beds and by the application of water sprays under the grate.

WG

closed fuel cycle. A type of nuclear *fuel cycle in which spent fuel from the reactor is reprocessed in order to recover the uranium and plutonium, which may then be recycled as fuel. *See* reprocessing.

PH

closed system. One which is isolated from the surrounding world, without exchange of materials, energy or information. Though such a system never truly exists, the concept is useful for many forms of analysis. For example, the operation of a *heat engine can be readily analysed by assuming it to be a closed system. The world as a whole can be regarded as a closed system for economic purposes if one ignores the solar radiation falling upon it. Indeed, if the world's economy reached a point where it was driven largely by *inherited energy resources and did not use to any great extent solar energy, then it would approximate to a closed system. Under such circumstances it may be considered to have a fixed amount of energy, whose quality was being slowly but inevitably degraded, implying an eventual cessation of activity. *See also* natural resources.

MS

cloud point test. A standard test used for fuels which contain dissolved wax. The test determines the temperature (cloud point) at which a cloud of wax crystals appears in a volume of oil cooled in a test tube. The information is important because oil with a high cloud point will not flow easily in cold weather with clay filters. At such a temperature blockage may result when the oil is passed through a fine orifice or a filter.

WG

Club of Rome. An organization founded in April 1968, when 30 individuals from 11 countries met together in Rome at the instigation of Dr Alexander King, then of OECD in Paris. Out of this meeting grew an informal organization with an invited membership of approximately 70 persons, with Aurelio Peccei, the Italian industrialist as its president. Its purpose was 'to foster understanding of the varied but independent components — economic, political, natural and social — that make up the global system in which we all live; to bring that new understanding to the attention of policy-makers and the public world-wide; and in this way to promote new policy initiatives and action'.

The first undertaking of the Club was to sponsor the Project on the Predicament of Mankind, and in 1972 its first report, a *systems analysis study of the world entitled *Limits to Growth* was published. It had a profound influence on world thinking and drew both acclaim and scorn. The second report *Mankind at the Turning Point* took the previous study further, and attempted to overcome some of the criticisms. It has subsequently published many studies. The present president is Alexander King.

MS

C_n fraction. The average number of carbon atoms per molecule in a petroleum fraction. It is of value only where the number is known with a fair degree of accuracy in low molecular weight fractions, i.e. for values of n between 1 and 8. For example, C_4 hydrocarbons include n-butane, iso-butane and the butenes.

WG

coal. An organic sedimentary rock formed from vegetation which has been altered by the combined effects of microbial action, pressure, heat and time, and consolidated between strata of non-organic rocks to form coal seams. The estimated age of such coal deposits ranges from 20 to 250 million years, the greatest proportion being laid down during the *Carboniferous period of geological time.

It is considered that the vegetation subsequently converted to coal flourished in swampy regions in a hot and humid climate, and when the atmospheric *carbon dioxide level was higher than at present. Dead vegetation collapsed into the swamps and was subjected to only slow microbial action due to the acid conditions. *Peat was formed and was eventually covered by silts and other sedimentary material due to land movement and the weathering of adjacent rock masses. The process of *coalification continued under this inorganic cover.

Coal as mined consists of organic matter associated with varying proportions of mineral matter. The latter arises from soluble salts in the peaty waters, from intimately intermixed silts and from rock accidentally included with the coal in the mining process. When the coal is burned such mineral matter forms *ash.

The organic matter is composed of carbon, oxygen, nitrogen and sulphur, the proportions depending on the stage reached in the coalification process. As it progresses, oxygen is slowly eliminated and the ratio of carbon to hydrogen increases. The end product would be a rock containing only carbon, i.e. graphite. These changes are paralleled by changes in the burning and *coking characteristics of the coal, and thus determine the type of use to which it will be put.

The extent of coalification determines the *rank of a coal. Coals of low rank, e.g. brown coal, contain higher proportions of oxygen and have a low carbon content, have high proportions of volatile matter and a low calorific value. High-rank coals, e.g. *anthracite, contain little volatile matter.

Coal has been superseded by oil and gas as a fuel for small combustion units. Thus it is little used for domestic purposes or for industrial steam generation. However, in time it may become necessary to use it again for such applications. It is used primarily in the very large combustion units of the electricity supply industry and in the manufacture of *coke for the metallurgical industries. Coal liquification by *hydrogenation is a technologically difficult process, but it is likely to be further developed as crude oil becomes scarce.

WG

coal equivalent. *See* tonne of coal equivalent.

coal field. A natural strata of coal within the ground. It may be near the surface, in which case it can be exploited by *opencast mining (US: strip mining) or by underground mines.

MS

coal gas. The mixture of permanent gases produced by the *carbonization of a coal. The composition of the mixture depends on the carbonizing temperature and the proportion of steam injected into the retort. Low-temperature carbonization, e.g. at 700°C, gives a relatively low yield (140 m^3/tonne) of gas containing high percentages of hydrocarbons and thus having a high *calorific value. High-temperature carbonization increases the yield of gas (330 m^3/tonne) but reduces the hydrocarbon content and the calorific value. The injection of steam increases the proportions of hydrogen and carbon monoxide in the gas and further lowers the calorific value. After treatment to remove suspended droplets of tar, ammonia and sulphur compounds, the gas may be piped to industrial and domestic premises. It is then called town gas.

WG

coalification. The combination of natural chemical and physical processes by which *peat is converted into *bituminous coal.

WG

Coalite. A proprietory solid smokeless fuel prepared by carbonizing coal in cast-iron retorts at 600°C. The product is a semi-coke, containing about 12% volatile matter, and may be burned with the minimum production of smoke on an open domestic grate or in a closed stove. It has a gross *calorific value of 28.5MJ/kg.

WG

coal liquefaction. Any process which converts coal into a liquid fuel. Since the ratio of hydrogen to carbon in the latter is considerably greater than that of coal, the process must involve the addition of hydrogen or the removal of carbon. Various *hydrogenation processes have been studied but none appear to have been economically feasible to date. Carbon can be removed as coke in a *carbonization process and then gasified to carbon monoxide in synthesis gas. Attempts have also been made to dissolve coal in a suitable solvent and to filter off the ash-forming constituents. In the Consul synthetic fuel process coal is heated in an oil solvent to about 400°C to cause its dissolution. The solvent is itself treated with hydrogen gas and in turn donates it to the coal to aid its conversion into liquid hydrocarbons. This and similar types of reactive solvent extraction processes being developed by Exxon in USA and by British Coal in the UK produce a liquid which is essentially a synthetic crude oil (syncrude). It is however sufficiently different from natural crude petroleums to require the development of specialized refinery processes to make it possible to obtain from it gasoline and diesel fuels of the required octane and cetane numbers.

Coal liquefaction processes such as the Bergius and the H-Coal operate by mixing pulverized coal with tar or a recycled heavy oil and a solid catalyst in powder or pellet form. The paste so formed is treated with hydrogen gas at 400–450°C and 200–250 bar. The reacted mixture is distilled to recover gaseous and liquid hydrocarbons from a heavy oil which is recycled, and the solid residue of unreacted coal and ash components.

During the Second World War Germany obtained liquid fuels from coal by the *Fischer–Tropsch process.

WG

coal tar. A brown–black viscous liquid product of the *carbonization of coal in gas-making retorts or coke ovens. The tar has a relative density in the range 1.1–1.2 and the yield is normally 32–56 litres per tonne of coal. Distillation of the tar yields a number of fractions ranging from a light spirit (benzole), which may be used as a motor spirit, to oils containing tar acids. The distillation residue is pitch.

WG

coal washing. A technique for the reduction of the ash content of coals. A number of processes have been developed in which particles of coal are suspended in water and agitated. This may cause the separation within the equipment of particles of different density. Thus mineral-matter particles and high-ash coal particles which have a higher density can be caused to separate at one part of the plant, and low-ash coal with a low density to separate at another point. A suspension of fine particles of iron oxide in water can be caused to behave like a high-density liquid in which coal particles float and mineral particles sink and thus give the required separation.

WG

cobalt-60. A radioactive form of the element cobalt, produced in nuclear reactors by irradiation of naturally occurring cobalt. It has a *half-life of 5.26 years and is extensively used as a source of radiation in medicine, science and industry.

DB

coefficient of performance (COP). A measure of the efficiency of *heat pumps and refrigerators. It is the ratio of useful heat delivered in a heat pump, or removed in a refrigerator, to the energy required to operate the mechanism. The useful heat is greater than the energy required to power the devices, and so the COP is usually considerably greater than one. If the higher temperature is T_A and the cooler temperature is T_B for the heat pump or refrigerator source and sink, then, by treating the systems as reversed *Carnot cycles, the theoretical maximum COP for a heat pump is

$$T_A/(T_A - T_B)$$

and for a refrigerator

$$T_B/(T_A - T_B)$$

Note therefore that the COP varies with the operating temperatures and definition used. Values of COP of about 3 in practical devices are common. *See also* second law of thermodynamics.

JWT

cogeneration. *See* combined heat and power.

coke. The solid residue which remains when coal has been carbonized in a retort or oven out of contact with air. The volatile constituents of the coal are thus distilled off and subsequently cooled to separate them into a mixture of permanent gases and a liquid tar. Coke prepared in this way at temperatures in excess of 1000 °C has a residual volatile matter content of about 2%. It is a hard porous material which is difficult to ignite on an open fire. It may be burned in a closed stove without production of smoke if the air flow through the grate is increased by a fan or by a tall chimney.

It is used largely as a fuel and a source of carbon monoxide in the blast furnace, in which iron ore is reduced to molten metallic iron. By blowing air through red-hot coke a mixture of carbon monoxide and nitrogen, called producer gas, is formed; by blowing steam through red-hot coke water gas, a mixture of carbon monoxide and hydrogen is formed. In selecting coke for the blast furnace its physical properties, e.g. crushing strength, are generally as important as its chemical characteristics.

WG

coke breeze. Finely divided coke from a screening plant. The particles are generally less than 12 mm in diameter.

WG

coke oven. A rectangular chamber, constructed from a refractory material such as silica brick, in which crushed coal is carbonized while at rest. The length of the chamber is about 12 metres, height about 4 m and width 300–450 mm. The coal is fed into the oven through several openings in the oven top, and removeable doors are provided at both ends to enable the hot coke to be pushed out of the oven with a ram at the end of the carbonization period. Pipes are fitted to the top of the oven to enable gas and tarry vapours to be collected. A large number of similar ovens, with gas heating flues between, are arranged in a battery. Producer gas or coke oven gas is burned with air in the flues to give a temperature of about 1350°C. Carbonization time depends on the width of the oven and is about one hour per 2.5cm of width.

WG

coke oven gas. The mixture of permanent gases produced by the *carbonization of coal in a coke oven at temperatures in excess of 1000°C. It has a composition similar to that of coal gas from a gasworks retort when steam is not injected into the coke. The gas mixture is used as a fuel to heat the coke ovens in which it was generated and to heat furnaces in an associated steelworks.

WG

coking characteristics. The behaviour of a coal when carbonized in a retort or coke oven and the properties of the coke produced. The extent to which swelling takes place during coke formation and the hardness and strength of the coke are important, together with its porosity and reactivity at elevated temperatures to carbon dioxide and water vapour.

WG

coking coal. A coal which forms a coherent mass of *coke when subjected to carbonizing temperatures in excess of 600–700 °C. WG

collective effective dose equivalent [collective dose]. *See* radiation dose.

collector (solar). A device or system that traps solar radiation. It usually incorporates a black absorbing surface with mirrors and/or transparent covers to improve efficiency. *See also* absorber. JWT

collusion. Agreement between legally separate companies or nations to co-operate in some way in order to increase their collective bargaining power in the market. The form taken by the cooperation varies from an informal agreement to share information to a formal agreement to establish a *cartel. Collusion is to be expected where *oligopoly is the market structure, as it is in the oil industry typically. Thus OPEC is a cartel, and it is frequently alleged that the major international oil companies act collusively in many areas of their operations. Collusion is generally taken to be against the interests of the customers of the colluding parties. MC

combined heat and power [cogeneration]. The supply and distribution, from central plant, not only of electrical power but also of the heat energy which is a byproduct of the electricity generating process and which would otherwise go to waste. In *district heating schemes, the waste heat from electricity generation is distributed throughout the surrounding industrial, commercial and domestic community for use in space heating and hot water supply. TM

combustion. The rapid *exothermic chemical reaction of a *fuel with oxygen. Energy is released in the form of heat and light, and during the reaction a *flame may form. With a carbonaceous fuel containing carbon, hydrogen and sulphur, combustion in air will produce a mixture of hot gases including carbon dioxide, water vapour, sulphur dioxide and nitrogen, and (as in an internal combustion engine) nitrogen oxides. WG

combustion energy. The energy released by *combustion. It is normally taken to mean the energy released when substances react with oxygen. This may be very fast, as in burning, or slow, as in *aerobic digestion. Though the heat releasable from any chemical reaction between a substance and oxygen may be precisely computed from *thermodynamic tables, the actual temperature achieved depends on whether combustion is with air or pure oxygen, and on the method of combustion. *See also* calorific value. MS

Combustion Institute. A publishing and symposium-organizing society.

Address: 986 Union Trust Building, Pittsburgh, Pa, USA. WG

comfort. *See* thermal comfort.

Comitato Nazionale per l'Energia Nucleare (CNEN). The national administrative authority of Italy for research and development in nuclear energy, safety, control and supervision of nuclear power plant and international cooperation in nuclear energy. In 1982 it broadened its remit to consider *renewable and *alternative energy sources and has been renamed Comitato Nazionale per la Ricerca e Suiluppo dell' Energia Nucleare e Energia Alternativa – ENEA.

Address: 125 Viale Regina Margherita, 00198, Rome, Italy. DB

commercial energy. An energy source or type which is traded. The term is used in contrast to energy sources, usually natural, like wood or dung, which are produced and consumed locally and appear in no recognized market. MS

commissioning. The process of bringing a new industrial plant into operation. MS

Commissariat à l'Energie Atomique (CEA). The national agency of France for the development of nuclear energy. It has nuclear laboratories at Saclay, Grenoble, Fontenay-aux-Roses and Cadarache and a nuclear-fuel processing plant at Cap de la Hague.

Address: 31–33 rue de la Fédération, Paris 15, France.

DB

commodity. The term is used in two senses in economics. Generally, to refer to any *good or *service produced in any economy. Specifically, for a raw material that is traded worldwide in an active market with *spot prices and *future prices determined on the basis of the interaction between supply and demand and *expectations about future changes in such. Examples of commodities in this sense are: copper, tin, sugar, rubber, oil. The context will usually identify the sense in which the term is being used. Thus, individuals have preferences, represented by *utility functions, as between the various commodities in the general sense of produced goods and services. A *commodity agreement refers to a commodity in the specific raw materials sense, as does the term *commodity market.

MC

commodity agreement. An international agreement to organize intervention in trade in some *commodity so as to stabilize its price. A typical form of intervention is the creation of a *buffer stock of the commodity.

MC

commodity market. A market in which a *commodity, i.e. a raw material, is traded. A feature of such markets is that they typically include a *futures market, as well as involving transactions to be effected immediately. *See also* forward price; spot price.

MC

common property resource. A *natural resource which is not owned by a single individual or company, and where rights to use are held in common by many individuals or companies. Examples are the atmosphere, the seas, large lakes and public lands. The market cannot regulate the use of common property resources since rights to use are not tradeable *assets. The result is that in the absence of non-market regulation of use such resources will be over-used and degraded.

A single oil field accessed by a number of oil-well operators is a further example of the common property resource problem. The total amount of oil which can be taken from a field decreases as the rate of extraction increases. A single owner of the field would take account of this trade-off in determining his *depletion programme. Where there are several operators exploiting the field, no one operator can be sure of capturing the benefits of a slower extraction rate for himself, so none of the operators will deplete at the rate required to produce the depletion programme which would be adopted by a single owner. The result is that the field is depleted too rapidly. In the USA recognition of this problem has led to legislation to control the spacing of and production from oil wells accessing a single field, and to legislation permitting the well-owners to collude in order to run the field as if it were owned by a single company. *See also* Tragedy of the Commons.

MC

commutator. A device for reversing the direction of an electric current.

TM

comparative advantage. The fundamental basis on which *commodities are internationally traded. Even if a country can produce more of everything than another (so having an absolute advantage), it will still trade if, for each unit of one commodity it gives up producing, it can produce relatively more of the other commodity than can the country with whom it trades. By specializing more in this commodity in which it has a comparative advantage and trading, it can have more of both commodities for its residents. The principle holds with more than two countries or two commodities.

MC

compensation principle. A principle to be used in considering the desirability of some proposed policy or project. It follows from the objective of achieving economic *efficiency in the solution to the *allocation problem. If those who will gain from the policy or project could fully compensate those who will lose, and still be better off than they are without the policy or project, then it should be undertaken. If such compensation is not possible then the policy or project should not be undertaken. This principle takes no account of the *distribution problem: it would accept, for example, a project which would make some rich people much better off after they had fully compensated poor people for their losses on account of the project. *See also* cost–benefit analysis.

MC

competition. *See* imperfect competition; monopoly; oligopoly; perfect competition.

complementarity. A property of two *commodities which have a 'cross price' *elasticity of demand which is negative so that (with all other prices remaining constant) a proportionate change in the price of commodity one leads to a proportionate change of the opposite sign in the quantity demanded of commodity two. Thus a lower price for commodity one leads to an increase in demand for commodity two. Standard examples of complementary commodities are coffee and sugar, tea and sugar, electricity and refrigerators, motor cars and petrol (gasoline). Generally, it would be expected that a fuel-using piece of equipment and the fuel concerned would exhibit complementarity. *See also* substitute.

MC

compounding. The process whereby interest earned on a loan is re-loaned at the same *interest rate. If the amount P is loaned for one year at the interest rate r, the amount to be repaid is

$$P(1 + r)$$

If this full amount is left on loan for a further year, the amount to be repaid at the end of the second year is

$$P(1 + r)(1 + r) = P(1 + r)^2$$

Generally, after t years of compounding, the amount to be repaid is

$$P(1 + r)^t$$

The proportionate change in the amount to be repaid from year t to year $t+1$ is r, which is the *rate of return on money loaned at r and compounded. This is the rate of return with which that on other *assets, such as oil in an oil field, or investment projects, such as an electricity generating plant, is to be compared. *See also* project appraisal.

MC

compound nucleus. The atomic nucleus formed when a *neutron is absorbed (*see* absorption) into a nucleus. For example, in the interaction of sodium-23 and a neutron the resulting sodium-24 is the compound nucleus. A compound nucleus is formed at an excited state of energy, the degree of excitation depending on the kinetic energy of the incident neutron and its *binding energy in the compound nucleus. Due to its excitation, the compound nucleus decays immediately to its ground state by the emission of *gamma radiation or of a particle such as a neutron, or both. The product of this decay distinguishes between neutron capture and *scattering reactions.

DB

compression ratio. In an internal combustion engine, the ratio of the volume of gas within the cylinder before compression to that after the gas has been fully compressed by the piston. In a modern automobile engine the compression ratio is about 9 : 1. Since the thermal efficiency of the engine increases with compression ratio, designers of engines are ever seeking to increase the value of the ratio. However, a limit is set by the petrol (gasoline) being used. If the compression ratio is too high, knocking occurs, i.e. combustion does not take place progressively through the cylinder. The mixture of petrol vapour and air in the cylinder distant from the spark plug detonates before the flame front reaches it, causing loss of power and overheating of the engine. *See also* octane rating.

WG

Compton scattering. One of the processes whereby *gamma radiation interacts with matter. The energy of each gamma *photon is transferred to an orbital electron in an atom, and a photon of reduced energy is subsequently emitted. The process can be regarded as scattering, as it has the effect of reducing the energy of the gamma radiation without absorbing it.

DB

concentration (economic). The extent to which an industry is dominated by a few companies, to which wealth is held by few individuals, to which oil reserves are located in a few countries, etc. Several ways of measuring concentration are used in economics. Let S_i denote the market share of company i or the share of oil reserves located in country i, where i runs from 1 through to n. Then some typical measures are:

(a) the Herfindahl index,

$$H = \sum_{i=1}^{n} S_i^2$$

$H = 1$ means a single company or oil-owning nation; $H = 1/n$ means equal shares;

(b) the entropy coefficient,

$$E = \sum_{i=1}^{n} S_i \log (1/5S_i)$$

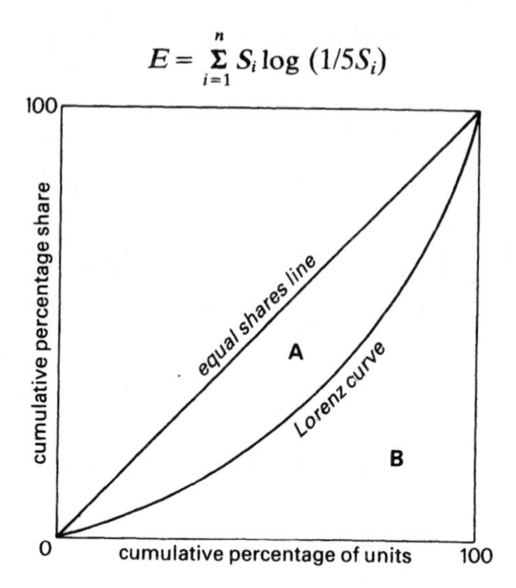

$E = \log n$ corresponds to equal shares;

(c) the Gini coefficient,

$$G = \frac{\text{area } A}{\text{area } A + \text{area } B}$$

where the areas refer to the *Lorenz curve shown in the diagram. $G = 1$ corresponds to equal shares by all n units and $G = 0$ corresponds to one unit taking everything.

It should be noted that the measures H, E and G will not all produce the same rankings of a given set of industries, for example.

MC

concentration factor. The ratio of the concentration of an element in an organism to its concentration in the surrounding water. *See also* biological concentration.

PH

concentrator. A device or structure used to concentrate power into a generator or energy supply system. Thus a mirror may be a concentrator for a solar energy device and a funnel may be a concentrator for an *aerogenerator.

JWT

concrete. A common structural material. It has important uses in nuclear engineering as the *biological shield for nuclear reactors, and as the structural and shielding material for storage tanks and processing plant handling radioactive material such as *irradiated fuel elements. *See also* prestressed concrete.

DB

condensation. The phase change occurring in a convective process during which a vapour gathers on a solid surface in the form of either a continuous film of liquid or a large number of liquid droplets. In buildings the phenomenon of condensation can occur, with harmful effects, when the methods of environmental control are inappropriate to the building-fabric's response to climatic variation. It can manifest itself as surface condensation or, more damagingly, as interstitial condensation between the leaves of multilayered wall or roof constructions.

TM

conductance (Symbol: G). That property of a material which permits the flow of electric current when an *electromotive force is applied to it. It is measured in *siemens (S). With a direct current, the inverse of conductance is *resistance; with an alternating current, the inverse of conductance is the resistance divided by the square of the *impedance. Each material, then, has a property known as electrical conductivity (σ) which determines, in association with the cross-sectional area of the conductor and the *potential difference, the actual flow of current. Electrical conductivity is measured in siemens per metre.

TM

conduction. The mode of *heat transfer in a solid material occasioned by a temperature difference between different parts of the material. Conduction also occurs in liquids and gases but is then generally associated with *convection and possibly, in the case of gases, with *radiation. Conduction within a solid is a transfer of internal energy, i.e. the energy of motion of the constituent molecules, atoms and particles of which the material consists. Each material has, then, a property known as its thermal conductivity (k) which determines, in association with the area of flow and the *temperature gradient, the actual rate of conductive heat flow. Thermal conductivity is measured in watts per metre per Kelvin. *See also* conductance.

TM

conductivity. (1) (electrical) *See* conductance. (2) (thermal) *See* conduction.

configuration factor. *See* form factor.

conservation of energy. *See* first law of thermodynamics.

conservation of mass. A fundamental law of nature which states that matter can neither be created nor destroyed. Thus if, in the process of making a machine, iron ore is turned into iron, then sheet metal and finally into a machine, which when worn out is discarded, there is neither more nor less iron in the world at any stage of these processes. The law permits one to make mass balances over any system of production so that what enters the system must either add to the system stock or leave the system. The law however is not strictly correct: whenever there is an increase in the *energy content of a system, *Einstein's equation tells us that there is a loss of mass. However, the relation between mass and energy is such that a great deal of energy can be released by an infinitesimal loss of mass. Thus, except for nuclear reactions, the law of conservation of mass may be taken as essentially correct.

MS

constant price series [real price series]. A series of figures for the movements in expenditure over time on some commodity or group of commodities, where the expenditures on the commodities are measured by multiplying the quantities bought by the prices ruling in some base period rather than by the prices actually paid. The purpose is to reveal changes in the total quantity bought. Thus, for example, given expenditure on energy at actual prices as

year	1	2	3	4	5
$	10	20	30	40	50

and the *index number for the price of energy as

1	2	3	4	5

then the constant price series for energy expenditure is

$	10	10	10	10	10

revealing that the quantity of energy bought did not change.

The use of constant price series is necessary to remove the effects of *inflation from *national income accounts, and in using economic data to calculate price and income *elasticities of demand for energy and other commodities. MC

Consul process. *See* coal liquefaction.

consumer. (1) In the context of the energy industry, any individual, household, industry or other agent which uses energy, either in its primary form (e.g. by burning coal or wood), in the form of electricity or by purchasing a commercial product, the manufacture of which involved the input of a significant amount of energy.

(2) In an *ecosystem, any animal which lives by feeding off other animals or off plant material. Primary consumers are those which feed off the primary producers, the green plants and algae. Secondary consumers feed off the primary consumers. In many ecosystems still higher levels of feeding are present. PH

consumption. In economic usage, the act by individuals of using goods and services to satisfy wants, which is the end purpose of economic activity. Due to the existence of *durable consumer goods, consumption in any period is not exactly measured by expenditure on that part of *final product destined for consumption, since durables are used over a number of periods. The difference between income and consumption is saving. Oil used by an individual for heating his house represents consumption: oil used by a firm in producing, say, oil-burning house-heating appliances is an *intermediate product. In popular usage, both ways of using oil contribute to total oil 'consumption'. In respect of a finite resource like oil, consumption also represents depletion. MC

containment. In a *nuclear reactor, the steel or concrete structure which encloses the reactor and which is designed to prevent the escape of radiation and radioactive materials and to maintain the pressure of the system both during normal operation and under accident conditions.

In a *fusion reactor, containment (or confinement) refers to the means by which the ionized *plasma of *deuterium and *tritium, at a reaction temperature of several million degrees, is prevented from coming in contact with (and vaporizing) the walls of the reaction vessel. This containment of the plasma may be achieved by applying strong external magnetic fields which combine with the magnetic fields due to the electric current in the plasma itself. Two shapes of magnetic containment are possible using the *torus and the magnetic mirror. In the torus, which is a doughnut-shaped tube, the magnetic field is continuous around the tube. In the magnetic mirror the field strength is greater at the ends of the straight containment tube than at its centre, so that the plasma is reflected from the ends towards the centre.

Inertial containment is an entirely different method of plasma confinement. In this system deuterium and tritium, possibly in the form of a pellet of a suitable compound such as lithium deuteride, are bombarded by a laser of very high intensity. The resultant sudden heating and compression of the pellet allows the fusion reaction to proceed for a very short time before it disperses.

DB

continental shelf. The land between the shore line and a depth of 100 fathoms (183 m). Most off-shore oil-recovery systems operate on the continental shelf. By international agreement the rights to the minerals below the shelf are owned by the adjoining territory.

MS

control rods. *See* control systems.

control systems. The means by which a complex system is controlled. In *nuclear reactors these are designed to control the *reactivity so that the reactor may be started up, shut down, undergo power changes and be operated at constant power for long periods. The most common type of control system consists of rods (control rods) containing materials such as *boron, *cadmium or *hafnium which have high neutron capture *cross-sections. As these rods are withdrawn from the core of the reactor, the reactivity increases to make the reactor supercritical and allow it to start up or increase power; as the rods are inserted into the core, reactivity decreases and the reactor becomes subcritical, resulting in power decrease or shut down. Control systems may include different groups of rods for different purposes. For example, a group of safety rods may be held out of the reactor at all times during normal operation, to be dropped into the core in the event of an emergency or malfunction in order to shut down the reactor rapidly. *See also* burnable poison.

DB

convection. The mode of *heat transfer in a fluid, involving gross motion of the fluid itself. Natural convection occurs at the interface between the two: fluid motion is due to the natural buoyancy forces occasioned by the changing density of the fluid in the vicinity of the interface. Forced convection requires applied motion of the fluid. The convection coeffecient, h, is a measure of the heat transfer over unit surface in response to unit temperature change, expressed in watts per metre Kelvin.

MS

convector. A device which promotes convective heat flow from a surface to a fluid. Specifically, a room-heating device forming part of a *central heating system or operating on electricity.

TM

conventions. Though the field of energy is on the whole a precise one, there are nevertheless a number of conventions which are based on custom, usage or convenience. For example, 0°C is the temperature at which ice melts at one bar pressure, though this has no particular thermodynamic relevance on an absolute scale. The compilation of *energy statistics requires certain conventions but these differ considerably, for example, as between OECD and the United Nations Organization. *See also* energy analysis.

MS

conversion ratio. An alternative term for *breeding ratio, usually applied to nuclear reactors whose breeding ratios are less than one.

DB

conversion tables. Tables have been included in the preface for the interconversion of units of energy, heat and work (table 1) and of units of power, i.e. energy per unit time (table 2). To convert a unit in the top row of either table into a unit in the first column, multiply by the value given at the intersection of the relevant row and column.

convertor. In nuclear engineering, a *nuclear reactor which has a *breeding (or conversion) ratio of less than one. All existing thermal reactors can be classified as convertor reactors. In contrast, *breeder reactors have a breeding ratio greater than one.

DB

coolant. A fluid circulated through any thermal system, such as the internal combustion engine of a car, for cooling purposes. In a *nuclear reactor it is circulated through the core to maintain the fuel at a safe temperature and transport the fission energy to heat exchangers in which steam is generated. Coolants used in current reactors include water, heavy water, liquid sodium and the gases carbon dioxide and helium.

DB

cooling load. The quantity of heat which must be extracted from an occupied space in order to maintain acceptable environmental conditions. This quantity varies over time and is a function on the one hand of climate, especially incident solar radiation, and on the other of the pattern of use of the building, especially lighting and occupancy levels. The maximum loading will dictate the sizing of cooling plant; the aggregation of the load over the cooling season will determine the energy consumption.

TM

cooling pond. A large tank of water in which the intensely radioactive *irradiated fuel elements unloaded from the *core of a nuclear reactor are stored for some time prior to reprocessing. The water in the tank acts as a radiation shield for the fuel elements and (possibly with the assistance of circulating pumps and heat exchangers) removes the *fission-product decay heat so as to keep them cool.

DB

cooling tower. A structure to promote the exchange of heat from a fluid, typically water, to the environment. The traditional pattern of tower, used to extract heat from the cooling water circulating in an electricity generating station, is some 50m high and of circular cross-section. A waist, some two-thirds up the structure, provides a venturi effect which causes air to be drawn in at the foot of the tower and expelled at the top; as it is drawn up within the tower it passes over the water which cascades down within the tower, thus effecting a convective and evaporative heat exchange. A more modestly sized cooling tower may be seen on the roof of buildings; such a tower lowers the temperature of the condenser fluid before it re-enters a *vapour compression chiller, evaporatively and convectively cooling it by contact with the air.

TM

coppicing. The cutting down of trees periodically so that new shoots may grow, thus obviating the need to replant following each harvest. This is an important factor when selecting species for an *energy farm. As most conifers (*softwoods) do not coppice, *hardwoods such as the *alder, *eucalyptus, poplar, sycamore and willow are preferred. Fast-growing species may be harvested every 3–5 years in *short rotation forestry to give high fuelwood productivities per unit land area.

CL

core (of a nuclear reactor). The central part of the reactor, containing the *fuel in which *fission takes place and, in the case of thermal reactors, the *moderator. The *coolant is circulated through the core, while the *cladding and structural components are required to give it strength and integrity.

DB

cosmic radiation. High-energy particles, principally *protons and *alpha particles which originate in outer space and bombard the earth's atmosphere. Interactions between these high-energy particles and the nuclei of oxygen and nitrogen in the atmosphere produce secondary cosmic radiation such as *gamma radiation which is incident on the earth's surface and is a component of the earth's *background radiation. The intensity of cosmic radiation in the atmosphere increases with altitude due to the fact that the atmosphere to some extent shields the earth from this radiation.

DB

cost. The total of the values (i.e. price times quantity) of the inputs to some economic activity; it will vary with the level of the activity. For a given level of the activity, cost is a measure of what must be given up in order to have the activity at that level. It should be compared with the *benefit of the activity in order to determine the desired level of the activity as that which maximizes the excess of benefit over cost. Where there is no *market failure affecting any input, cost is the same as expenditure on inputs; thus private and *social cost are the same and there is no *external cost.

In *project appraisal there is a stream of costs extending over time; these have to be reduced to present value terms for comparison with benefits at their present value, using the *discount rate. If the activity arises in the *public sector, or if market failure affects any input, then *cost–benefit analysis is called for. The construction of a dam for hydroelectric generation will mean the destruction of a natural environment, the loss of which is an input to the project which will not have a price put on it by markets, but which should be included in cost, in a cost benefit analysis, as a measure of what has to be given up to have the dam and hence the electricity.

MC

cost-benefit analysis. Most generally, any evaluation of an economic activity which considers *social costs and social *benefits as well as private costs and benefits. Typically, the term is used in a narrower sense to refer to a method of *project appraisal for investment decision arising in the *public sector, or for such arising in the private sector but which have effects where *market failure operates. The essential idea is that of the *compensation principle: the project should go ahead only if the gainers could compensate the losers and still be better off. It should go ahead only if the benefit exceeds the cost, with social benefits and costs included.

With benefit and cost streams extending over time, it is necessary to reduce each to a *present value, using a *discount rate, before comparing them. The project's net present value is the present value of benefit minus the present value of cost, and the decision rule is to undertake the project if its net present value is positive. Projects can be ranked by their net present value, and should be undertaken in the ranking order. With B_t as benefit at time t, C_t as cost at t, ρ as the discount rate, and T as the project lifetime, net present value is

$$\sum_{t=0}^{T} (B_t - C_t)/(1 + \rho)^t$$

Where market prices are either unavailable or inappropriate, inputs and outputs have to be priced, to get the B_t and C_t series, using *shadow prices, the derivation of which is a major difficulty in the application of cost–benefit analysis. Also, it is frequently difficult to enumerate and quantify all of the consequences, and hence inputs to and outputs from, a project. A further problem, the principles of which are disputed among economists, is the selection of the appropriate discount rate, ρ. Alternative candidates are the average *rate of return available on private sector projects, and the social rate of *time preference. The *distribution problem also gives rise to controversy. Typically, all contemporaneous cost and benefit items are weighted equally, irrespective of income levels of the individuals to whom they accrue. Some economists argue that extra (less) weight should be attached to benefits (costs) accruing to the less well off.

The decision on constructing a dam, for hydroelectric generation, in a national park illustrates the issues arising. The principal item in the benefit stream is the quantity of fossil fuel not burned in each year of the project's lifetime, this quantity being relatively easy to estimate. With a project lifetime of (say) 50 years, considerable uncertainty will, however, attach to the prices at which the quantities should be valued. Benefit items in terms of recreational opportunities for sailing and fishing will also arise, but estimating the quantity and value of such is extremely difficult. Construction costs arise early in the project lifetime and should not be difficult to estimate, ignoring, as is required throughout the analysis, the effects of *inflation on *absolute prices and costs. However, the cost stream arising from the recreational opportunities lost to national park visitors are extremely difficult to quantify and value: this stream may well effectively persist for ever, the effects of construction being essentially permanent and irreversible. This problem interacts with the choice of discount rate, since high rates mean that effects far into the future carry very small weight with respect to a decision based on present values. The significance of the distribution problem can be seen from two polar possibilities. In one, all electricity consumers, boat users and fishermen are rich, whereas all national park visitors are poor. In the other, it is the national park visitors who are rich and the project beneficiaries who are poor.

MC

cost effectiveness analysis. A special case of *cost–benefit analysis where the benefit stream is fixed and known to be worth having, the problem being the identification of the particular project, from among several feasible alternatives, which will deliver that benefit stream at least cost. The analysis has been used in assessing alternative weapons systems designed to serve the same military purpose. A choice between two types of nuclear reactor, given a firm decision to build one of them, would be a cost effectiveness analysis.

MC

cost insurance freight (cif). A contract for the shipment of goods overseas where the shipper includes in the price the insurance and freight charges, as well as the cost of the goods. In interpreting, for example, data on imported oil prices it is important to know whether these are reported on a cif basis or an fob (*free on board) basis. Movements in cif prices will include the effects of any variations in insurance and freight costs, as well as variations in the price of oil as such.

MC

cost-in-use. The total expenditure, including *capital costs and *recurring costs, incurred over the life cycle of equipment, plant, buildings, etc. In order to decide between alternative acquisitions, the costs-in-use of the alternatives may be predicted by assuming the life-span of each and the rates of interest which will prevail over the life-cycles. Costs-in-use may be expressed as a single 'present worth value' or as an 'annual equivalent' value.

TM

coulomb (Symbol: C). The SI unit of electric *charge, defined as the quantity of electricity transported per second by a *current of one *ampere.

TM

Council for Mutual Economic Assistance (COMECON). An intergovernmental council established in 1949 to improve and coordinate the planned economic development and integration of members: effectively a *customs union involving the USSR and eastern European states, with special features arising from the non-market basis on which the members' economies are largely organized. The founder members were Bulgaria, Czechoslovakia, Hungary, Poland, Romania and the USSR. Subsequent additions have been Albania in 1949 (left in 1961), German Democratic Republic in 1950, Mongolia in 1962, Cuba in 1972 and Vietnam in 1978. In 1964 Yugoslavia became an associate member.

In energy terms trade within COM-

ECON principally involves the movement of oil and natural gas from the USSR to eastern Europe: the east European states have participated in energy extraction developments in the USSR and in the construction of an oil pipeline to supply eastern Europe from the USSR. COMECON trade in energy with non-member states is small, but there are indications that this situation may change in the future.

MC

cracking. The breakdown of higher molecular weight hydrocarbons, by the action of heat and catalysts, into hydrocarbons of lower molecular weight. Hydrogen and solid carbon are usually produced as well. *See also* catalytic cracking; thermal cracking.

WG

critical assembly. *See* criticality.

critical group. A group of members of the public who receive more radiation than any other group as a result of the discharge of a particular *radionuclide to the environment from a particular nuclear plant. For example, the critical group for the discharge of the radionuclide caesium-137 from a nuclear reprocessing plant would be the group of persons who eat locally caught fish and shell-fish.

PH

criticality. The necessary condition for a self-sustaining fission *chain reaction to be established. A critical assembly of *fissile and other materials is one in which a chain reaction, once started, will continue at a steady rate, giving a constant power output. A subcritical assembly is one which does not have enough nuclear fuel for criticality, and cannot sustain a chain reaction. A supercritical assembly is one with more fuel than that necessary for criticality and in which a chain reaction takes place at an ever-increasing rate, giving an increasing power output. The critical size is the size of system which, for a given composition, leads to criticality. *See also* critical mass; multiplication factor.

DB

critical mass. The amount of nuclear *fuel required for the *criticality of a nuclear reactor. The critical mass is not a unique value but depends on the following factors: the composition of the reactor, i.e. the materials chosen for fuel, *moderator, etc.; the shape of the reactor; the percentage of *fissile isotopes in the fuel.

The variation in critical mass between different types of *nuclear reactor may be quite large. For example, the critical mass of natural uranium in a graphite-moderated thermal reactor is much larger than that of mixed uranium-plutonium fuel in a fast breeder reactor.

DB

cross-section (nuclear). A means of quantifying the rates at which *neutron and *gamma-ray interactions take place with atomic nuclei and electrons. It is an extremely important concept in the design and operation of *nuclear reactors.

The interaction rate per unit volume of any nuclide is equal to $\sigma N \varphi$, where N is the number of atoms per unit volume, σ is the microscopic cross-section (or the cross-section per atom) and φ is the neutron or gamma *flux. The product σN can be regarded as the total cross-section of all the atoms in unit volume of material; it is called the macroscopic cross-section, Σ, for neutron interactions or the absorption coefficient, μ, for gamma-ray interactions.

Different types of interaction are characterized by different cross-sections. In the case of neutrons, for example, the rate of *scattering per unit volume is expressed as $\sigma_s N\varphi$ or $\Sigma_s\varphi$, while the rate of *absorption per unit volume is $\sigma_a N\varphi$ or $\Sigma_a\varphi$; σ_s (Σ_s) and σ_a (Σ_a) are the scattering and absorption cross-sections respectively.

DB

crowding-out effect. A shorthand term for a variety of arguments to the effect that increased government expenditure must be at the expense of private-sector expenditure on either *consumption or *investment. The arguments concern the method by which the government expenditure is financed. If, for example, the government borrows to finance expenditure on a pipeline to collect gas from offshore oilfields, the argument is that the borrowing will push up *interest rates, so that private-sector investment falls as fewer projects pass the net *present value test at the higher interest rate. Some of the private-sector projects crowded out in this way may well have been in the energy supply or conservation field. Not all economists accept that the crowding-out effect is always operative. Some argue that, anyway, the public-sector project to be financed may show a higher rate of return than the private-sector projects crowded out. *See also* project appraisal.

MC

crude oil. *Petroleum from an underground reservoir which has been stabilized by the removal of the greater part of the hydrocarbon gases possibly associated with it. Crudes may vary from pale mobile liquids to dark viscous semisolid substances.

WG

C_3 plants. *See* 3-carbon plants.

cube factor [energy pattern factor]. A number used to relate wind speed to the power in the wind. This power, *P* (energy passing per unit time), is proportional to the cube of the wind speed, *v*, i.e.

$$P = a\,v^3$$

where *a* is a constant. However, wind speeds fluctuate greatly even over short periods, so the average power in the wind is given by

$$P = a(\bar{v^3})$$

where $(\bar{v^3})$ is the average of the cubed speed. $(\bar{v^3})$ is not equal to $(\bar{v})^3$, the average speed cubed. Their ratio

$$(\bar{v^3}) \div (\bar{v})^3$$

is called the cube factor or energy pattern factor. Commonly the value is nearly two. *See also* Betz theory.

JWT

Culham. A laboratory which is the centre of the United Kingdom Atomic Energy Authority's research in nuclear *fusion, plasma physics and associated technology. The Joint European Torus (JET), which is a cooperative project with other EEC countries is being built on an adjacent site.

Address: Culham Laboratory, Abingdon, Oxfordshire, England.

DB

curie (Symbol: Ci). A unit which is named after one of the pioneer scientists in the discovery of *radioactivity, Marie Curie, and was until recently the standard unit of radioactivity. Its value is based on early measurements of the activity of one gram of radium-226, and is defined as

$$1 \text{ curie} = 3.7 \times 10^{10} \text{ disintegrations/second}$$

It is now being superseded by the SI unit of radioactivity, the *becquerel, but it is likely that the curie will continue to be used for some time.

DB

currency. The notes and coins in circulation in a country, which constitute only a small proportion of the country's *money supply. The term is also used to refer to the particular named unit in which a country's money supply is measured, so that the currency of the UK is the £ sterling whereas that of the USA is the $.

MC

current. (1) (electric; symbol: *I*) A flow of electricity within a conductor. Current is measured in *amperes (A).

(2) (mass) A flow of fluids along a conduit.

TM

current account. (1) A subsection of the *balance of payments in which a country's overseas trade in goods and services is recorded.

(2) A type of *bank account on which cheques can be drawn and into which deposits can be paid. Interest is not usually paid when such an account is in credit, but it is usually charged if the account is overdrawn (i.e. in debit). In the USA current-account deposits are called demand deposits.

MC

customs union. A grouping of countries with no trading or customs barriers between them but who usually have an agreed common external tariff to non-members. The EEC is a customs union.

MC

cut-in speed. The lowest wind speed at which a rotating *turbine, usually linked to an electric generator, is connected to produce useful power. The speed is particularly important for *aerogenerators connected to an electrical *grid since at wind speeds below the cut-in wind speed the machine could draw power from the grid and operate as an electric fan. Cut-in speed for many aerogenerators is about 43 m/s (about 8 mph), usually specified at hub height or the standard meteorological measuring height of 10 m.

JWT

cyclotron. A particle *accelerator in which the charged particles are accelerated to high energies in a spiral orbit by means of intense electric and magnetic fields.

DB

cycloparaffins. *See* naphthenes.

D

dam. A wall or other form of *barrier constructed to retain water in a reservoir. Earth dams are largely composed of soil and rock obtained from the nearby surroundings and faced by a solid slope to prevent erosion by the water. Highly stressed sections of a dam are made of reinforced concrete.

JWT

damper. A moveable shutter arranged between the combustion chamber of a furnace and the chimney to control the rate of flow of the exiting hot gases.

WG

dark respiration. An oxidation process, mainly of carbohydrates, by plants to provide energy. It occurs primarily in the *mitochondria of plant cells. In order to survive and grow, plants need to respire using oxygen from the atmosphere, liberating carbon dioxide and water in the process. The quantitative difference between the carbon dioxide assimilated in *photosynthesis and that lost in respiration is called the net photosynthesis. Dark respiration occurs as rapidly in the light as in the dark. As much as 70% of the gross carbon dioxide fixed in photosynthesis may be used up in the process, greatly reducing potential *biomass yields.

The respiratory pathway may take two routes, one of which supplies only one-third as much energy to the plant as does the other. The elimination of the latter (which constitutes 15–20% of total respiration) by techniques of *genetic engineering would result in substantial increases in *primary productivity. In addition to dark respiration, plants also evolve carbon dioxide in an apparently wasteful light-stimulated process known as *photorespiration.

CL

Darrieus wind turbine. A vertical-axis wind machine with usually two thin blades in an egg-beater shape meeting at the top and the lower end of the axis. The blades or airfoils are shaped to develop *lift for high-frequency rotation.

JWT

data sources. *See* statistical sources.

daughter product. The derived product(s) of a radioactive decay process produced from the decay of an original radioactive substance. For example, in the decay of the radioactive *fission product tellurium-135:

$$^{135}_{52}\mathrm{Te} \rightarrow {}^{135}_{53}\mathrm{I} \rightarrow {}^{135}_{54}\mathrm{Xe}$$
$$\rightarrow {}^{135}_{55}\mathrm{Cs} \rightarrow {}^{135}_{56}\mathrm{Ba}\ \text{(stable)},$$

the nuclides iodine-135, xenon-135, caesium-135 and barium-135 are all daughter products of tellurium.

DB

daylight. The illumination available from the sky vault. The distribution of daylight throughout the year depends on the latitude of the locality: at the poles daylight is available continuously over half the year only, whereas at the equator daylight is available for about 12 hours daily throughout the year. Although the quantity of available daylight is affected by weather, for building-design purposes an international standard level of illumination from a uniformly overcast sky has been proposed and agreed as the basis for calculation.

TM

debentures. *See* capital.

decay. *See* radioactivity.

decay constant (Symbol: λ). A measure of the rate at which a radioactive substance decays (*see* radioactivity). Radioactive decay is a random process and the decay constant is defined for a single atom as the probability per unit time that it will decay. For a sample of material containing a large number (N) of atoms, the rate of decay is proportional to the number of atoms still existing, i.e.

$$dN/dt = -\lambda N$$

This is the basic equation for the rate of radioactive decay. dN/dt is the rate of decay; the negative sign denotes that N decreases with time.

DB

decay heat. *See* fission product decay heat.

decentralized energy. Energy supplies generated in dispersed locations and used locally, thus maintaining a low *energy flux from generation to supply. For example, the widespread use of *solar energy for domestic water heating from roof top collectors would result in decentralized energy. The term is often used for renewable energy supplies since these harness the energy flows of the natural environment, which are predominantly dispersed and have relatively low energy flux. In contrast, the centralized energy supplies of large-scale fossil and nuclear sources produce large energy fluxes which are most economically used in a concentrated manner.

JWT

decommissioning. The permanent shutdown of a nuclear facility at the end of its useful life. There are three principal methods of decommissioning in use: mothballing, entombment and dismantlement.

Mothballing is a temporary measure to allow radioactive decay to occur before final action is taken. It involves the removal of all easily removable radioactive objects from within the facility. Strict security measures are necessary, such as locked or welded doors and guards on constant duty, to prevent access of unauthorized persons during the mothball period. At a later date dismantlement or entombment is required.

Entombment involves making the contaminated parts of the facility permanently inaccessible by use of demolition techniques and by covering the remains with reinforced concrete. In dismantlement all contaminated parts are removed piece by piece, using remote-controlled equipment and heavy shielding. Then all surfaces are scoured with remotely operated equipment using water, steam and various chemicals. It may be possible to completely demolish a facility with explosives and demolition balls but this depends on local population density, cost, etc.

The total cost of decommissioning depends on the method utilized and on the length of time allowed for radioactive decay before permanent decommissioning measures start. For example mothballing a reactor for several years or decades can greatly simplify decommissioning operations. Reliable methods of estimating the cost of decommissioning have not yet been developed.

PH

decontamination. The process whereby a dangerous, toxic or radioactive spillage or leak is removed, rendering the immediate environment once more safe.

MS

decrement. The diminution of heat flow through a barrier, such as the wall of a building, due to its *thermal capacity. Decrement factors (*f*) for different densities and thicknesses of wall construction have been calculated by making the simplifying assumption that diurnal variations in climate are sinusoidal. These factors, used with equivalent factors for *thermal lag, allow the dynamic thermal behaviour of a space within a building to be approximated.

TM

dedicated. Describing or relating to a machine or system which is operated for a single particular purpose or in a prescribed position, e.g. electric *batteries used solely for charging by an *aerogenerator.

JWT

deforestation. The removal of mature trees from an area of land at a rate faster than young trees can replace them. It often results in the inability of the land to retain soil *nutrients and water. This in turn can lead to an often irreversible decline in the soil's capacity to support plant life, hastening the onset of *desertification. Deforestation often occurs because of an over-dependence on wood as a source of energy, particularly in developing countries. For example, the Sahel region of north Africa and also northern India and Nepal have been severely hit by what has become known as the firewood crisis, and so the environment has suffered accordingly. The clearing of forests to make way for farming land or for other purposes also contributes to deforestation, and it is estimated that 245,000 km^2 (approximately the size of the British Isles) of tropical forests are lost in this way each year.

CL

dehydration. The partial or complete removal of water from a substance by heating or some other physical process. Dehydration also describes the removal from one or more compounds of the elements hydrogen and oxygen, which together form water, by a chemical reaction. For example, ethyl alcohol may be dehydrated to ethylene:

$$C_2H_5OH = C_2H_4 + H_2O$$

See also ester.

WG

dehydrogenation. A chemical reaction which removes hydrogen from a molecule. In the *reforming of petroleum fractions, *paraffins are dehydrogenated to olefins and *naphthenes are dehydrogenated to aromatics.

WG

dehydrosulphurization. *See* hydrodesulphurization.

delayed neutron precursors. The radioactive *fission products which lead in the course of their decay to the production of *delayed neutrons. An example is the fission product bromine-87, which decays with a *half-life of 55 seconds (*mean life of 80 seconds) to form krypton-87. In 2% of these decays the krypton-87 is formed at an excited state of energy from which it decays instantaneously to produce krypton-86 and a neutron. This process, therefore, produces delayed neutrons with a mean delay time of 80 seconds.

DB

delayed neutrons. Although most *fission products and their *daughter products decay by the emission of *beta particles, a very few decay by the emission of *neutrons. These neutrons, produced some time after the originating fission event, are called delayed neutrons. The delay time is characterized by the *mean life of the *delayed neutron precursors. The fraction of delayed neutrons in the total of fission-produced neutrons is 1% or less, but they are crucially important in the control of a nuclear reactor undergoing a power increase. In the absence of delayed neutrons, safe reactor control would be very difficult or impossible.

DB

delivered energy. The actual amount of energy available or consumed at point of use. The concept recognizes that in order to have a unit of economically usable energy, there are prior exploration, production and delivery systems, each of which detracts from the net amount of energy delivered to the point of use.

MS

Delphi method. A method for technological forecasting, based on an intuitive consensus of expert opinion, in which a panel of experts are questioned individually with regard to their personal expectations for a series of hypothetical future events. It was developed in the early 1960s by Olaf Helmer and Norman Dalkey and their colleagues under the auspices of the RAND Corporation.

The method involves a carefully designed programme of sequential individual interrogations, usually conducted by questionnaire, interspersed with information and opinion feedback. Since group discussion, or presentation of position papers to a group, is avoided, the method eliminates the influence of interpersonal relationships which may bias the group consensus. After the first set of questionnaires is returned from the experts the results are collated and analysed statistically. At the beginning of the second round the participants are shown the results of the first round with a list of anonymous comments, and then asked to submit their revised estimates and comments. In the third (and later rounds) this procedure is repeated with further commentary and impersonal debate. Ideally, if the method works, the experts clarify and revise their thinking with a resulting convergence and narrowing of the range of estimates. The forecaster is provided with a valuable indication of the degree of unanimity and range of opinion.

PH

demand. *See* demand function.

demand curve. *See* demand function.

demand deposit. *See* current account.

demand function. The relationship between the demand for some commodity and the levels of those variables which determine demand for the commodity. By demand is meant the quantity which customers plan to purchase at the ruling price. The variables which determine demand are incomes and prices. For a particular commodity, the particular form for its demand function will arise from preferences between this and other commodities, as represented in household *utility functions. The form of the demand function for a commodity fixes the various *elasticities of demand.

It is part of the work of *econometrics to attempt to fix the numerical values of the parameters of demand functions, and hence of elasticities, from available data on incomes and prices and on quantities purchased. It is then possible to forecast future demand conditional upon income and price levels. Such conditional forecasts, treated with due caution in view of the inherent uncertainties, are then, for example, an important input to studies of likely future developments in energy markets.

A demand curve is the relationship between the demand for a commodity and the price of the commodity, when all other prices and incomes are held constant. Demand curves are popular because they can be represented graphically: they generally, but not universally, slope downwards, with quantity demanded falling as price rises.

MC

denominator. The divisor or bottom line in a division calculation. For example, in the calculation a/b, b is the denominator.

MS

deoxyribonucleic acid. *See* DNA.

Department of Energy (UK). The Ministry responsible for national policies in relation to all forms of energy, including those relating to the efficiency with which energy is used and the development of new sources of energy. It is also responsible for the international aspects of energy policy and the government's relationship with those energy industries which are in the *public sector.

Address: Thames House South, Millbank, Lon-don, SW1P 4QJ

MC

Department of Energy (US). The cabinet department of the US government, responsible for unifying and coordinating administration of US national energy policy and programmes. It combines the jurisdictions, programmes and personnel of several former federal agencies, including ERDA, FPC and FEA, in the areas of energy research and development, conservation, data collection, regulation and policy development.

Address: Washington, D.C. 20545, USA.

PH

depleted uranium. *See* uranium.

depletion allowance. An offsetting allowance against taxable income arising in mineral extraction. Since mineral deposits are non-renewable *natural resources, extraction necessarily reduces the value of the *asset, which is the mine or well. Hence, for mines and wells a depletion allowance is essentially the same as a *depreciation allowance such as is generally available to *firms in respect of their liability for profits tax (in recognition of the fact that the value of the firm's capital equipment is reduced in the act of generating profit). A firm operating an oil well in respect of which a depletion allowance is operative will typically also get a depreciation allowance in respect of the equipment being used to extract the oil. Relative to the situation without a depletion allowance in operation, it is generally argued that the effect of a depletion allowance is to speed up the rate of extraction so that the mine or well is more quickly exhausted.

MC

depletion programme. The plan over time according to which the owner of a non-renewable *natural resource deposit, such as a coal mine or an oil well, proposes to extract the resource and eventually exhaust the deposit. The nature of the depletion programme for a given resource deposit is determined by many factors, and will be revised as such factors change. Among the determining factors are: extraction costs and their variation with the rate, and cumulated past rates of extraction; the expected future behaviour of the price of the resource; the situation regarding the taxation of income and profits arising from extraction, especially whether or not there exists a *depletion allowance; whether or not the owner is the sole owner of the deposit (*see* common property resources); the *interest rate. Thus, for example, an increase in the rate of interest will, other things remaining constant, cause a revision of the depletion plan in favour of more rapid depletion, since it raises the *rate of return on holding money as compared with that on holding stocks of the unextracted resource.

MC

deposit gauge. A device which is used to measure the rate of deposition of *particulate material from the atmosphere around power stations and industrial plants.

PH

depreciation. (1) A measure of the decrease in the value of an *asset over a period of time. The decrease in value arises from wear and tear in use and/or obsolescence, and the sum of depreciation over the useful life of the asset should equal the asset's original value. Usually, firms compute depreciation per year as a constant amount given by the asset's original value and its assumed useful life, with depreciation being added to other costs in a year for computing *profit for the year. Typically, depreciation allowances can be offset against *gross profit in assessing liability for profits taxation. However, the rules which the taxation authorities specify for the determination of such tax offsets are not necessarily the same as the accounting conventions adopted by companies.

The accounting conventions of, and tax-allowance rules affecting, companies in the business of extraction from mineral deposits are especially important in affecting their *depletion programmes. Hence, for example, depreciation and its treatment are crucially important in the oil industry, where special *depletion allowances for tax purposes are common. The matter of depreciation is also important in the compilation of the *national income accounts, where, for example, national income can be measured gross or net of depreciation. It is the latter which is the more appropriate measure of a country's output for most purposes, since it indicates what is available for *consumption or *investment after making allowance for that part of current output needed to maintain the country's existing *capital stock intact.

(2) A decrease, under a system of flexible *exchange rates, in the value of a nation's *currency against other currencies. *See also* appreciation. MC

derived working limit. *See* radiological protection standards.

derv. Acronym for *d*iesel-*e*ngined *r*oad *v*ehicle. Derv fuel is a high-quality material burnt in high-speed *diesel engines. According to British Standards requirements it should have a minimum *cetane number of 50. In practice the value for the commercial material lies normally between 60 and 65. WG

desertification. The natural or man-made increase of desert areas. A common cause is decrease of trees or overgrazing by animals. Since trees are removed for wood fuel, there is a close relationship between energy demand that increases the need for *firewood and desertification. JWT

design rating. The anticipated output of an industrial plant, such as an electrical generation plant, based on the design calculations. MS

desulphurization. The removal of sulphur from a fuel. It is carried out to avoid pollution problems that would normally result when untreated fuels are burned and sulphur oxides are formed. WG

detector. *See* radiation detector.

detention time. *See* retention time.

detonating explosives. *See* high explosives.

deuterium. An isotope of *hydrogen, having an abundance of 0.015% in naturally occurring hydrogen. Deuterium is twice as heavy as hydrogen, its nucleus containing one proton and one neutron, but chemically they are identical. The separation of deuterium from hydrogen is a very expensive process, making use of the small physical differences between the properties of hydrogen compounds and their deuterium equivalents. Consequently, pure deuterium and its compounds are very costly. The most important commercial separation method is the Girdler sulphide process, which depends on the variation with temperature of the concentration of deuterium in water and hydrogen sulphide.

Deuterium is important as one of the fuels in *fusion reactions Two deuterium atoms can combine to give hydrogen, *tritium (the radioactive isotope of hydrogen) and a large energy release. They can also react to yield helium, a neutron and a large energy release.

These reactions are *exothermic, so deuterium represents a very large source of energy once the technology of fusion power has been developed.

Deuterium is also important in nuclear energy by virtue of its low mass number and its low neutron capture *cross-section: the compound *heavy water (deuterium oxide) is used as a *moderator in some types of thermal reactor.

Symbol: D or $^{2}_{1}H$; mass number: 2; atomic mass: 2.041.

See also heavy water; nuclear reactor (CANDU).

DB

Deutsche Gesellschaft für Mineralölwissenschaft und Kohlechemie e. V. A publishing society for petroleum science and coal chemistry.

Address: 2 Hamburg 1, Nordkanalstr. 28, Federal Republic of Germany.

WG

devaluation. If a country sets an agreed *exchange rate (referred to as either a parity rate, as under the Bretton Woods Agreement, or a central rate, as under the Smithsonian Agreement) which is expressed as the number of home currency units per dollar (or some agreed *numeraire) then a devaluation denotes a rise in this rate; a *revaluation denotes a fall in this rate. Note, if the exchange rate is expressed as dollars per unit of home currency (e.g. the sterling exchange rate) then the movement is the opposite to that just stated. *See also* appreciation; depreciation.

MC

development threshold. The level of development at which a country achieves literacy, health care, diet and infant mortality rates characteristic of developed countries. It has been observed that these indicators *inter alia* are approximately similar across most developed countries, while they are at much poorer levels in less developed countries. A commonly used criterion of development threshold is a certain level of per capita *gross national product, but a more reliable and less inflation-prone criterion may be one based on national energy flux, where this is defined as national energy consumption divided by national territorial area.

PH

dewaxing. The removal of wax from a petroleum fraction by cooling and then separating it as solid crystals, or by contacting the fraction with a selective solvent which dissolves the non-wax hydrocarbons and precipitates the wax as fine crystals.

WG

dew point. The temperature at which air or other gas is so saturated with vapour that condensation takes place. Since the concentration of water vapour that will saturate air at various temperatures is known accurately, and has been tabulated, a measurement of dew point temperature may be used to determine air humidity, i.e. the concentration of water vapour in air for a given set of conditions.

WG

dielectric. A substance containing few, if any, free electrons. It therefore resists the passage of an electric current. Dielectric materials are used in the insulation of electric cables.

TM

diesel cycle. *See* heat engine.

diesel engine. An internal combustion engine in which ignition is caused by compression. The temperature of air in the cylinder is thus raised above the spontaneous ignition temperature of the *diesel fuel that is sprayed into the cylinder head. The *compression ratio in a diesel engine may be as high as 22:1, thus it is more strongly built and heavier than a petrol (US: gasoline) engine. Since there is no high-tension spark circuit, there is less to go wrong with the engine under adverse conditions. It has a longer life and lower maintenance cost than a petrol engine. Diesel engines range in size from tiny units used in model aircraft to very large units used in oil-tanker vessels and power stations.

Diesel engines operate on a four-stroke cycle in which air is drawn into the cylinder and compressed. Just before the point of maximum compression the fuel is injected. The fuel ignites and drives the piston down in a power stroke. The piston then returns pushing waste gases to exhaust. To ensure vaporization and complete combustion of the fuel, the engine is designed to take in a large excess of air. The products of combustion thus contain little or no carbon monoxide. WG

diesel fuel. A liquid fuel for an engine in which ignition is achieved by air compression. Air so compressed is raised to a temperature of 500°C in a few rotations of the crankshaft. When the fuel is injected into the hot cylinder, it vaporizes and is raised to its *autoignition temperature. For successful operation this should be at least 30°C below that of the air. Among the hydrocarbons, alkanes have the lowest autoignition temperatures and are preferred in a diesel fuel. Fuels with low *octane ratings are thus most suitable as diesel fuels. For high-speed diesel engines, such as are used in road vehicles, a petroleum fraction (derv) with a boiling range of 200 – 300°C is used. Large stationary or marine diesel engines with better cooling facilities can use fuels of lower volatility and less well refined. The ignition quality of a diesel fuel is measured in a standard engine and compared with the reference hydrocarbon, cetane, and thus given a *cetane number. WG

diesel index. A measure of the ignition quality of a *diesel fuel which does not require it to be tested in a standard engine. The *API gravity and *aniline point are determined, and the diesel index calculated from a formula incorporating the results of these tests. There is good agreement between the values of *cetane number and diesel index for the same fuel. WG

diffuse radiation. Solar radiation arriving not from the disc of the sun, but from the sky and clouds or reflected from neighbouring objects. JWT

diffusion. (1) The distribution of light rays in numerous directions during reflection or transmission.

(2) The random motion of atoms and molecules in a gas or liquid. TM

diffusion coefficient. *See* Fick's law.

diffusion length. An important parameter in nuclear-reactor theory which characterizes the distance travelled by neutrons in a material before being absorbed. For three important *moderators used in thermal reactors, the values of the diffusion length for thermal (low-energy) neutrons are:

water	2.85 cm
heavy water	170 cm
graphite	51 cm

These figures provide one reason why, in order to minimize neutron leakage, the cores of graphite-moderated reactors are larger than those of water-moderated reactors. DB

diffusion process. *See* enrichment.

diminishing marginal returns ['law' of diminishing returns]. The marginal return to an input to production is the additional output arising when the level of the input in question is increased by a small amount, the level of other inputs being held constant. The law of diminishing marginal returns states that with successive increases in the level of an input, the marginal return to it will eventually decrease. The law is, in economics, an empirical generalization rather than a logical necessity following from the basic laws of nature. A good example relating to energy concerns the use of fertilizer in agriculture, fertilizer production being energy-intensive. With a fixed amount of land, successive increases in fertilizer application, after a certain level has been reached, yield decreasing crop gains. Since the existing level of fertilizer input is, typically, much higher in the UK than in India, for example, the gains to increased input would, typically, be higher in India than in the UK. MC

diode. An electrical device which permits current to pass in one direction only. Thus alternating electric current may be transformed to direct current by connecting a diode in series with the electrical supply.

JWT

direct current (dc). An electric *current which flows in one direction only. *See also* alternating current.

JWT

direct cycle. A term used to indicate that the *coolant in a *nuclear reactor serves also as the working fluid in the thermodynamic power cycle, and passes from the reactor to the turbine without the need for an intermediate heat exchanger. There are considerable thermodynamic and economic advantages in this arrangement. The boiling-water reactor is the only current example of a direct-cycle nuclear power plant (*see* diagram). However, the high-temperature gas-cooled reactor may be developed for a direct-cycle gas-turbine power plant, with helium gas as the reactor coolant and working fluid.

DB

direct energy. Fuel consumed in the operation of a manufacturing, agricultural or other activity. It is in contrast with indirect energy, which is the energy used in the manufacture of the inputs used in the above activities. The sum of the direct and indirect energies leads to the assessment of *energy requirement and *gross energy requirement.

MS

direct radiation. *See* *beam radiation.

discounted cash flow. The cash flow series arising from a project expressed in *present value terms. Thus, with R_t and E_t as receipts and expenditures arising at time t, where the project life is T, the net cash flow is

$$(R_1-E_1), (R_2-E_2), \dots (R_T-E_T)$$

and the discounted cash flow is

$$\frac{(R_1-E_1)}{1+\rho}, \frac{(R_2-E_2)}{(1+\rho)^2}, \dots \frac{(R_T-E_T)}{(1+\rho)^T}$$

where ρ is the appropriate *discount rate. *See also* internal rate of return; project appraisal.

MC

discount rate. in *cost–benefit analysis, the conversion factor by which money values arising at different points in time are made commensurable at a single point in time. Thus the stream of current money values arising at times 1, 2, 3, . . . T, and denoted by V_1, V_2, V_3, . . . V_T,

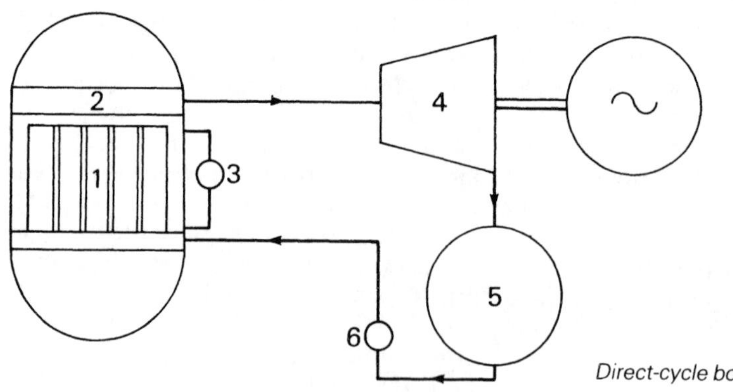

Direct-cycle boiling-water reactor

becomes a stream of *present values at time 0, given by

$$\frac{V_t}{1+\rho}, \frac{V_2}{(1+\rho)^2}, \frac{V_3}{(1+\rho)^3}, \ldots, \frac{V_\tau}{(1+\rho)^\tau},$$

where ρ is the discount rate. In *project appraisal projects, the discount rate to be used is the appropriate *interest rate on the funds to be used to finance the project. In cost–benefit analysis the appropriate discount rate to use is less clearcut and the matter of some dispute. Where cost–benefit analysis is being applied to a *public sector project financed by government, it is often the case that the government fixes the discount rate value to be used, which rate is sometimes referred to as the test discount rate. Decisions on projects are more sensitive to changes in the discount rate the longer the life of the project. Thus many energy supply investment decisions, involving long construction periods as well as long project lifetimes, as for example in the case of nuclear power stations, are sensitive to the value used for the discount rate.

MC

discriminating monopolist. A monopolist whose customers can be separated into two or more separate markets in which different prices can be charged. Thus, for electricity supply the industrial market can be separated from the domestic market, since purchasers in one market cannot resell their purchases in the other market. Where possible, discrimination enables a monopolist to increase total revenue and profit so long as the *elasticities of demand are different in the various separate markets.

MC

disequilibrium. The condition in an economy of not being in an *equilibrium state, so that disequilibrium analysis is concerned with the forces making for, or not, a return to an equilibrium state. For example, the 1973/74 oil price shock put the energy and related markets into disequilibrium. Subsequent changes in the prices of other fuels were one of the forces making for a return to, a new, equilibrium in energy and related markets. *See also* general equilibrium analysis; market equilibrium; partial equilibrium analysis.

MC

disintegration. *See* radioactivity.

dispersants (for oil spills). Chemicals which break down oil into small droplets which are then rapidly dispersed in the sea and cannot recombine to form a slick. They accelerate the natural process of dispersion so that *biodegradation of the oil is enhanced. Dispersants are made up of two chemical components: a surface active agent (surfactant or detergent) and a suitable solvent. The precise nature and proportions of the two components vary from product to product but they can in general be classified into two categories: those based on hydrocarbon solvents and those based on alcohol solvents.

Dispersants have their limitations when used on oil at sea. They are generally capable of dispersing most liquid oils and liquid oil-in-water emulsions, but are not suitable for dealing with stable water-in-oil emulsions (*chocolate mousse). In the 1960s over-enthusiastic use of dispersants caused considerable damage to coastal ecology as a result of poor application techniques and the toxicity of the hydrocarbons used as solvents.

PH

disposal (of radioactive wastes). *See* radioactive waste management.

dissipation, energy. A precise term for

energy consumption, since, according to the *first law of thermodynamics, energy can neither be lost nor gained. It is therefore in strict scientific terms incorrect to talk of energy production or consumption. Energy dissipation is thus often used to refer to energy consumption, inferring correctly a process of degradation to a lower temperature.

MS

dissociation. The reversible decomposition of a molecule into one or more fragments; these may themselves be relatively stable molecules, atoms, radicals or ions. In combustion processes above 1500 °C, there is significant dissociation of the carbon dioxide and water vapour. Thus the temperature of the gases does not rise in proportion to the heat absorbed by them, since dissociation is usually an *endothermic reaction.

WG

distillate fuel. A light fuel oil containing only relatively volatile material. A fuel oil which when distilled leaves no high-boiling residue.

WG

distillation. A process in which a liquid is boiled and the vapours collected and condensed. It provides a convenient means of separating components of liquid mixtures of differing volatility. Distillation may be a batch process as in whisky preparation, or continuous, as in an oil refinery.

WG

distillation column. An industrial device used in *distillation. Often called a *fractionating column in a petroleum refinery, it is a tall metal tower of a height sufficient to achieve the desired separation of the components of a liquid mixture and of a diameter adequate to handle the desired output of distillate. It is normally lagged to prevent heat loss, and may be continuous or batch operated.

WG

distillation cut. A distillate collected from a *fractionating column between two predetermined temperatures: a liquid with a narrow specified boiling range.

WG

distribution problem. The question of fairness or justice in the allocation of resources. Competitive market forces may, in ideal conditions, allocate resources in such a manner as to produce *efficiency in allocation. There is no presumption that the outcome, in the absence of government intervention via direct or *indirect taxation or *subsidies, will be generally regarded as fair. While rising energy prices act to conserve energy, they also cause hardship for low-income households. It has been argued that the government should act to protect low-income households from the full effects of increased energy prices by way of subsidies or special allowances.

MC

district heating. The supply and distribution of heat energy to domestic, commercial and industrial consumers throughout a neighbourhood or district. The heat may be raised specifically for the purposes of district heating or may be co-produced with electricity generation, as in a *combined heat and power system.

TM

diversity. (1) In general, the range of sources or facilities.

(2) The ratio of the maximum load likely to be experienced by an energy supply and distribution system to the system's total connected load.

TM

DNA (deoxyribonucleic acid). The genetic material which controls the structure and function of all living organisms. It is found within the nuclei of cells in structures called chromosomes.
PH

domestic heating. The system employed within a house to provide *thermal comfort. Domestic heating systems take a wide variety of forms: open-fire combustion of fossil fuels, closed-fire combustion of fossil fuels to heat circulating water or air, radiant conductive or convective use of electrical energy, *district heating. According to some sources, the maintenance of thermal comfort in buildings, in Europe, accounts for over 50% of the total primary energy used; the bulk of this amount of heat energy is used in domestic premises. TM

domestic heating oil. Liquid fuel used in domestic central-heating installations. It may be a grade of *paraffin (US: kerosene) used with vaporizing *burners or gas oil used with pressure-jet burners.
WG

domestic income. The sum of the payments to *factors of production in an economy, and a measure of the total incomes generated by the production of *domestic product. It differs from *national income by the amount of income received from abroad. *See also* national income accounts.
MC

domestic product. The total output *value produced in an economy in terms of goods and services to meet the demand for *final product, as distinct from goods and services which are *intermediate products used only in the production of other goods and services. If measured *net, as opposed to *gross, it is the value of domestically produced output potentially available for *consumption without running down the nation's stock of *capital. It is a widely used indicator of the level of economic activity. Cross-country comparisons of energy-use levels are most usually presented in terms of figures for each country for the ratio of *per capita energy use to per capita domestic product, thus revealing differences in the energy intensity of economic activity. *See also* national income accounts.
MC

Doppler broadening. When the temperature of the fuel in a *nuclear reactor increases, as happens when its power increases, the shape of the *resonances in the neutron *cross-sections of the fuel also change. The resonance peaks become broader, an effect known as Doppler broadening, and the absorption of resonance-energy neutrons increases. This increase in the neutron absorption rate affects the *reactivity of the reactor. In a thermal reactor the principal effect is to increase neutron capture in uranium-238, and reactivity is decreased. In a fast reactor neutron capture in uranium-238 and fission in plutonium-239 both increase and, depending on the relative amounts of these two nuclides, reactivity may either increase or decrease. It is desirable from the point of view of reactor safety that the reactivity should decrease following a power and temperature rise. DB

Doppler coefficient. The measure of the change in *reactivity of a *nuclear reactor per degree change in the temperature of the fuel due to the effect of *Doppler broadening. It is important from the point of view of reactor safety that this coefficient should be negative, denoting a decrease of reactivity following a rise in the fuel temperature.
DB

dose. *See* radiation dose.

dose equivalent. *See* radiation dose.

double duct system. *See* dual duct system.

double glazing. *See* glazing.

doubles. Coal with an approximately uniform lump size, the pieces of which are about 50 mm in diameter.
WG

doubling time (nuclear). The time taken for a *breeder reactor to produce a surplus of fissile material which is equal to the original amount loaded into it. The doubling time is inversely proportional to the *rating of the reactor and is proportional to the *breeding ratio minus one. Doubling times for typical breeder reactors are of the order of 20 to 30 years.
DB

Dounreay. A research establishment of the UK Atomic Energy Authority in Caithness, Scotland which is the centre for fast breeder reactor research and development in the UK. The Dounreay Fast Reactor (DFR) was commissioned in 1963 and operated as a research and materials testing reactor until 1978. More recently the Prototype Fast Reactor (PFR) has been built at Dounreay and is now operating at a power of 600 MW(th) and is supplying 250 MW of electricity to the national grid as well as providing a fast breeder reactor experimental facility. Fuel reprocessing work at Dounreay is aimed at completing all stages of the plutonium fuel cycle for fast reactors. It is the proposed site of the European Demonstration Reprocessing Plant [EDRP].

Address: Dounreay, Thurso, Caithness, Scotland.
DB

downwind. On the side of a device facing away from the wind. Relative to another position, the position where wind arrives after some delay.
JWT

drag. The force acting on an object in a fluid (liquid or gas) that pulls the object parallel to the direction of relative motion. The component of drag force is perpendicular to the complementary component of *lift force, as shown in the diagram, where an elementary *airfoil *blade section is drawn.
JWT

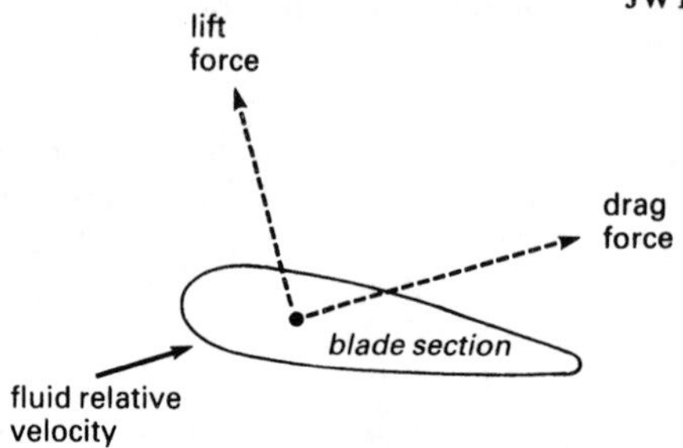

drag-type turbines. Machines rotated by the *drag force of a moving fluid. The standard meteorological cup *anemometer has a drag-type turbine. The maximum speed of a drag-type turbine blade cannot be greater than the speed of the fluid, in contrast to a *lift-type machine.
JWT

draught. A term usually restricted to furnace technology and describing the pressure difference causing a flow of gas through the furnace. It is the difference in density between the hot gases within and the cool air outside the chimney which causes the pressure difference.
WG

draught animals. Animals used for pulling vehicles, machines or tools.
JWT

drill. A device for boring holes. When applied to oil technology, it refers to a complex system able to drill holes of 10 cm or more in diameter up to several thousand metres into the earth in search of oil or gas.
MS

dross. Coal particles which are sufficiently small to pass through a screen with a 25 mm (1 inch) diameter mesh size. It is thus a mixture of particles ranging from about 20 mm diameter to dust.
WG

drouth. Literally dryness, referring to the atmospheric environment; quantitatively 100 minus *relative humidity. MS

dry ash-free (DAF). A basis for calculating and reporting the analyses of coal after the content of moisture and ash have been subtracted from the total. WG

dry basis. A basis for calculating and reporting the analysis of a fuel after the moisture content has been subtracted from the total. For example, if a coal sample contains $A\%$ ash and $M\%$ moisture, the ash content on a dry basis is:

$$\frac{100A}{100-A}\%$$

WG

dry bottom furnace. A pulverized fuel combustion chamber in which the lower section is cooled so that droplets of molten ash solidify and collect as a dry powder.

WG

dry coal. Coal which has been heated in a laboratory oven to 100 °C until its weight is constant. *See also* moisture content.

WG

dry gas. *Flue gas containing no water vapour or *natural gas containing no condensable hydrocarbons under operating conditions. WG

dry mineral-matter free (dmmf). A basis for calculating and reporting the analyses of coal after the moisture and mineral content have been subtracted from the total. WG

dual duct system [double duct system]. A system of *air conditioning for a building in which dehumidified air is circulated throughout the building via two parallel ducts from a centrally located plant room. Hot air circulates within one duct, cold air within the other. The proportion of hot air and cold air delivered to any room within the building can be thermostatically controlled.

TM

dual pressure cycle. A thermodynamic power cycle in which steam is generated in the heat exchangers at two pressures in order to increase the thermal efficiency of the plant. Such a cycle is associated with Magnox *nuclear reactors.

DB

ducted rotors. *Wind turbines with conical-shaped ducts or funnels to enhance the flow of wind onto the rotors and *airfoils (*see* figure).

JWT

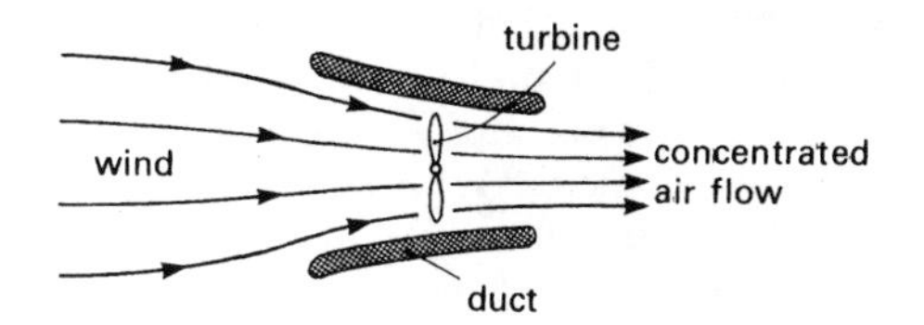

Ducted wind turbine

ductwork. A system of passageways through which run the engineering services between and within buildings. Ductwork between buildings takes the form of covered trenches which may provide passageway for a range of services – electricity, gas, *district heating, drainage and sewage. Within the building fabric ductwork runs horizontally and vertically to facilitate distribution of the engineering services to all parts of the building. Certain services, e.g. conditioned air, will themselves be ducted, these ducts running either within or outside the more general ductwork system.

TM

dumping. The selling of a commodity abroad at a lower price than is charged at home. It is a form of price discrimination. It can arise in the *short run when producers wish to unload unwanted stocks. When it arises as a means of securing an increased market share (predatory pricing) then there may be justification for antidumping duties, as allowed under the rules of the *General Agreement on Tariffs and Trade.

MC

Dumping at Sea Act 1974 (UK). The national legal authority for the implementation of the *Oslo and *London Dumping Conventions of 1972. Ships without a licence from the British government are prohibited from dumping substances or articles within British territorial seas and, if the dumping is from a British ship, within the high seas.

PH

duopoly. A market in which there are only two producers and (usually) a large number of purchasers. Each producer has some control over his price and output, but must take account of the reactions of his competitor. Different economic *models can be formulated depending on the type of reaction assumed.

MC

durable [durable commodity]. A commodity that yields a flow of services over a number of time periods, rather than being used up in a single period. The commodities comprising a stock of *capital are all durable in this sense, the services provided in a period of time being *depreciation in that period. Consumer durables are durable commodities bought by households. A household's central heating system is a consumer durable, whereas the fuel which it uses is not. Patterns of expenditure on durables are less quickly responsive to changes in relative prices than patterns of expenditure on non-durables. The durability of fuel-using equipment can slow down the rate at which fuel-use patterns respond to changes in the *relative prices of fuels.

MC

durain. A dull greyish-black constituent of coal, recognizable when viewed in the lump. It is hard and difficult to break up and is seen under the microscope to be composed of plant fragments and spores. It is generally interbanded with clay and is thought to originate from vegetable mush collected in swamps during the Carboniferous period of geological time.

WG

dynamic viscosity (coefficient of). *See* viscosity.

dynamometer. Any of a class of instruments which use the principles of *electromagnetism to measure electric current, voltage and power in a circuit. One group of instruments measures torque and hence the power of a prime mover or electric motor.

TM

dyne. A unit of *force, now generally superceded by the SI unit, the *newton (N). One dyne applied to one gram mass during one second will produce a velocity of one cm per second. One dyne is equal to 10^{-5} N.

TM

E

eco-energetics. A type of *energy analysis in which natural energy flows are included. A basic philosophy of eco-energetics is that an economically useful activity is one in which the amount of non-renewable fuel consumed is less than the effective solar energy thereby captured. In this way consumption of fossil fuels can lead to an expansion of the energy supply to the world.

MS

ecology. The study of the relationship of living systems with each other and with their environment. Human ecologist view mankind as a living system which must exist in ecological balance with its environment and other living systems if its life is to be sustained. They suggest that the ecological basis of mankind has been little appreciated by engineering technologies which may over-consume the earth's natural resources. Energy-consuming activities of mankind that affect such ecological balances include air pollution from fossil-fuel power stations, radioactive effluent from nuclear power stations, the destruction of natural woodland and forests for fuelwood, and the submerging of land in hydro-power reservoirs.

JWT

econometrics. The application of the methods of statistical inference to historical data for the purpose of testing and quantifying hypotheses about economic behaviour. A widely used econometric technique is *regression analysis. A typical application is the study of the demand for, say, oil. Economic theory suggests that the quantity of oil demanded falls as the price of oil rises, and increases with increases in gross *domestic product. Most studies have shown the data to be consistent with these hypotheses, and have provided numerical values for the various *elasticities of demand for oil. These values can be used for forecasting the future demand for oil, conditional upon knowing future oil prices and levels of gross domestic product. The methods used also indicate the degree of error to be expected in the conditional forecast.

MC

economic growth. An increasing level of *national income, or equivalently *national product, measured *net and in terms of *constant price series. Some economists argue that measured in this way, the rate of economic growth overstates the rate of increase in the well-being of the average citizen due to the neglect of *external costs. It has also been argued that in the future economic growth will be more difficult to achieve due to the depletion of *natural resources, especially energy resources. Most economists take the view that economic growth will remain a desirable and feasible objective of economic policy for the foreseeable future, with resource constraints being overcome by the accumulation of *capital and the effects of *technical progress. *See also* national income accounts.

MC

economics. A social science concerned with the way resources are allocated among alternative uses in order to satisfy human wants. It deals with the production of goods and services, how they are distributed and the price at which they are exchanged. Basically it deals with the relationship between ends and scarce means which have alternative uses. Positive economics deals with economic statements which can be falsified by appealing to real world observations; normative economics deals with questions which cannot be falsified by empirical observation alone because they involve statements about what ought to be. Economics is subdivided into *microeconomics and *macroeconomics.

MC

economies of scale. The decrease in a firm's average costs in the *long run as output rises. They can arise from many sources, e.g. as a result of output increasing in greater proportion to a proportionate increase in all inputs (i.e. increasing returns to scale); a reduction in the costs of inventory holding; marketing economies; managerial economies. Internal economies of scale arise within the firm as a result of its own actions, while external economies of scale arise from the expansion of the industry as a whole. Economies of scale have been strongly evident in the electricity supply industry and in the oil transport industry, showing up in the historical trends towards larger generating sets and tankers respectively.

MC

economizer. A heat exchanger linked to a steam boiler in which heat from the flue gases is transferred to the entering feedwater. The heat loss is thus reduced and the thermal efficiency of the boiler plant increased.

WG

ecosystem. A term first used by A. G. Tansley in 1935 to describe a community of living organisms and their environment. An ecosystem is any area with a boundary through which the input and output of energy and materials can be measured and related to some unifying environmental factor. On land the major ecosystems, usually called biomes, are the forests, grasslands, savannas, tundra and deserts. Each of these can be divided into smaller units, such as oak-wood, rain forest and swamp. All of the various ecosystems are interconnected and together form the planetary ecosystem or *biosphere.

PH

effective dose equivalent. *See* radiation dose.

effective temperature. *See* temperature.

efficiency. (1) In economics, production or action with the minimum of expense, waste and effort, i.e. the best use of resources to produce something. Technical efficiency is the maximum output for a given set of inputs or the minimum input requirements for a given output. Economic efficiency refers to the least cost method of producing some level of output. Thus technical efficiency does not involve the price of inputs while economic efficiency does.

Allocative efficiency is the best allocation that can be achieved with given resources, where best means that it is not possible to alter the allocation so as to make one individual feel better off except at the cost of making some other individual(s) feel worse off. Allocative efficiency is also referred to as Pareto optimality. It does not imply that the allocation concerned is fair or just. The fact that, for example, the electrical supply industry is operated in a way that is technically efficient does not itself imply that its operations are economically efficient, or that they are promoting allocative efficiency. While technical and economic efficiency are necessary for allocative efficiency, they do not guarantee it.

(2) In terms of energy, efficiency may be considered in two ways: absolute and relative. The absolute efficiency is the ratio of energy (as *heat or *work) transferred from a source to a function or purpose. For example, a *heat exchanger can be designed so that almost 100% of the heat supplied as a hot stream is transferred to the colder stream. By contrast if work is to be done, the energy that can be transferred as work is thermodynamically limited and can never reach 100%. The theoretical maximum efficiency is then calculated from the *Carnot cycle. Where heat transfer is concerned, *see* first law efficiency; where work is concerned, *see* second law efficiency.

The relative efficiency is the ratio of the actual energy transferred (as heat or work) relative to the maximum that could be transferred if the device was operating at theoretical efficiency, calculated from the Carnot cycle. Thus if a heat source at 1000 kelvin is used to generate electricity by heat transfer to a steam in a *boiler, which then feeds steam to a *turbine to produce electricity, the efficiency of heat transfer in the first stage may be as high as 90% (first law efficiency) while in the second step only 40% of the heat is turned into electricity (work). However it may be shown from the Carnot cycle that the maximum theoretical efficiency for the transfer of work can only be

$$(T - T_0)/T$$

In the above example, T is 1000 K and T_0, the *ambient temperature, is say 10°C or 283 K. Thus the theoretical or maximum possible efficiency is

$$\frac{1000-283}{1000} = 71.7\%$$

and the relative efficiency is

$$\frac{.4 \times .9}{.717} = 50.2\%$$

Similarly a wind generator may convert 20% of the energy in the wind to electricity. However it has been shown by the *Betz theory that 59% is theoretically available. The relative efficiency is then

$$\frac{.20}{.59} = 34\%$$

In popular usage efficiency is often confused with effectiveness.

MC, MS

effluent. Any waste released from an industrial or agricultural plant to the surrounding environment. For example, a nuclear fuel reprocessing plant releases a gaseous effluent into the atmosphere and a liquid effluent which must be carefully disposed of. *See* radioactive waste management.

PH

Einstein's equation. An equation which expresses the relationship between mass and energy and is valid for any process or reaction in nature in which energy is released or absorbed and the mass of the reactants changes. It states that

$$E = mc^2$$

where E is the energy released or absorbed in any reaction, m is the mass decrease or increase and c is the velocity of light. In the SI system c is equal to 3×10^8 m/s, and the equivalence between mass and energy is:

1 kg mass is equivalent to 9×10^{16} J of energy

This is equivalent to the heat of combustion of two million kg of oil.

In an *exothermic reaction energy is released and the mass of the reactants decreases. In an *endothermic reaction energy is absorbed and the mass of the reactants increases. As a consequence, the classical laws of conservation of mass and conservation of energy are not separately valid for such processes, but are replaced by the statement that:

$$\text{energy} + (\text{mass} \times c^2) = \text{constant}$$

DB

elasticity. A dimensionless measure of the responsiveness of one economic variable to a change in some other economic variable. The concept is most widely used in the context of the *demand function. Where the term is used without qualification it refers to the 'own price elasticity of demand', i.e. the proportionate change in the quantity demanded, divided by the proportionate change in the price of the commodity concerned, it being understood that all other prices and incomes remain unchanged. If, for example, with everything else constant, a 1% rise in the price of oil leads to a 2% reduction in the quantity of oil demanded, the own price elasticity of demand is –2. When reporting elasticities of demand it is conventional to omit the minus sign, negativity being understood.

The value of the elasticity of demand for a commodity will depend largely on the ease with which it can be replaced in use by other commodities, and it is not a number constant over all circumstances and all times. Apples have a relatively high elasticity of demand, being generally replaceable by other fruits. Oil has a relatively low elasticity of demand, with quantity falling little as price rises due to the difficulty of replacing it in most uses. As the price of oil rises, so the gain to overcoming the difficulties will increase, and the elasticity of demand would be expected to increase. Oil use can be reduced by installing improved control equipment, and for any given proportionate increase in the price of oil, the gain from so doing will be the greater the higher the initial price of oil. The role of time is important. In the *short run oil users are committed to particular *durable items of oil using equipment so that the response to oil price rises is constrained. In the *long run equipment can be changed: the full effect on demand of an oil price rise may take many years to work itself out.

The 'cross price elasticity' of demand is defined for pairs of commodities, being the proportionate change in the quantity demanded of commodity one, divided by the proportionate change in price of commodity two. Where a cross price elasticity is negative, the commodities exhibit *complementarity; where it is positive, the commodities are *substitutes.

The 'income elasticity of demand' is the proportionate change in quantity demanded, divided by the proportionate change in income, all prices being held constant. For most commodities the income elasticity of demand is positive. If the income elasticity of demand for energy is less than unity, as it appears to be for developed economies, energy demand will grow less fast than *gross domestic product, other things being equal. If the own price elasticity of demand is low, as it appears generally to be, in the short run at least, energy demand will not be very responsive to changes in the price of energy.

MC

elastic scattering A form of neutron *scattering in which the total kinetic energy of the neutron and nucleus is conserved. It is analogous to the collision between two perfectly elastic spheres. Elastic scattering is of particular importance in thermal reactors (*see* nuclear reactor) in which repeated scattering collisions between neutrons and the nuclei of the *moderator reduce the energy of the neutrons to thermal energy.

DB

electric fire. A space heating device powered by electricity. An electric current passes through a filament of high *resistance, causing it to become incandescent and radiate heat.

TM

electricity. The form of energy derived from either moving or stationary electric charge, usually *electrons, and the science, technology and applications associated with this energy form. The movement of electric charge results in an electric current. Magnetic forces arise from these movements and can be utilized in *electric motors and *generators. Lighting and electronic effects can also be produced by an electric current. Electrical energy can always be dissipated as heat.

The factor stimulating the movement of electric charge is the *electromotive force (emf or voltage), which is measured in *volts. Most countries have a mains electric standard voltage of about 220 V; North America has a 110 V standard. Current can be *direct (dc) or *alternating (ac) and is measured in *amperes.

Electricity is a high-quality form of energy in that it can be used to do *work with little *transformation loss, this loss having already been expended in the production of the electricity.

MS

Electricity Council. A watch-dog committee to safeguard consumers' interest in respect of electricity provision and pricing in England and Wales.

Address: 30 Millbank, London, SW1P 4RD

MS

electric motor. A machine for transforming electrical energy into mechanical energy. The tranformation is effected, according to the principles of *electromagnetism, by passing a current through a coil in such a way that the magnetic effects induced cause rotation of the *armature and shaft of the motor. Electric motors exist in designs of considerable variety: they can run off direct current or off single-phase or polyphase alternating current supplies.

TM

electrode boiler. A device which raises the temperature of water within a steel tank by means of an alternating current passed between electrodes submersed in the water. Resistance to the flow of electricity through the water raises its temperature. The conversion of electrical energy to heat energy is of the order of 98%.

TM

elephant grass. *See* napier grass.

embodied energy. A popular term that seeks to embrace the notion that when a good or service is produced and energy is dissipated in so doing, that energy use is described as the embodied energy. Although it may be calculated with precision, the value arrived at will depend on the conventions adopted. For example the *IFIAS convention requires that the calculation be based on the total amount of all the various primary energies in the ground consumed.

MS

emergency core cooling system (ECCS). A standby cooling system installed in pressurized water reactors (*see* nuclear reactor) and designed to operate in the event of a fault or accident resulting in depressurization of the water in the core and loss of flow. In these circumstances, even with the reactor shut down, the *fission product decay heat of the fuel in the core would cause overheating and fuel melting unless the core continues to be cooled. The ECCS consists of one or a number of high-pressure water systems which automatically inject water into the core if the reactor pressure drops below its normal operating level.

DB

electrolysis. The separation of chemical components by the passage of electricity. For example, the term is used when *direct current is passed through water (perhaps containing some acid), so producing hydrogen and oxygen. Hydrogen is evolved at the negative electrode (the cathode) and oxygen at the positive electrode (the anode). Electrolysis is important in many processes, for instance the manufacture of aluminium from the ore bauxite. Such processes demand considerable electrical power, and are economical only when supplied by cheap *hydropower.

JWT

electromagnetic radiation. A disturbance travelling at the speed of light when energy radiates from one place to another through empty space, air or some other transparent medium. The wavelength and frequency of the radiation is distinctive, so defining particular bands of the total *electromagnetic spectrum such as visible light, infrared and radio regions. *See also* radiation.

JWT

electromagnetic spectrum. The division of *electromagnetic radiation into regions by wavelength and frequency. In order of increasing wavelength (i.e. decreasing frequency) the named sections of the spectrum are the gamma ray, X-ray, ultraviolet, visible, infrared, microwave and radio wave regions. The distribution of energy in the spectrum from a heated surface is given by *Planck's radiation distribution law and the total emitted energy by the *Stefan–Boltzmann law.

JWT

electromagnetism. The science of the properties of and relationships between magnetism and electric current. An electromagnet is a core of iron or steel partly surrounded by a coil through which an electric current is passed; the electromagnet, while the current flows, behaves with the properties of a permanent magnet. Conversely, if the magnetic flux close to an electric circuit is changed, an *electromotive force will be induced in the circuit with the consequent flow of electricity.

The phenomenon of electromagnetism governs the operation of the *electric motor and the electricity *generator.

TM

electromotive force (emf). The *force which tends to cause flow of an electric current in a circuit by producing a *potential difference between parts of the circuit.

TM

electron. The negatively charged particle which is one of the constituents of the atom. Its mass is 0.00055 *atomic mass units (u), and its charge is 1.602×10^{-19} coulombs. Atomic electrons are bound in orbits around the *nucleus of the atom and the number of electrons in a neutral atom (which is equal to the number of *protons in the nucleus) determines its *atomic number, Z, and its chemical properties.

Free electrons can be produced by the *ionization of an atom, or by certain types of radioactive decay process in which high-energy electrons, known as *beta particles, are produced. *See also* electricity.

DB

electronic charge. The charge of the *electron, which is negative in value, or the *proton, which is positive in value. The magnitude of the charge in each case is 1.602×10^{-19} coulomb.

DB

electronvolt (Symbol: eV). The unit of energy used in atomic and nuclear physics. It is defined as the energy gained by a particle of unit *electronic charge when it passes through a *potential difference of 1 volt:

$$1\,\mathrm{eV} = 1.602 \times 10^{-19}\,\mathrm{J}$$

A commonly used multiple of the electronvolt is the megaelectronvolt, MeV, equal to a million (10^6) electronvolts.

DB

electrostatic precipitator. A device for collecting *particulate material from waste gases such as those released by fossil-fuel power stations. The basic principle of operation is based on the fact that particulates, moving through a region of high electrostatic potential, tend to become charged and are then attracted to an oppositely charged electrode where they can be collected and removed.

PH

electrostatics. The science and study of electric charges at rest.

TM

Embden – Meyerhof – Parnas pathway. *See* glycolysis.

embodied energy. The energy associated with the production of a *good or service. It is a widely used term with many different interpretations. It can imply energy requirement, as in *gross energy requirement, fuel consumption per unit of output, the sum of direct and indirect energies consumed, or the renewable and non-renewable energy dissipated in the course of a production process.

MS

emissions taxes. A response to *pollution problems which involves the levying of a tax per unit of discharge into the environment of residuals arising in production, as happens, for example, with the burning of coal to produce electricity. Typically, the receiving environment is a *common property resource, so that there is *market failure and the emissions are in excess of the level called for by the requirement of *efficiency in allocation. Following the *polluter pays principle, emissions taxation leads to a reduction in residuals discharge and a reduction in *external costs. If the rate of taxation is set so as to attempt to achieve efficiency in allocation, the emissions tax is a *Pigovian tax. It is extremely difficult to compute the tax rates required for Pigovian taxes, due to the difficulty of measuring the cost of pollution-induced damage; 'arbitrary' tax rates are typically the practical recommendation.

MC

emissivity [emittance]. (Symbol: ε) The property of the surface of a body which determines, in relation to its temperature, the quantity of heat or light radiation it emits. Emissivity is measured as the ratio of the *radiation emitted by the body to the radiation emitted by a *black body at the same temperature.

TM

emittance. *See* emissivity.

emphysema. A medical condition common among coal miners in which the air sacs (alveoli) in the lungs become uneven and distended due to destruction of the alveolar walls. The disease is characterized by extreme shortness of breath, especially after exercise. It is one of the diseases of the respiratory system the incidence of which is related to air pollution and which is common in persons suffering from *pneumoconiosis.

PH

emulsifier (for oil spills). *See* dispersants.

endogenous. Coming from within or emanating from within the system under examination.

MS

endothermic reaction. A chemical or nuclear reaction in which energy is absorbed and the mass of the reactants increases. This mass change is of very small order, particularly in the case of chemical reactions. *See also* Einstein's equation.

DB

energy. In common parlance, a generic term used to cover sources of heat and power without specifying what sort and without regard to quality. Thus one speaks of energy resources, energy wastage, energy efficiency. As commonly used, these words have no precise meaning nor can they be quantified. The use of the word is confused by the utilization of 'energetic' and 'energy' to signify vigour.

In a strict thermodynamic sense, energy is a concept invented to overcome the fact that though *work can be converted, *joule for joule, into *heat, the reverse is not the case. When heat or work is removed or put into an isolated system, that system changes and ends up in a different state. The property which accounts for this change of state is called the *energy content of the system. It is an inherent property of the system: every system for a given set of conditions has a certain energy content. Thus, if the initial energy content is E_1, and the final value is E_f, then the change in energy, ΔE, is given by

$$\Delta E = E_f - E_1 = (J \times Q) - W$$

where Q is heat, W is work and J is the *mechanical equivalent of heat.

This concept of energy can take account of the fact that work is required to change an ore into a metal, to raise a heavy load, to produce a unit of electricity. For the lay mind, confusion is created through the fact that both heat and work are expressed in the same units, normally joules or some multiple thereof.

Energy plays a key role in economic development, since by the utilization of energy in appropriate machines, man can multiply many times his own unaided effort.

The word comes from the Greek, *en* (in) and *ergon* (work).

MS

energy accounting. A term, now little used, referring to the process of calculating the *energy requirement of goods and services. The term *energy analysis is more usual.

MS

energy analysis. The methodology whereby the energy required to manufacture a *good or create a service may be computed, taking into account both *direct and *indirect energy use. Energy analysis recognizes several *system boundaries for computation, leading to concepts such as *gross energy requirement, *process energy requirement, direct energy, indirect energy and *energy requirement for energy. The term also applies to the utilization of such data for subsequent energy analysis or economic analysis. *See also* IFIAS.

MS

energy audit. An assessment of the energy flows in a production process, usually with a view to establishing where economies can be made.

MS

energy balance. The accounting of energy inputs and outputs in any process. On the basis of the *first law of thermodynamics, heat may neither be lost nor gained. An energy balance, therefore, simply sets out to show that all input energies are matched by either changes in the state of the system or as output energies. An energy balance is always preceded by a *mass balance.

MS

energy capital. The value of the stock of capital equipment in the energy supply or distribution industries. It can also imply the energy used in creating such capital. *See also* capital energy requirement.

MS

energy carrier [energy vector]. A portable substance which can release energy when desired. For example, electricity cannot be stored readily but it can be converted into substances which can subsequently release energy, such as hydrogen or ammonia. Fossil fuels, especially in their refined state, are considered to be energy carriers.

MS

energy coefficient. *See* energy elasticity.

energy content. The intrinsic energy of a substance, whether as gas, liquid or solid, in an environment of given pressure and temperature (with respect to a datum set of conditions). Any change of the environment can create a change of the state of the substance with a resulting change in energy content. Such a concept is essential for the purpose of calculations involving use of *heat to do *work. However, the phrase is also loosely used to imply *energy requirement of production of the substance. The two resulting numbers are quite unrelated. On occasion, the term is also used to imply the fraction of cost of production represented by the use of *direct energy.

MS

energy cost. The price paid for energy. However, the term sometimes infers the amount of energy dissipated in the production of a good or service. A more rigorous expression of this idea is *energy requirement.

MS

energy density. A term used with different meanings, but usually relating to the energy which can be obtained from an energy store by combustion, electrical output, heat transfer, etc. The table gives approximate values. Energy density is usually expressed in units of energy per unit volume or unit mass when applied to such systems.

JWT

energy dissipation. *See* dissipation, energy.

energy economics. The study of the way economic systems use energy, especially energy from *fossil fuels, and possible replacements (nuclear energy, wave energy, wind energy, etc.) for these sources. Energy is taken as meriting spe-

Energy Store	*Energy Density: dry matter* MJ/kg	GJ/m^3 (atmospheric pressure)
oil	42	37
coal	32	42
wood, dried (approx)	15	4.5
charcoal	28	6.2
methanol	21	17
ethanol	28	22
biogas (65% CH_4)	20	0.02
hydrogen gas	120	0.01
wood gas (approx)	17	0.02
hydrogen liquid	120	8.7
battery, lead-acid (approx)	0.1	0.5
Hydro (100 m head)	0.001	0.001
water, $\Delta T = 70°C$	0.3	0.3
straw	18	5
cattle dung	14	4

cial attention for two main reasons. First, energy use is all-pervasive in economic activity in that every production process involves some input of energy, if only in the form of human labour input. In a modern economy, most production processes use energy derived from fossil fuels directly, and all use such energy indirectly. This pervasiveness of energy use has come to be recognized as important, especially in the recent past with energy prices rising relative to other prices. Second, energy production and use involves several problems for economic analysis and policy. No one of these problems is peculiar to energy, but the several problems come together in a unique way in the case of energy, and are given special force by the dependence of a modern economy on energy use.

Fossil fuels are non-renewable *natural resources: more use now means less use in the future, making the problem of fairness in the pattern of use over time acute. The extraction, transportation and use of fossil fuels gives rise to many detrimental *externality situations, resulting in water and air pollution and in the reduction of the level of amenity provided by many natural environments. This is also true of the so-called 'alternative' energy sources, especially in respect of amenity loss with, for example, wind or wave energy extraction installations.

Projects for energy supply facilities typically involve long development and construction periods, large inputs of *capital, and are attended by *risk and *uncertainty. In such circumstances, *project appraisal and *cost benefit analysis are difficult problems within the context of which highly controversial issues can arise which are not capable of resolution in technical terms. This is currently the case with projects for electricity generation using *nuclear reactors.

MC

energy elasticity. The proportionate change in a nation's energy consumption, divided by the associated proportionate change in the level of economic activity. The level of economic activity is usually taken as the level of *gross national product; energy consumption has been given by useful energy, *delivered energy or *primary energy. Measured energy elasticity coefficients vary widely across nations, as might be expected. They also vary over time for any one nation. The usefulness of energy elasticity calculations is in revealing these variations and so prompting enquiry into whether systematic explanations for them can be found. Some of the observed variations can be explained in terms of climatic conditions, the price of energy relative to other commodities, industrial structures, or differing energy supply systems.

MC

energy farm. An area of land or water (fresh or salt) devoted to growing specific plants for their ability to furnish energy. Such plants range from aquatic algae to fast-growing trees. *Mixed cropping is a common method of attaining increased *primary productivity over that obtained from an energy farm monoculture. It is important that photosynthetic efficiency is high to reduce land area requirements.

The cultivated biomass can be used directly as a fuel through *combustion (of wood, for example) or indirectly via *pyrolysis or gasification to produce a range of solid, liquid and gaseous fuels. Crops containing a high moisture content are better fermented in the aqueous state, as in the fermentation of sugar from *sugar-cane to ethanol and the anaerobic digestion of *algae to methane.

Energy crop plantations can store energy for use at will, are renewable, are dependent on available technology with minimum capital inputs, can be developed with current manpower and material resources, are not expensive and are ecologically benign if well managed. Associated problems are land use competition, the comparatively large areas required, fertilizer and water requirements and the variable lag times before meaningful and renewable quantities can be harvested.

Large energy farms exist in Brazil, growing mostly sugar-cane and cassava for fermentation to ethanol. Silviculture plantations, based on hybrid poplars and other *coppicing species, are being developed in Sweden, Canada and the USA, principally for wood-fired power stations. The production of ethanol from sugar-cane, molasses, maize, wheat and sugar-beet crops is occurring in the US and Japan. Many other countries, such as Australia, New Zealand and Ireland, are undertaking feasibility studies on various types of energy farming systems, while among the developing countries reafforestation programmes are being caried out in Nepal, Kenya, China, South Korea and many more nations.

CL

energy indexing. The assessing of *energy qualities, and expressing them relative to some arbitary energy type.

MS

energy flux. The energy passing through or incident upon a given area of surface, measured in terms of energy per unit area per unit time, i.e. joules per square metre per second or watts per square metre.

MS

energy intensity. The energy required to create a unit of product, valued in money units. It can be expressed, for example, in megajoules per dollar. Data may be obtained from energy related input-output tables or from census of production (UK).

MS

energy level. *See* nuclear energy levels.

energy manager. One whose task is to improve the efficiency of energy use, whether in industry, homes or else-where.

MS

energy park. An industrial complex devoted to energy delivery, together with ancillary activities such as waste treatment, recycling and fuel preparation. It is usually thought of in terms of nuclear energy, but can refer to a solar energy park.

MS

energy pattern factor. *See* cube factor.

energy policy. A process of evaluation and decision about energy use and supply. It is normally considered at the national level, in which a government decides on incentives and restraints on energy supply, development, support for new technologies, consumption taxes and other fiscal devices aimed at meeting the nation's energy needs. Policies frequently change as circumstances change, particularly in respect of the understanding of how energy enters the economy, and in response to the world scene. International bodies such as the *International Energy Agency or the European Commission also offer analysis aimed at leading to appropriate energy policies. *See also* depletion programme; fiscal policy.

MS

energy profit ratio. A little-used concept analogous to the *energy requirement for energy, but expressed as the output of energy from a system as a proportion of the input energy.

energy production. The extraction and subsequent transformation of fossil or fissile energy sources, and their delivery to the economic system. The term is inaccurate because according to the *first law of thermodynamics, energy, can neither be destroyed nor created and hence cannot be produced; it is arguably applicable to the processes of nuclear fission and fusion, where matter is turned into energy. National energy production should be interpreted as the rate at which energy is extracted, transformed and delivered within the national system. *See also* system boundary.

MS

energy quality. The ability of energy to do *work is related to its temperature above the ambient environment. Since various fuels generate heat at various temperatures, some have greater ability to do work, even if they produce the same amount of heat, and hence they have different energy quality. It has proved difficult to bring this important concept into economic analysis, not least because the effectiveness of energy is sensitive to the technology of use, while the calculations are sensitive to the *system boundary chosen for the calculations. Some economists use a ratio of 2.6:1 to compare the quality of electricity to coal, with correspondingly better ratios for oil and gas but poorer ones for lignite, peat or wood.

The question of energy quality is largely ignored in official energy statistics, which may thus be very misleading. For example, both in 1968 and in 1973 UK consumption of coal and oil amounted to 7.85×10^6 TJ, while economic output had markedly increased. It might be thought that this was a sign that energy was being more efficiently used. However, in 1968 56% of that energy came from coal, while in 1973 it was 44.8%. Without taking into account energy quality, it is impossible to say what improvement took place.

MS

energy ratio. (1) In economics, the ratio of energy use by a country (national economy) per unit of *gross domestic product (GDP). It therefore has the dimensions of energy per unit money, and is a changing value whose computation is made upon the basis of annual national statistics. There are two inherent sources of error. Firstly, the energy use may be computed for each fuel at different system boundaries (*see* energy statistics) and secondly, the GDP is a financial measure and hence not necessarily a measure of physical output.

Comparison of energy ratio between countries for any one year must therefore take into account *exchange rates, while for any country a comparison of one year with another must take into account *index numbers of output. The *energy elasticity coefficient equals the marginal energy ratio divided by the average energy ratio. *See also* purchasing power parity.

(2) In technology, where the output of a process has an *energy content which is valued as such, and requires energy in its production, the energy ratio is the ratio of the two. There is no firm convention for which is the divisor. The term is most often found in analysis of food production, where it is usually expressed as units of metabolizable (i.e. food) energy produced for so many units of *non-renewable energy consumed. Thus low-intensity agriculture will have a high energy ratio – significantly greater than unity – while high-intensity agriculture, as found in the industrialized world, has a ratio often less than unity.

MS

energy requirement. The energy required to produce and bring a *good or service to a particular point in the market place. It is usually expressed in terms of so many units of energy per item or per mass. The term has been adopted in an effort to overcome the ambiguity of the expression 'energy cost', which is sometimes used in the same sense but also implies an expenditure of money. The definition is not in itself complete, since it tells one nothing about the *energy quality nor the *system boundary chosen for the summing process. The latter can be resolved by computing the *gross energy requirement. *See also* process energy requirement.

MS

energy requirement of energy (ERE). The sum of all primary energies dissipated to yield one unit of *delivered energy. Since no primary energy source can be delivered without prior energy expenditure on exploration, production, refining and delivery, ERE is always greater than unity. As energy resources become more inaccessible, ERE can be expected to rise. ERE for the world in 1981 was approximately 1.35. The concept provides a rigorous way of comparing various energy sources and forms of delivered energy. For example, the ERE of electricity from coal is in the range 3.5–4.00; the ERE of gasoline from Middle East oil (in the Middle East) is as low as 1.05 while from the North Sea (delivered in Europe) it is close to 1.15.

MS

Energy Research Support Unit (ERSU). A unit established by the Rutherford Laboratory of the UK Science and Engineering Research Council. The unit liases with universities and its personnel act in an advisory and assessment role for many projects, including the energy requirements of buildings and *wind power.

Address: Rutherford and Appleton Laboratories, Chilton, Didcot, England.

JWT

energy slave. Work done by energy on behalf of a person. The concept simply makes the point that most of the world's energy consumption goes into doing work of one kind or another. One can compute that amount of work in terms of how many able-bodied human beings would be necessary to carry out the same work, and so compute the number of energy slaves serving an individual. For example, a tentative estimate would be that the average American lives with the help of 50 energy slaves, but that in India there is less than one per person.

MS

energy statistics. A compilation of data on energy use, usually by type of energy and on a national basis. Energy statistics are subject to conventions which differ from one country to another, and one international compilation (e.g. United Nations) to another. These conventions arise because there is no unique and unequivocal measure of energy that readily lends itself to compilation by statisticians. Some statistics show electricity in terms of its heat energy, some in terms of the fossil energy needed to make the electricity (some three to four times greater). Normally energy statistics ignore the energy dissipated in delivering energy to a national frontier, so that, for example, oil from the Middle East delivered to Rotterdam is only counted in terms of the oil unloaded, not the additional energy expended in transportation or exploitation. The variability of these conventions means that one cannot compare one set of statistics with another without establishing their basis, and then recomputing. For example, UK energy consumption apparently differed at one time by 17%, depending on whether the data was listed in *Eurostats or in the UK Department of Energy publications, simply by virtue of different conventions. The figure (*overleaf*) shows the different system boundaries used in compiling Eurostats.

MS

energy storage. Energy retained in a form where it may be readily recovered. There are many forms of energy storage. In thermal storage, energy is retained in a heated mass (e.g. a *rock bed) or in the latent heat of a material changing phase (e.g. solid to liquid salt). In chemical storage, energy is retained in reaction products and may be released by the reversed reaction (e.g. nitrogen and hydrogen becoming ammonia). *Batteries and *accumulators producing and storing electricity are a form of chemical storage. *Fuel cells may be classified as chemical storage producing electricity. Mechanical storage occurs when energy is retained as kinetic energy (e.g. a flywheel) or potential energy (e.g. com-

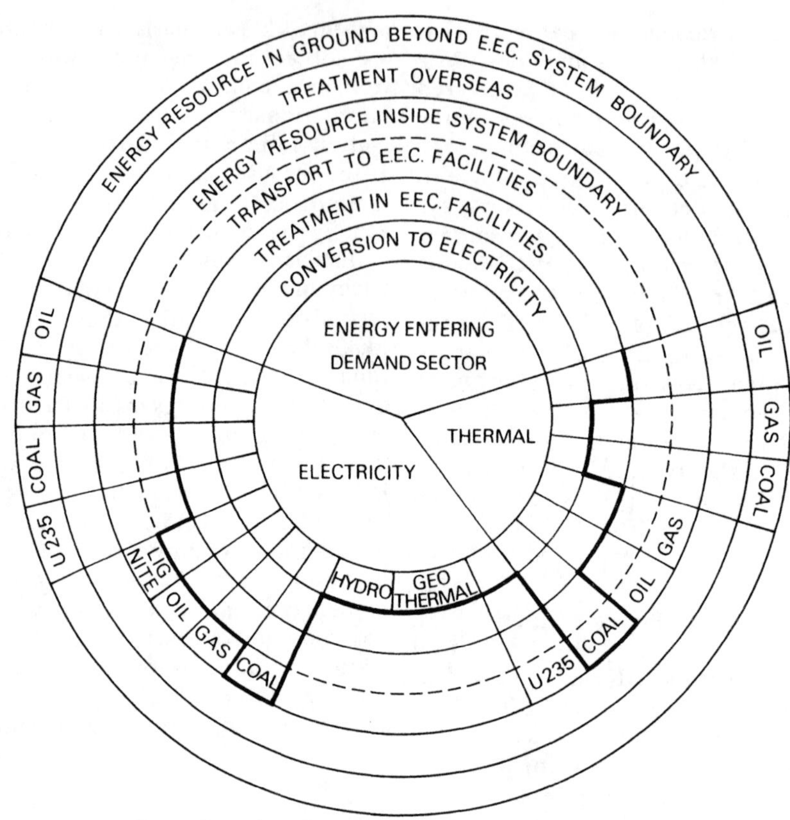

System boundary diagram (EEC Energy Transformation System)

*heavy line indicates the *system boundary at which *Eurostat data is recorded*

Energy Technology Support Unit (ETSU). A unit of the UK Department of Energy with expertise in all forms of energy, especially *alternative energy supplies other than nuclear power. A series of publications have been produced on specialist options for UK energy supplies, such as *wave power. The unit acts in an advisory and assessment role to industry, research and development organizations, and other similar bodies. Grants are awarded for research and demonstration projects. The unit relates to the energy programmes of the European Economic Community.

Address: AERE Harwell, Oxfordshire, England.

JWT

energy transformation system. A physical as opposed to an economic concept of the energy industries. Without exception, no energy resource can be used without some processes of extraction, purification, treatment and delivery. The concept describes the network of industries which carry through the entire process from search and exploitation to delivery to the customer.

MS

energy vector. *See* energy carrier.

Engel curve. The relationship between the proportion of the houshold budget spent on a commodity, or group of commodities, and household income. Originally Engel's law referred to food, stating that the proportion of income spent on food fell as income rose. Numerous studies have confirmed this in many countries at many times. The slope of an Engel curve is the *elasticity of expenditure with respect to income.

MC

enhanced recovery. The initial recovery of crude petroleum from a bore hole is determined by the pressure in the underground reservoir due to water below it or gas dissolved in the petroleum. This pressure may reduce to a low level and recovery virtually cease when as little as 25% of the petroleum in the reservoir has been brought to the surface. Secondary recovery methods such as the injection of methane, carbon dioxide or water into the reservoir may enhance the recovery of petroleum to as much as 50% of that originally in place. Tertiary recovery involves use of heat (e.g. steam) or detergents to liberate the oil from the reservoir. Enhanced recovery reduces the *net energy of the fuel source.

WG

engineering services. The generic term commonly given to the system of supply and distribution of a variety of facilities within a building or industrial plant. These can include hot water, cold water, lighting, electric power, space heating, elevators etc. Typically, they can account for between 35% and 65% of the *capital cost of a finished building.

TM

enriched uranium. *See* uranium.

enrichment. A process of enrichment that, when applied to natural *uranium, is designed to increase the concentration of the *fissile isotope uranium-235. Natural uranium contains only 0.715% of fissile uranium-235 while most power reactors require enriched uranium with 2–3% uranium-235 for *criticality. Enrichment is also necessary to produce weapons-grade uranium with about 90% uranium-235.

Two processes are at present used commercially for enrichment: the gaseous diffusion process and the centrifuge process. Both make use of the fact that the gaseous compound uranium hexafluoride, UF_6, consists of a mixture of $^{238}UF_6$ and $^{235}UF_6$, which have slightly different densities. In the gaseous diffusion process, in which the UF_6 gas is pumped through a series of semipermeable membranes, the lighter $^{235}UF_6$ molecules diffuse slightly faster than the denser $^{238}UF_6$ molecules. Thus there is an enrichment of the uranium, but several hundred stages are required to produce a significant increase in the uranium-235 content. The process requires very large quantities of energy and consequently the cost of enriched uranium from diffusion is very high.

In the centrifuge process the UF_6 gas, contained in cylinders, is rotated at very high speed. Under the effect of strong centrifugal force, the $^{238}UF_6$ tends to be more concentrated near the perimeter and the $^{235}UF_6$ more concentrated near the centre of the cylinder. Thus gas withdrawn from the centre of the cylinder is enriched in uranium-235. The centrifuge process requires considerably less energy than gaseous diffusion.

A third experimental process involving the use of lasers has been proposed for separation of uranium-235 and uranium-238. Laser enrichment depends on the fact that the two isotopes are excited by light of different wavelengths. A precisely tuned laser beam can excite the uranium-235, which becomes ionized and may be separated from the unexcited uranium-238 in an electric or magnetic field. This process known as atomic vapour laser enrichment (AVLIS) is being developed in the USA and UK, but has not yet (1987) reached commercial development.

DB

enthalpy (Symbol: H). A concept which is best expressed in terms of change of enthalpy between one state of a system and another. It is the increment (decrement) in the total heat content of the system, equal to the heat absorbed (lost) by the system during the process, carried out at constant pressure, when the only work done is that of expansion or compression. Thus from the first law of thermodynamics the change in enthalpy is given by

$$\Delta H = \Delta(U + pV)$$

where U is the *internal energy, p the pressure and V the volume. When a chemical reaction such as combustion occurs, there is an enthalpy of reaction. *See also* combustion energy; gross calorific value.

MS

entropy (Symbol: S). A vital concept in thermodynamics which allows one to determine the potential of any system to do useful work. It is best understood by way of example.

Every system, if left to itself, will change slowly or rapidly to some final stable state; for example, a mechanical clock winds down, a cup of coffee cools down. As these systems move towards their final equilibrium state, they lose ability for spontaneous change. *Work can only be extracted from a system which enjoys the capacity for spontaneous change. The further a system is from equilibrium, the greater the amount of useful work it can carry out. The passage towards equilibrium does not necessarily imply that the system is losing energy. It may in fact remain constant. What is being lost is the availability of its energy to do useful work. In other words, the entropy of the system is increasing. Changes in entropy are quantitatively measured in a reversible process from the relation

$$dS = \frac{dQ}{T}$$

where dS is the change in system entropy and dQ is the heat absorbed at temperature T. *See also* second law of thermodynamics; third law of thermodynamics.

MS

entropy coefficient. A measure of economic *concentration.

MC

environment. The surrounding of a specified system. Environmental factors are the physical, chemical and possibly biological properties of these surroundings which may be influenced by biological or man-made activities. Thus air is part of man's environment which itself may be affected by industrial *pollution.

All energy supply and consumption affects the local environment to some extent. Of particular importance has been air and water pollution from fossil fuel use, and thermal heating of rivers and estuaries from nuclear power stations. CO_2 emissions from fossil fuel and forest burning have been sufficient to change significantly the global concentration of *CO_2 in the atmosphere.

*Renewable energy supplies draw energy from the environment, but do not in themselves cause air or water pollution. Thus renewable energy generators (e.g. *wind turbines, *hydropower) cause aesthetic interference of the visual environment rather than physical or chemical damage.

JWT

environmental impact statement. A form of *technology assessment introduced in the US by the National Environmental Policy Act (NEPA) of 1969. Section 102 (2)(c) of this act requires that before any major federal project proceeds, the applicant or agency responsible must file a draft environmental impact statement which must be circulated and made public for comment prior to final Environmental Protection Agency decision. It should describe:

(a) the environmental impact of the proposed action;
(b) any adverse environmental effects which cannot be avoided as a result of the proposed action;
(c) alternatives to the proposed action;
(d) any irreversible and irretrievable commitments of resources which would be involved.

The applicant is required to consult with and obtain the comments of any other Federal, State or local agencies which have jurisdiction by law or special expertise, or are authorized to enforce environmental standards. The applicant must also make copies of the statement and comments available to the public under the Freedom of Information Act.

PH

Environmental Protection Agency (EPA). An independent agency of the US federal government. *Inter alia*, its role is to minimize the environmental impact of energy production and consumption. It has many regulatory and monitoring functions: it issues radiation protection guidelines to all federal agencies, establishes radiation protection standards, monitors environmental radiation levels, establishes water quality standards and auto emission standards, issues waste water discharge permits, etc.

Address: 401 M St., S.W., Washington, D.C. 20460, USA.

PH

enzyme. A protein biological catalyst produced by living cells to speed up the rate of a biochemical reaction occurring within or outside the cell. Intracellular enzymes operate inside the cell whereas extracellular enzymes are secreted outside to act in the transformation of a substrate into another compound(s). Enzymes may sometimes be induced by the cell to initiate a chemical reaction, and thus are formed only when a particular type of substrate is available. In this way hydrolytic cellulase and amylase enzymes produced by moulds can be used to break down cellulose and starch molecules respectively to fermentable sugars in the manufacture of *ethanol fuel. Other enzymes important in energy production and conversion include *hydrogenase, *nitrogenase and those in *glycolysis, the *Krebs cycle and the *Calvin–Benson cycle.

CL

equilibrium. In economics, a situation where the plans of economic agents are mutually consistent, with the result that there is no tendency for change in the absence of external shocks. Thus *market equilibrium exists when demand equals supply in the sense that buyers actually buy the amount they planned to buy at the ruling price and sellers actually sell the amount they planned to sell at the ruling price. Much of economic theory is concerned with establishing the conditions under which a unique equilibrium exists, the factors which determine the nature of the equilibrium, and the stability properties of the system. A stable equilibrium is one which will be restored by *disequilibrium behaviour following some external shock.

The concern with the analysis of equilibria is justified by the contention that it concentrates on the underlying determinants of the state towards which the system or market generally tends to move, and avoids the complications which arise in disequilibrium. Thus analysis of the oil market is more fruitful if it considers the *demand and *supply functions and the resulting market equilibrium rather than focusing on, for example, temporary interruptions to supply as with an embargo by oil-exporting countries. *See also* general equilibrium analysis; partial equilibrium analysis.

MC

equilibrium price. *See* market equilibrium.

equities. *See* shares.

ere. *See* energy requirement of energy.

erg. A unit of work or energy which has been generally superceded by the SI unit, the *joule (J). One erg is equal to a force of one *dyne acting through a distance of one centimetre. One erg is equal to 10^{-7} J.

TM

ester. The product of the chemical reaction of an organic acid with an *alcohol, e.g.

$$\underset{\text{acid}}{RCOOH} + \underset{\text{alcohol}}{R'OH} \rightarrow \underset{\text{ester}}{RCOOR'} + \underset{\text{water}}{H_2O}$$

where R and R′ are any organic radical. Esters are stable substances, most having characteristic pleasant odours. They are used as solvents and in polymer manufacture.

WG

ethane. An odourless flammable gaseous *aliphatic hydrocarbon, C_2H_6. It occurs in natural gas and in refinery gases. Its gross calorific value is 52.02 MJ/kg.

WG

ethanol. The systematic name for ethyl alcohol, C_2H_5OH. It is a colourless volatile flammable liquid. Apart from its value as a solvent and as a raw material in many chemical processes, it may be employed as a fuel for spark-initiated internal combustion engines both alone and mixed with gasoline. Although the *calorific value (GCV = 30.15 MJ/kg) is lower than that of gasoline, the higher latent heat of vaporization of ethanol compensates for this in lowering the cylinder temperture and thus allowing a higher charge density.

WG

ethene. *See* ethylene.

ethylene [ethene]. An odourless flammable gaseous hydrocarbon, C_2H_4. The molecule has one double valency bond; it is thus the simplest to refine. It is produced in the petroleum refinery by the *thermal cracking of mixtures of paraffin hydrocarbons and steam. Owing to the double bond it is a very reactive substance and is employed as a petrochemical feedstock, and in the formation of polymers such as polyethylene and polystyrene. It is thus too valuable to be burned as a fuel. Its gross calorific value is 50.4 MJ/kg.

WG

ethyne. The systematic name for *acetylene.

eucalyptus. A tree which provides an excellent source of *firewood both in the tropical highlands and in the semi-arid regions of the world. *Eucalyptus globulus* is the most extensively planted eucalypt in the world, with an estimated world area of some 8000 km^2, mostly in mild temperate climes and in cool tropical highlands. Its wood has a specific gravity of 0.8–1.0 and a calorific value (air-dried) of 20 GJ/t. It burns well and leaves little *ash. In India the species is widely cultivated for both firewood and charcoal manufacture, while its dense and widespread root system is valuable for erosion control and land reclamation. The trees *coppice vigorously at least twice, but yields usually fall off in the third coppice. Annual wood productions of 10–30 m^3/ha are reported in countries as far apart as Italy and Peru, giving energy yields of 160–300 GJ/ha.y.

The most popular eucalypt species for growth on poor soils in dry climates is *Eucalyptus camaldulensis.* Its wood has a specific gravity of 0.6, a calorific value of 20 GJ/t and makes an outstanding fuel. Some examples coppice well for over six rotations and annual yields per hectare of 20–25 m^3 (equivalent to 240–300 GJ) have been reported from Argentina, 25–30 m^3 (300–360 GJ) from Turkey and 30 m^3 (360 GJ) from Israel. However, where the soil is of very poor quality, productivity can be as low as 2–11 m^3/ha.y (25–130 GJ/ha.y).

CL

Euphorbia. A latex-producing plant, many species of which grow well on arid land. It contains hydrocarbons of high molecular weight (in the range 10,000 to 20,000). It has been postulated that this latex could be converted into light hydrocarbon fractions as in crude oil. This would provide a *bioenergy source without the need for the elaborate *biological energy conversions associated in converting say, cellulose to ethanol. *Euphorbia lathyrus* and *Euphorbia tirucalli* have already undergone trials in California, showing annual yields equivalent to 20-25 barrels of crude oil per hectare. There are prospects for advancement by the application of *genetic engineering techniques to produce improved varieties.

CL

Euratom. An agency of the European Economic Community which exists to promote and provide assistance for nuclear research in EEC member states, and foster research in the Community's own nuclear laboratories which are at Ispra, Italy; Petten, Netherlands; Karlsruhe, Germany FR; Geel, Belgium. It operates an inspectorate based in Luxembourg.

Address: 200 Rue de la Loi, 1040, Brussels, Belgium.

DB

eurocurrency. Claims to a country's *currency held by non-residents of the country. It was originally known as eurodollars because the market for such claims emerged in the late 1950s when it was dominated by the lending and borrowing of US dollar balances through London and other European centres. This is now much less the case and gives rise to a terminological difficulty as the prefix euro- and the term dollar no longer accurately describe the situation. Thus, for example, nowadays a claim on Japanese Yen held by an American bank would now be covered by the term eurocurrency. For this reason the eurocurrency market is now sometimes referred to as the 'xeno currency' market.

Since the oil price rises of 1973–74, the market has become very important in financing international trade related to oil, with oil-importing countries borrowing in it to finance *balance of payments deficits on *current account, and with oil exporting countries (especially OPEC members with large balance of payments surpluses) lending in the market. *See also* petrocurrency.

MC

eurodollars. *See* eurocurrency.

European Economic Community (EEC). A *customs union founded by the Treaty of Rome in 1957, signed by France, West Germany, Italy, the Netherlands, Belgium and Luxembourg (the 'Six'). In 1973 the UK, Eire and Denmark joined under the Treaty of Accession. Greece joined in 1980 and Spain and Portugal in 1986. Its aim is to promote free economic activity and increased efficiency by the removal of trade barriers and other forms of restrictions between member states. It also aims at integrating economic policies on agriculture, transport and industry. It has further attempted to coordinate its monetary policies by means of the European Monetary System. The formal treaties of the EEC do not specifically refer to energy policy, but there are periodic meetings of the energy ministers of member states and some agreed objectives for energy policy. Thus, for example, it is agreed that the EEC as a whole should seek to reduce its dependence on imported energy.

MC

European Free Trade Association (EFTA). An association of countries formed by the Stockholm Treaty signed in 1959 by Austria, Denmark, Norway, Portugal, Sweden, Switzerland and the UK. Finland became an associate member in 1961 and Iceland attained full membership in 1970. In 1973 Denmark and the UK left upon their accession to the *European Economic Community. The association is concerned only with trade and aims at reducing trade restrictions between members, but each country in the association is free to impose its own *tariffs and *quotas on non-members.

MC

Eurostat. Statistics, including energy statistics, produced by the statistical office of the European Commission for all countries of the European Economic Community.

MS

eutrophication. The depletion of oxygen in a body of water caused by excessive growth of vegetation due to *nutrients in the water, chiefly nitrates and phosphates. Typically such vegetation may be *algae or water weeds. This situation can arise when nutrient-bearing wastes are released by farms or industries such as fertilizer manufacture or food processing. In the case of an aquatic *energy farm where the *biomass is frequently harvested this could be deliberate policy. Otherwise, under uncontrolled conditions, the *environment becomes further polluted and incapable of supporting *aerobic forms of life.

CL

evaporation. The process by which a liquid, through the application of heat or a reduction of pressure, changes phase and becomes a vapour.

TM

event tree. The outcome of an analysis of the possible events following an initial incident. It is used in accident analysis. An event tree is constructed by postulating an initial failure within a system and then endeavouring to imagine the sequence of all possible events which might follow. *See also* fault tree.

PH

excess air. Air in excess of the amount required theoretically to burn completely a unit mass of a fuel. It is normally calculated and reported as a percentage of the *theoretical air. In practice most fuels require more than the theoretical air to prevent the production of smoke. The excess depends both on the nature of the fuel and the design of the combustion unit. For efficient operation it should be kept to a minimum. Typical values are 50% excess air for coal-fired, 20% for oil-fired and 10% for gas-fired installations.

WG

exchange controls. The legal restriction by a country of the freedom of its citizens to engage in transactions involving the exchange of its *currency for other currencies. Such restrictions are imposed for *balance of payments reasons and as a means of managing the *exchange rate.

MC

exchange rate. The price of a *currency in terms of another currency, expressed either as the number of domestic currency units per unit of foreign currency (the usual definition) or the number of foreign currency units per unit of domestic currency (the UK usage). The spot exchange rate is the price for the immediate delivery of a currency (*see* spot price). The forward exchange rate is the price set now for the future delivery or purchase of a currency; a forward rate exists for each forward contract, e.g. 1 month, 3 months, 6 months, etc., the most usual being the 3 month rate (*see* forward price). The parity exchange rate was the rate agreed between a country and the *International Monetary Fund under the *Bretton Woods Agreement. If the exchange rate is defined in terms of the home (foreign) currency then a rise (fall) in the parity rate denotes a *devaluation, and conversely for a *revaluation. The effective exchange rate is an *index number which is a weighted average of a country's exchange rates with the currencies of the countries with which it trades.

With floating exchange rates *balance of payments surpluses and deficits lead to exchange rate changes. Particularly since the oil price rises of 1973–74, this means that a country's exchange rate is much affected by its pattern of trade in energy products, especially oil. Thus, the UK's exchange rate is now higher than it would otherwise be by virtue of North Sea oil. A higher (lower) exchange rate means better (worse) *terms of trade and lower exports, other things being equal. Thus North Sea oil improves the UK's terms of trade and works to reduce UK exports. *See also* appreciation; depreciation. MC

excise tax. An *indirect tax levied on goods and services, the amount paid being based on the quantity purchased rather than on the value of the quantity purchased as is the case with *ad valorem taxes. Consequently the tax-inclusive price of commodities subject to excise taxation does not move in line with *inflation, so that their *real tax-inclusive price falls with inflation, unless the tax rates are revised upwards. Common excise taxes are those on tobacco, beer, wines and spirits, petrol and diesel fuel. MC

excitation (nuclear). The raising of the energy state of an atom or *nucleus above its normal ground state. For example, an atomic nuclear can be excited by absorbing a *neutron, the degree of excitation corresponding to the *binding energy of that neutron into the nucleus. *See also* absorption. DB

exergy. *See* negentropy.

exhaust gas. Gaseous products of combustion from an internal combustion engine. WG

exogenous. Coming from outside; external to the system under examination. MS

exothermic reaction. A chemical or nuclear reaction in which energy is released and the mass of the reactants decreases. The mass decrease is of very small order, particularly in the case of chemical reaction. All combustion reactions release energy. *See also* Einstein's equation. DB

expectations. The views held about future economic conditions as they affect decisions to be taken currently. For example, in deciding between ordering nuclear and coal-fired electricity generating plant, the expectation of the relative prices of coal and uranium 20 years hence is a crucial input to the decision. Also, wage bargains typically hold for 12 months, so that for workers seeking to maintain or improve the *real wage, the increase in the wage rate to be sought depends on the expected rate of *inflation for the 12 months following the agreement. The way people form their expectations on the basis of their past experience is there fore important in affecting the way the economy behaves, and is currently the subject of active debate and research in economics. MC

explosion limits. *See* flammability limits.

explosive. A material which can be caused to release over an exceedingly short period of time a large amount of chemical energy in the form of heat and physical energy in the form of a shock wave. There are two main types of explosive both of which consist of highly combustible fuel material and an agent to supply the oxygen for its combustion.

Low or deflagrating explosives, when suitably confined and ignited, burn rapidly to produce large volumes of hot gas causing a pressure increase and a bursting effect. Blasting explosives used in mining are of this type and contain a finely divided carbonaceous fuel and a solid oxidizing agent.

High or detonating explosives require a detonator to set them off properly. Speed of burning is so rapid that a shock wave is created which precedes the combustion through the exploding mass. They thus cause a shattering effect rather than the bursting effect of the deflagrating explosive. A high explosive like nit-

roglycerin has both the fuel component and the oxygen necessary for its reaction contained in the one molecule.
WG

export. Something produced domestically and sold to an ultimate buyer overseas. If it is a service rendered to a foreign resident, it is an invisible export, as for example with banking and insurance services. One country's export is another country's *import. *See also* balance of payments; invisible trade; visible trade.
MC

exposure. Of the many meanings, in the energy context it refers to the degree to which a building is exposed to the external environment, and the nature of that environment. For example, a building on headland by the sea would be more exposed than one situated inland in a hollow.
MS

external benefit. The difference between the social *benefit of some economic activity and the benefit as perceived by the operator of the activity. The divergence between social and private valuation is an example of *market failure, and the market outcome in respect of activities which give rise to external benefits is such that they are conducted at lower levels than is required for *efficiency in allocation, assuming that there are no *external costs involved. An example of an external benefit would be where the flooding of a river valley for the purpose of hydroelectric generation created new recreational facilities in respect of which users were not charged. To the extent, in this example, that flooding the valley also deprived others of the recreational use of the unflooded valley for which deprivation the firm operating the hydroelectric facility did not have to pay, there would be also arising external costs. *See also* externality.
MC

external combustion engine. *See* heat engine.

external cost. The difference between the social cost of some economic activity and the private cost as perceived by the operator of the activity. It is the damage involved in a detrimental *externality situation, as with air pollution arising from the burning of coal to produce electricity. The principal pollution effects are damage to human health, increased corrosion, and damage to plants and animals. The external cost of the electricity produced is the money value of all such damage. It is extremely difficult to measure. It is however necessary to attempt the measurement in order to consider how much it is worth paying, in terms of resources to be used or electricity output foregone, in order to reduce the pollution. As in this particular example, external costs are generally associated with the overuse of *common property resources (here the atmosphere). One instrument for reducing the external costs associated with environmental pollution is an *emissions tax. In some circumstances regulation and direct control of the external-cost generating activity may be preferred. *See also* social cost.
MC

externality. A side effect of production or consumption for which no payment is made. It arises in a situation where an individual's *utility function or a firm's *production function includes a variable, the level of which is set by some other individual(s) or firm(s) without regard to the effect on the individual or firm. In particular, the term refers to situations where the interdependence is not the subject of market transactions, and externalities are examples of *market failure. Externalities may be beneficial or harmful to the affected party. An example of the former type is that of on the job training of an individual by a firm, where the individual subsequently moves to another firm without the payment of any kind of transfer fee. An example of a detrimental externality situation would be incidental pollution arising from energy use, e.g. exhaust fumes from motor cars. *See also* external cost.
MC

extraterrestrial disposal. *See* radioactive waste management.

F

factor of production. A primary input used in the process of production. In economics it has been conventional to divide these into *labour, *capital and *land, with the term land referring to all *natural resources. This usage is now changing, with natural resource inputs being increasingly referred to as such. Also, the different classes and types of natural resources are now being recognized as distinct factors of production. For example, energy is now often treated as a factor of production to be distinguished from all others.

MC

Fahrenheit scale. A scale for measuring temperature on which 212 marks the boiling point of water at standard atmospheric pressure and 32 marks the freezing point of water. Thus the scale between these two points can be divided into (212–32) i.e. 180 equal divisions, or degrees Fahrenheit (°F). To convert a temperature in degrees Fahrenheit (F) to a temperature in degrees centigrade (C) the following equation is used:

$$(F-32)\,5/9=C$$

PH

fail-safe. Refers to a system which, when it malfunctions, does so in a manner that results in increased safety, e.g. *shut-down.

MS

fall-out. The deposition onto the earth's surface of airborne particles. Typical sources are volcanic eruptions, dust storms, and nuclear explosions, whether from bombs, bomb testing or accidents to commercial reactors, such as in the *Chernobyl incident. Such particulates can remain suspended in the atmosphere for long periods and are usually brought to earth by rainfall.

MS

fan coil system. A system of *air conditioning a building in which hot or cold water is circulated throughout the building from a centrally located plant room. In each space to be heated or cooled, a fan unit is sited. The fan draws air from the room, blows it over the hot or cold water coil and returns it to the room. In summer cold water will be supplied, with a changeover to hot water in winter. Dehumidified air from central air-handling plant or fresh air from outside may also be supplied by the system.

TM

fantail. A small horizontal-axis wind turbine mounted sideways on a large horizontal-axis turbine so as to steer the large machine into the wind direction.

JWT

farad (Symbol: F). The SI unit of electrical *capacitance. A capacitor has a capacitance of one farad if, when charged by one *coulomb of electricity, a potential difference of one *volt exists across its plates.

TM

fast breeder reactor [fast reactor]. *See* nuclear reactor.

fast flux test reactor. A test facility designed to provide fast neutron environments for testing fuel and materials for fast breeder reactors (*see* nuclear reactor).

PH

fast neutrons. Neutrons with high kinetic energy, though below that of *fission neutrons, whose average energy is about 2 MeV. In a fast breeder reactor the spectrum of energy of fast neutrons extends from a few KeV up to about 2 MeV.

DB

fault tree. An analysis of the many possible events following an initial incident. Its purpose is to assess the probability of failure at each step along *event tree paths that are responsible for a particular accident within a system. A typical application would be the events and their probabilities following an emergency shut down of a nuclear reactor.

PH

feedback. An impact upon an agent of the result of that agent taking action. In principle, no act by any person, machine, plant or system can fail to create feedback, since any act must alter the state of the system. In reality, the feedback is often too trivial to be measurable, or occurs as a result of a long chain of events so that it cannot be identified. Nevertheless, in many situations there is detectable feedback. For example, if there is an increase in gasoline tax, this will affect gasoline consumption, which in turn will affect the amount of tax collected. Feedback can be positive or negative. An example of positive feedback is the process of giving birth. The greater the population, the greater the number of births and hence the greater the population. Negative feedback is reflected in deaths. The more deaths, the smaller the population and hence the smaller (eventually) the number of deaths. *See also* negative feedback.

MS

feedstock. A *petroleum fraction used as a raw material for a petrochemical process and fed to the processing plant for the production of chemicals, such as ethylene and toluene, and for the production of polymers, detergents, etc.

WG

fermentation. The transformation by micro-organisms of organic compounds, chiefly *carbohydrates, into end-products useful to man. It also provides energy for the micro-organisms and is carried out in the absence of oxygen. Though microbial cells such as bacteria and yeasts are usually present within the fermentation vessel, cell-free extracts containing the required *enzymes can also bring about the desired reaction. A large number of products can result, including methane, ethanol, acetic acid, acetone, higher organic acids and alcohols, and hydrogen, according to the micro-organisms and substrates chosen and the reaction conditions.

The most common fuel-producing fermentations are the *anaerobic digestion of organic matter to *biogas mediated by *methanogenic bacteria, and the yeast fermentation of sugars to *ethanol. Reaction rates are governed by various parameters such as temperature, pH, mixing, and loading rates. *Mesophilic fermentations occur at approximately 35 °C, while the faster *thermophilic fermentations operate at 55–60 °C. The fermentation vessel or fermenter may vary greatly in size and design and may function as either a closed or open system. A closed or batch system receives all the necessary nutrients for microbial growth before fermentation begins, while in an open or continuous system organisms and nutrients can continuously enter and leave the fermenter so that a steady state is set up. In its broadest sense fermentation now refers to *aerobic as well as to anaerobic reactions and the end-products can include enzymes, vitamins, hormones, antibiotics, amino acids, nucleic-acid-related compounds, single cell protein, polysaccharides, solvents and so on.

CL

fertile (nuclear). Denoting a susbtance which by nuclear transformation becomes a nuclear fuel. The isotopes uranium-238 and thorium-232 are referred to as fertile because when neutrons are captured in them the *fissile isotopes plutonium-239 and uranium-233 are produced. *See also* breeding.

DB

Fick's law. A law, formulated by Adolf Fick in 1885, which describes the way in which a substance such as a gas diffuses through another substance. In general terms it states that the rate of diffusion of one substance through another across a given cross-sectional area is proportional to both the area and the concentration gradient at that point:

$$\text{rate of diffusion} = D \times A \times G$$

where D is the diffusion coefficient, A the area and G the concentration gradient. The diffusion coefficient is a constant and its magnitude depends on the size and charge of the diffusing molecule, the nature of the material through which it diffuses and the absolute temperature.

PH

film badge. A radiation detector carried by persons who may be exposed to ionizing radiation in the course of their work. It consists of a small piece of film enclosed in a plastic holder which is attached to the clothing of the person at all times when he or she is in the radiation environment. The film is developed, normally after being in use for one month, and the degree of blackening is a measure of the accumulated *radiation dose to which thc person has been exposed during that month.

DB

final boiling point. The temperature at which 95% of a petrolem fraction, undergoing a test distillation, has distilled over.

final product. Something produced for *consumption, *investment, *export or for use by government. It is to be distinguished from an *intermediate product. The provision of final product for consumption is the end purpose of economic activity. Investment involves not consuming available output so as to add to the stock of *capital, making more available for consumption in the future. Exporting is necessary to make it possible to import, so exploiting *comparative advantage so as to increase consumption opportunities. Where energy is directly used by a household it is treated in economics as a final product, but where it is used by a firm to produce some commodity for use by households it is not.

MC

finance house. *See* bank.

fines. Finely divided solid fuel from a crushing plant; particles which pass through the finest mesh in a range of sieve sizes.

WG

Fingal process. A process developed by the UK Atomic Energy Authority at Harwell during the period 1958–68 for the solidification of high-level *radioactive waste.

PH

finite difference method. An approach to the approximate numerical solution of the partial differential equations such as used in *heat transfer calculations. The method replaces the derivatives in the partial differential equations with approximations in the form of finite-sized differences between values at particular locations in the solid, liquid or gas through which heat is flowing.

TM

firedamp. Low molecular weight *hydrocarbon gases, predominantly *methane, which occur naturally in coal seams and which are released as the coal is mined. Its accumulation in the coal pit may lead to a disastrous explosion. Owing to its low density the gas collects under the roof of the tunnels and thus may be sucked away to a collection point through suitably arranged ducts. The methane so 'drained' from a pit may be burned in a boiler plant at the pit head to provide steam for various services.

WG

firewood [fuelwood]. Wood used as a source of fuel. The dominant use of firewood is for cooking although obviously it is used for space heating and personal comfort heat. The total amount of energy produced from firewood on a world scale is about 30×10^{12} MJ per year (equivalent to about 1.5×10^{9} tonnes per year), which corresponds to about 6% of total world energy use. Since this use is primarily by two-thirds of the world population in the Third World, there is generally little appreciation of the importance of firewood in the industrialized countries. In Africa for instance firewood represents 60% of total energy supplies.

The calorific value of wood depends on the species and especially on the moisture content. In general dry wood has a calorific value of about 19 MJ/kg and a density of about 0.9 tonne/m^3.

JWT

firm. An entity which produces output for sale on a market. Thus, as the term is used in economics, a firm is not necessarily a limited liability company, and it is not necessarily in the private sector. An electricity generating utility, for example, is a firm irrespective of whether it is privately or publicly owned. Government agencies which provide services for which there is no direct charge are not firms.

MC

first law efficiency. A term that has come to imply the ratio of the amount of heat delivered for a certain duty or objective compared to the amount of heat supplied. The words 'first law' are intended to convey the notion based on the *first law of thermodynamics that the heat and rather than the work potential of the energy is being measured. Typically such a term is used to describe the efficiency of a furnace providing heat for, say, a central heating system. The heat supplied is that obtained from the combustion of the fuel. The heat delivered is the heat dissipated in the space to be heated. The ratio of the two is the first law efficiency. In a typical domestic installation this may be between 60% and 70%, depending on the fuel, the condition of the system and the *load, the balance being dissipated to the environment. By contrast the *second law efficiency of such a system may be as low as 8%.

MS

first law of thermodynamics. The law stating that energy cannot be destroyed or created. Thus the total of all forms of energy (e.g. heat, work, chemical energy) remains constant, i.e. is conserved, in a *closed system unless nuclear reactions occur. *See also* Einstein's equation.

JT

fiscal policy. The use of government expenditure and taxation to regulate the level and composition of total economic activity, and to redistribute income and wealth in the economy. The government, according to *Keynesian economists, can and should increase (decrease) the size of its budget deficit to stimulate (reduce) economic activity in order to reduce unemployment (inflation). The structure of taxation affects the composition of output and the way it is produced. Thus increased *indirect taxes on fuels would, other things remaining constant, raise the price of fuels relative to other commodities, so inducing some *substitution of non-energy for energy commodities in consumption and production.

MC

Fischer–Tropsch process. A process for the conversion of soil fuels to liquid fuels, which was developed in Germany prior to the Second World War and supplied a large proportion of German requirements for gasoline during that conflict. A mixture of hydrogen and carbon monoxide made by passing steam over red hot coke was, in the original process, passed through a catalyst containing nickel and cobalt oxides at 200°C and at atmospheric pressure. A mixture of alkanes was formed, ranging from propane to solid paraffin wax. A petrol fraction with a low octane rating could be separated. This was subjected to catalytic cracking to give a petrol with an octane rating of about 68. The value could then be increased as required by the addition of tetraethyl lead.

In recent times the process has been improved by using better catalysts and working under a pressure of 15 atmospheres. It is worked successfully in South Africa. Its success economically depends on a plentiful supply of cheap coal as the source of coke.

WG

fissile. Denoting or relating to isotopes which can undergo *fission with neutrons of very low (effectively zero) energy. The three most important isotopes in this category are uranium-233, uranium-235 and plutonium-239. *See also* fertile; fissionable.

DB

fission. The process whereby the *nucleus of a heavy atom splits. It may occur spontaneously but the probability of this is very small. The usual cause of fission is the excitation of a heavy nucleus when it absorbs a *neutron. The mechanism of fission is postulated as follows (with letters referring to the diagram):

(a) a neutron is absorbed into the heavy nucleus to form a *compound nucleus at an excited state (b), the excitation being due to the *binding energy of the neutron in the compound nucleus. The excited compound nucleus is deformed due to its excess internal energy (c) and splits into two nuclei of intermediate mass (d) with the release of a number of fission neutrons (usually two or three) and gamma radiation. In addition a large amount of energy is released in the process.

The fission products are the two nuclei of intermediate mass produced by the splitting of the compound nucleus. A large number of different pairs of fission products may be formed, all of them nuclei with mass numbers between 70 and 160 and all of them radioactive. The following process, involving the fission of uranium-235 and the production of lanthanum and bromine as the fission products, is just one of many possible fission events:

$$^{235}_{92}\mathrm{U} + ^{1}_{0}\mathrm{n} \rightarrow ^{236}_{92}\mathrm{U}\ \text{(excited)} \rightarrow$$
$$^{147}_{57}\mathrm{La} + ^{87}_{35}\mathrm{Br} + 2\,^{1}_{0}\mathrm{n}$$

The number of fission neutrons produced in each fission is variable, being in the range one to four. The average number of fission neutrons produced per fission, symbol ν, is an important nuclear parameter for any fissionable fuel. For

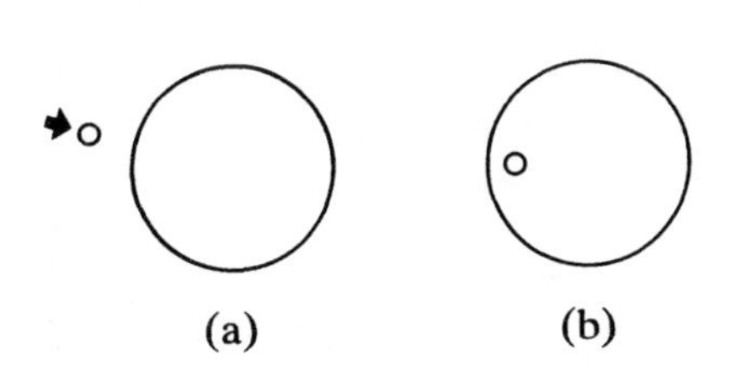

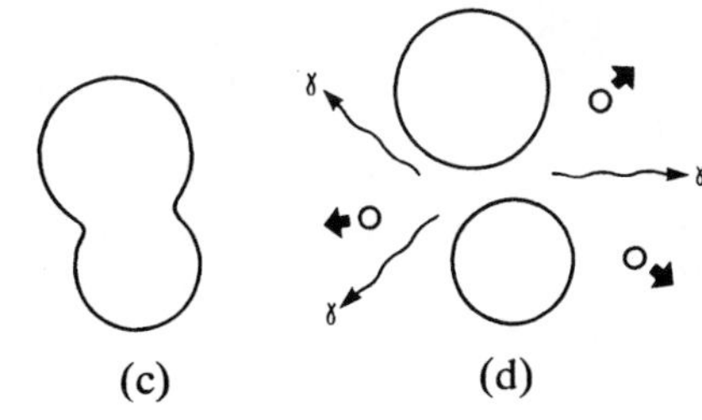

fission in uranium-235 caused by thermal neutrons, for example, its value is 2.44. The fission neutrons have a spectrum of energies up to about 8 MeV, the average energy being about 2 MeV.

The energy released per fission is about 200 MeV. Expressed in an alternative way, if 1 kg of uranium-235 could be completely fissioned, the energy released would be 8.2×10^{10} kJ, which is almost two million times the energy released by the combustion of the same mass of oil.

DB

fissionable. Denoting or relating to those elements which can undergo *fission with neutrons whose energy is in the range from zero to about 10 MeV. The most important isotopes in this category are uranium-233, uranium-235, uranium-238, thorium-232 and plutonium-239. The term fissile is restricted to those isotopes which can undergo fission with very low (effectively zero) energy neutrons. The most important isotopes in this latter category are uranium-233, uranium-235 and plutonium-239. The two isotopes uranium-238 and thorium-232, which are fissionable but not fissile, can only undergo fission with neutrons whose energy is greater than about 1 MeV. *See also* fertile.

DB

fission chamber. An instrument for detecting *thermal neutrons which consists of a *gas-filled radiation detector lined on its inside with uranium-235. Thermal neutrons entering the detector may cause *fission in the uranium-235 lining, and the resulting high-energy fission products then cause intense *ionization in the gas-filled space of the detector. This ionization can be detected and measured in the same way as in other gas-filled detectors.

DB

fission energy. *See* fission.

fission neutrons. *See* fission.

fission product decay heat. The decay of radioactive substances releases energy as heat. In a *nuclear reactor which has been operating for some time, the concentration of radioactive *fission products and the rate of energy release from their decay builds up to such a level that after shut down of the reactor significant quantities of heat are still released in the core. In order to prevent overheating and possible melting of the fuel, it is necessary to continue to circulate coolant through the core after shut down.

DB

fission products. *See* fission.

fixed bed. A solid fuel combustion chamber or gasifier in which the reacting solids do not move, while the air or other gas flows between the pieces. *See also* catalytic cracking; moving bed.

WG

fixed carbon. A constituent in the *proximate analysis of coals and cokes which is determined by difference, i.e. one hundred minus the sum of the percentages of moisture, ash and volatile matter. It is thus coke minus ash.

WG

flame. A volume of gas which is undergoing chemical reaction and emitting heat and light. Flames are produced in combustion processes by the burning in air of combustible gases and vapours rising from heated liquids and solids.

When an issuing jet of gas is ignited a diffusion flame is formed. This consists of a thin-walled envelope, called the flame front, containing unburned gas and surrounded by air. Since gas and oxygen molecules meet only on diffusing into the flame front, the reaction rate is necessarily slow and the flame large. It is shaped by the convection currents produced in the air around it. If the gas is a hydrocarbon, the molecules will be thermally cracked on approaching the flame front and before reacting with oxygen. Thus particles of carbon are deposited in the flame, become incandescent and make the flame luminous. Such a flame deposits soot on a cool surface with which it comes in contact.

An explosion flame is created when a source of ignition is introduced into a gas/air mixture with a composition within the *flammability limits. In this case the flame front moves rapidly through the mixture.

An issuing jet of gas/air mixture when ignited produces an aerated flame. Combustion is rapid so the flame is small, noisy and non-luminous. Most gas appliances are set to give aerated flames since they are small and compact and burn cleanly without creation of smoke or deposition of soot.

WG

flame speed. If a long horizontal tube is filled with a mixture of combustible gas and air and a flame applied to one end of the tube, the flame will travel from one end of the tube to the other. The rate of travel is said to be the flame speed. The speed is affected by variables such as temperature, pressure and mixture composition, and thus comparable results can be obtained only in a standard apparatus. Alkane/air flames (e.g. natural gas/air) have a flame speed about 30 cm/s while hydrogen has a speed under the same conditions of about 340 cm/s.

WG

flame temperature. A temperature calculated from the *calorific value of the fuel being burned, and the volume and *heat capacity of the gaseous products of combustion. If it is assumed that there are no heat losses from the system, a maximum value of temperature is obtained. In practice the flame loses heat by radiation and thus the actual value of flame temperature is likely to be a little less than the calculated value. When hydrocarbon fuels are burned in the theoretical amount of air, a flame temperature of about 1900 °C is obtainable.

WG

flammability limits [explosion limits]. When a flammable gas is introduced into a compartment filled with air in the presence of a source of ignition such as a glowing electric element, no flame is produced until a certain concentration of gas in air is reached. The concentration at which a flame spreads through the compartment is known as the lower limit of flammability. For hydrocarbon gases the lower limit is normally about 1–2% by volume in air at atmospheric pressure.

If air is introduced into a compartment filled with a flammable gas in the presence of a similar source of ignition, again no flame results until a certain concentration of air in gas is reached; this corresponds to the upper limit of flammability in air. For hydrocarbon gases the upper limit is somewhere in the range of 7–12% by volume in air.

Between these limits all mixtures of gas and air will explode if a source of ignition is introduced.

WG

flare. A permanent flame located at the top of a stack in a petroleum refinery or oil rig. It is essentially a safety feature and is used to dispose of hydrocarbon gas in excess of requirements. Although a seemingly wasteful practice, the cost of providing safe storage or pipeline facilities for such gas may far outweigh the economic value of the gas.

WG

flashing. The evolution of vapour when a rapid drop in the pressure of a liquid, whose temperature is just below the saturation temperature, results in boiling.

This effect is of particular importance in pressurized water reactors (*see* nuclear reactor) since any fault resulting in a sudden depressurization (such as a pipe break) may result in flashing, which in turn may leave the core denuded of coolant and liable to overheating, or even meltdown, unless emergency cooling is supplied. *See also* emergency core cooling system.

DB

flash point. The temperature to which a liquid fuel must be heated in a standard apparatus to produce sufficient vapour to give a flammable mixture with air. The fuel under test is heated in a small covered cup and a small standard flame introduced periodically into the air space above the liquid. When the flash point temperature is reached, a small flash of flame travels momentarily across the surface of the liquid. Minimum flash point requirements are stated for liquid fuels so that they may be transported, stored and utilized safely.

WG

flask. In the context of the nuclear industry, a container for holding, storing or transporting radioactive materials. Flasks have to be designed to high standards of strength and integrity, suitable for long periods of service and able to withstand conceivable accidents without damage.

DB

flat plate collector. A solar *collector that traps energy by absorption of radiation on a black flat surface. The surface passes heat to a fluid, gas or liquid, perhaps flowing within fixed tubes.

JWT

Fleming's rule. An empirical method for deducing the direction of the *electromotive force (emf) induced in a conductor in relation to the direction of the magnetic flux and the direction of motion of the conductor. The rule derives from Lenz's law, which states that the direction of an induced emf is always such that it tends to set up a current opposing the motion or the change of flux responsible for inducing that emf.

TM

Flowers Report. *See* Royal Commission on Environmental Pollution.

flue gas. The gaseous products of combustion from a furnace which pass to a flue or chimney. The mixture is likely to contain carbon dioxide, carbon monoxide, water vapour, oxygen and nitrogen together with small amounts of sulphur oxides.

WG

fluidization. The process by which a bed of solid particles is lifted by gas blown up through it until the particles are just separated from one another. The rate of gas flow should be sufficient to support the weight of the particles without carrying them away. The bed of particles takes on the appearance of a liquid and is said to be fluidized. Solid objects may be caused to float on top of the bed and move freely in it. Such a system has many advantages in the chemical and fuel industries since the motion of the particles ensures efficient mixing and good temperature control and transfer of heat. *See also* catalytic cracking.

WG

fluidized combustion. A method of burning small coal particles by injecting them into a fluidized bed of hot ash (*see* fluidization). Excellent heat transfer to boiler tubes in or above the bed is achieved. Combustion is virtually smokeless, and if limestone is added to the ash the lime formed absorbs any sulphur dioxide released from sulphur in the coal and thus the *flue gases contain little more than carbon dioxide and water vapour.

WG

fluorescence. Following the absorption of radiation by a body, the emission by that body of radiation of a different (usually longer) wavelength. The phenomenon is used in lighting technology: a gas-filled *lamp, internally coated with a powder that fluoresces under the action of electrical discharge, provides a shadowless white or coloured light. Lamps using the phenomenon of fluorescence are more efficient, in energy terms, than those using the phenomenon of *incandescence.

TM

flux. (1) A flow of material or energy across or through a surface. A particular flux will be defined by its units, energy per unit time per unit area (also called energy flux density or intensity).

(2) (nuclear) A flux of *neutrons (or gamma photons) is a measure of their number density and speed, being proportional to both these quantities. If a parallel beam of neutrons (or photons), all of the same energy, is considered, then the flux φ is equal to the number of neutrons (or photons) per unit volume n multiplied by their speed v. More generally, where neutrons (or photons) have a spectrum of speeds (or energies) and are moving in different directions, the flux is equal to the total track length per second of all neutrons (or photons) in unit volume of the material through which they are passing.

JWT, DB

fly ash. Small solid particles of inorganic matter which separate as ash on the burning of a fuel and which are lifted from the fuel bed and carried forward with the draught into the waste gas flues. They tend to deposit on the surfaces of heat recovery units situated in the flues and reduce their efficiency.

WG

flywheel. A rotating device of relatively large mass used to store kinetic energy and/or to smooth out fluctuations in the rotational machinery.

JWT

focusing collectors. Solar devices incorporating mirrors and/or lenses to concentrate direct sunshine onto an absorbing surface (*see* collector). Focusing collectors must be continuously controlled to point towards the sun.

JWT

foil activation. A technique for measuring neutron *flux in which a thin foil (or disc) of an element such as gold is irradiated by *neutrons. The foil material undergoes a capture reaction to produce a radioactive isotope. For example, the irradiation of a gold foil containing 100% gold-197 ($^{197}_{79}Au$) produces some radioactive gold-198 by the reaction

$$^{197}_{79}\mathrm{Au} + ^{1}_{0}\mathrm{n} \rightarrow ^{198}_{79}\mathrm{Au} + \gamma$$

The *radioactivity of the irradiation product (gold-198 in the above example) is proportional to the neutron flux in which the irradiation takes place, and this radioactivity can be measured by an instrument such as a *scintillation detector.

DB

Food and Agriculture Organization (FAO). An agency of the United Nations, established in 1945. It carries out research and provides technical assistance, especially to developing countries, in agriculture, forestry and fishing. It also helps in promoting international *commodity agreements and has sections devoted to questions of energy consumption in agriculture and fishing.

Address: Via delle Terme di Caracalla, Rome, Italy.

MC

food chain. A series of living organisms with interrelated feeding habits, each serving as food for the next in the chain; thus one organism provides for a higher organism. A set of interwoven food chains makes up a food web. A food chain starts with a primary producer, an *autotrophic organism such as a green plant which can produce matter directly from inorganic materials by *photosynthesis. The next link in the chain is the primary *consumer who lives off the primary producer; the primary consumer may in turn provide the food for a secondary consumer. A food web traces the path of *solar energy through an *ecosystem. Solar energy is stored by the primary producers in the chemical products of photosynthesis. When these organisms are consumed some of this stored energy is used in the synthesis of new biomass in the consumer, some is used to support the basic life processes of the consumer, while the rest is released as waste heat.

PH

food web. *See* food chain.

fool's gold. *See* pyrites.

force (Symbol: *F*). The influence exerted on a body causing an alteration to its state of rest or motion. The force required to effect such change in the state of rest or motion is proportional to the mass of the body and the acceleration which it experiences. The unit of force is the *newton (N).

TM

forced draught. Air blown into a combustion chamber or under a solid fuel grate by a fan in order to increase the rate of *combustion.

WG

forecasting. In economics, the prediction of the level to be taken by one or more variables, given assumptions about the levels to be taken by some other variables. Thus any forecast is conditional. The conditions on which a forecast is based cannot be controlled and so may not be realized, with the result that the forecast level is not the level which actually occurs. Thus, for example, the techniques of *econometrics may have been used on historical data so that the past value of the price *elasticity of demand for oil is known with some precision. This does not mean that accurate forecasts of the demand for oil can be made, since the future price of oil is not known. The situation is not essentially different from that in other sciences. In physics, for example, successful predictions are the result of closely controlled conditions. In meteorology, where conditions cannot be controlled, forecasts are often wrong. The lack of control over conditions is not the only source of error in economic forecasts. Incorrect or oversimplified *models of economic behaviour also contribute to forecasting error. *See also* technological forecasting.

MC

foreign exchange. The currency of another country. When importers buy goods or services from other countries they usually need to pay in terms of the currency of the exporting country. Foreigners will equally be demanding home currency in order to pay for home exports. Since there is no single international currency then foreign exchange is required for credits and debits in international trade. The foreign exchange market is the international market where foreign currencies are traded. The two main currencies used as foreign exchange are dollars and sterling.

MC

form factor [configuration factor]. A measure of the unobstructed 'view' which one object has of all others within some given three-dimensional geometric configuration. The measure is used in the prediction of how much light or how much radiant heat emitted from one object would impinge on another. For simple geometries the factor can be exactly determined; for more complex geometries it is approximated.

TM

forward market. *See* futures market.

forward price. In a contract to buy or sell

goods or securities in the future, the price agreed at the time the contract was made. There is a different forward price for each forward date. (In the foreign exchange market these are usually 3 and 6 months ahead.) The forward price is to be contrasted with the *spot price, which is the current price agreed for the immediate buying or selling of a good or security. *See also* hedging.

MC

fossil fuels. Fuels derived from living organisms which have been fossilized by being subjected to geological forces over extremely long periods of time. Such materials range from natural gas to petroleum, tar sands and oil shales; from peat through bituminous coal to anthracite. Although the process of fossilization is probably continuing to produce fuels, man's rate of use is such that deposits may be considered finite and subject to complete depletion in due course. In the end alternative sources of energy must be sought.

WG

4-carbon plants [C_4 plants]. Mostly tropical plants able to fix carbon dioxide from the air more efficiently than can temperate crops. This is achieved not only by way of the common *Calvin–Benson cycle, but also via a pathway involving 4-carbon compounds as intermediates. A distinguishing feature of these C_4 plants (which include *sugarcane, *maize, sorghum and many tropical grasses) is the possession of two distinct chloroplast types: mesophyll and bundle sheath. The carbon dioxide accepter molecule is phosphoenolpyruvate (PEP), a 3-carbon compound present in the mesophyll cells and so forming a 4-carbon molecule, oxaloacetic acid. The oxaloacetic acid is then reduced to malic acid, conveyed to the bundle sheath cells and decarboxylated. This generates a C_3 molecule, pyruvic acid, and releases a carbon dioxide molecule for fixation in the 3-carbon cycle. Meanwhile, the pyruvic acid is phosphorylated to regenerate PEP for accepting carbon dioxide molecules once more.

PEP is more reactive with carbon dioxide than is ribulose diphosphate, the carbon dioxide accepter molecule in the 3-carbon cycle. Additionally, C_4 plants can utilize carbon dioxide at lower concentrations than can C_3 plants and have a competitive advantage in conditions of high insolation and temperature and restricted water supply. Thus productivity (and hence *bioenergy potential) is greatest in the tropical and more arid regions of the world. Furthermore, C_4 plants exhibit lower rates of the wasteful *photorespiration process, thus making them even more productive and promising as crops grown specifically for energy.

CL

four-stroke cycle. Another name for the Otto cycle (*see* heat engine).

fraction. *See* petroleum fraction.

fractional distillation. A distillation process in which a mixture of liquid components is separated into fractions of different boiling points. A *fractionating column is used.

WG

fractionating column. A tall tower used to separate *petroleum fractions according to their boiling points. The vertical structure is separated into a number of compartments known as plates, and vapours rising from the base of the tower meet a descending trickle of condensed liquid. The plates are designed to promote contact between the two streams. The temperature of the tower falls as the vapours rise from bottom to top, and thus components with different boiling points will tend to condense at different levels and may be drawn off as a liquid sidestream at a convenient plate. The tower base is connected to a heated tank, known as a reboiler, from which undistilled residue may be removed, and the material to be fractionated is preheated and injected into the column at a suitable level.

WG

Francis turbine. A machine driven by a fluid, gas or liquid, which moves past *blades of a special configuration mounted on a rotor. The Francis turbine accepts fluid around a horizontal circumference and allows the flow to leave centrally down a vertical pipe.

JWT

free-burning. A term applied to non-coking coals. During combustion on a grate such coals do not fuse together. This causes a reduction in the surface area available for reaction and so reduces the capacity of a furnace.

WG

free energy. (1) In thermodynamics, that part of the energy content of a system which is free to do work. It is an important concept for it permits one to calculate the work available from a given quantity of energy at a given temperature, for any specified system. When the calculation is done for a system at constant temperature and pressure, such as occurs in most chemical processes and reactions, it is known as Gibbs free energy (G). The net change in G (ΔG) for a specified chemical reaction can define precisely the extent of the reaction possible. Most combustion reactions have very negative ΔG and so react almost 100%; many commercially important processes do not have high negative ΔG and so do not give high yields. An example is the formation of ammonia from its elements.

(2) In economics, energy at no cost.

free on board (fob). A contract for the shipment of goods overseas where the shipper includes in the price the costs of getting the goods onto the ship/plane/truck, but not the insurance and freight charges for shipment to the goods' overseas destination. In interpreting, for example, data on imported oil prices it is important to know whether such are reported fob or cif (*cost insurance freight).

MC

free trade. A situation where there are no restrictions, such as *tariffs or *quotas, on the trade between countries. If such a situation existed, trade would be based on the principle of *comparative advantage, and, in the absence of any sources of *market failure, the result would be one of allocative *efficiency for the world economy. In fact, a situation of complete free trade has never existed, countries erecting 'barriers to trade', i.e. tariffs and quotas, for both economic (domestic employment) and non-economic (strategic security) reasons. The gains to free trade which economic theory predicts do, however, form the basis for the philosophy of the *General Agreement on Tariffs and Trade (GATT), which seeks to reduce the extent of barriers to trade in the world economy.

MC

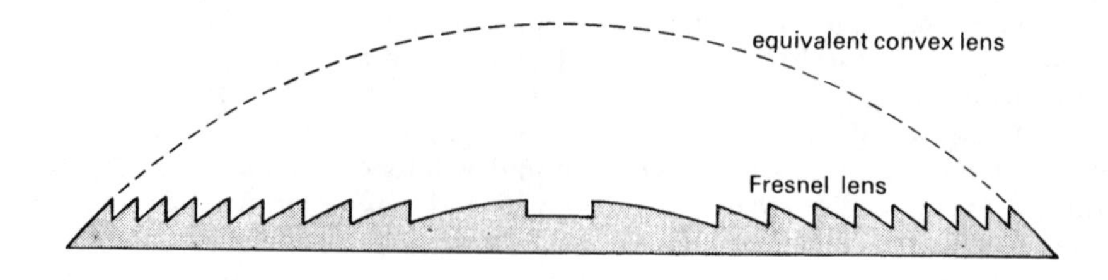

freezing point. The temperature at which a gas or liquid becomes a solid.
MS

Fresnel lens. A device for concentrating light, consisting of concentric rings of transparent glass or plastic, each having the top section of an equivalent convex lens (*see* diagram). Thus light concentration occurs from a thin device with little weight, which is appropriate for use in large or moving structures, e.g. solar *focusing collectors.
JWT

Friends of the Earth (FOE). An international organization committed to environmental conservation based on natural ecological principles. It frequently acts as a political pressure group, having much influence in planning and lobbying. It takes a strong attitude against the development of *nuclear energy and for *renewable energy.

Address: (UK) 9 Poland Street, London, W1; (US) 124 Spear Street, San Francisco, California 94105.
JWT

fuel. (1) (chemical) Any substance, solid, liquid or gaseous, which is available in commerical quantities and may be burned in the oxygen of the air to generate heat. A fuel can also be caused to react with the oxygen in an oxidizing agent, such as an inorganic nitrate or chlorate. The heat is then released explosively.

(2) (nuclear) An element which can readily *fission to produce energy; principally the three *fissile isotopes uranium-235, uranium-233 and plutonium-239. Of these, uranium-235 is the most important at present as it occurs in nature. In addition, the *fertile isotopes thorium-237 and uranium-238, which also occur in nature, can be regarded as nuclear fuels because they can be converted to fissile uranium-233 and plutonium-239 respectively.
WG, DB

fuel assembly. *See* fuel element; nuclear reactor.

fuel cell. A device that produces electrical energy directly from the controlled electrochemical oxidation of fuel. It does not contain an intermediate heat cycle as do other electrical generation techniques. *See also* battery.
TM

fuel consumption. Where fuel is used to carry out a function, the amount of fuel per unit function. In a car, for example, it is the kilometres achieved per unit mass or volume of the fuel.
MS

fuel cycle (nuclear). A cycle which comprises all the operations involved in the mining of uranium, *enrichment (if necessary), the manufacture of *fuel elements, their use in reactors, storage and reprocessing after use, the separation and disposal of *fission products and the recycling of uranium and plutonium for reuse in reactors (*see* diagram, *overleaf*). *See also* transportation of nuclear fuels.
DB

fuel element. The unit of nuclear *fuel fabricated for loading into the core of a *nuclear reactor. It consists of the fuel itself (possibly in the form of a compound such as uranium oxide) enclosed in *cladding, together with any structural parts and attachments necessary for handling the fuel element. Typical fuel elements used in existing nuclear reactors include natural uranium metal rods in finned *Magnox cladding, as in British gas-cooled reactors, and enriched uranium oxide pellets in stainless steel or *Zircaloy tubes, as in advanced gas-cooled reactors and pressurized-water reactors. In the latter case the individual tubes or fuel pins, which are about 1 cm diameter, are assembled in bundles known as fuel assemblies.
DB

fuel gas. A combustible gas which is available in sufficient quantity to be used commercially as a fuel. It may be burned

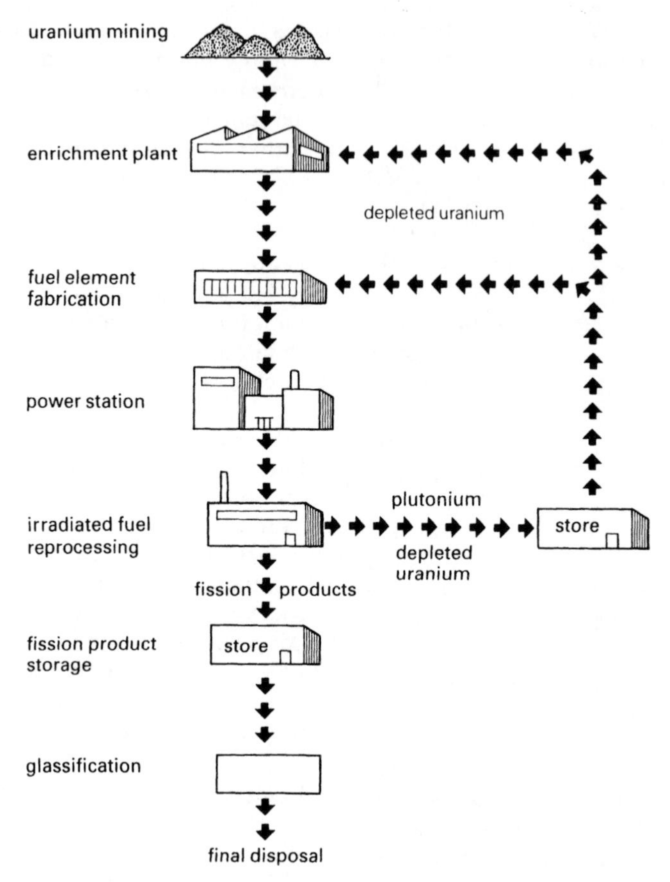

near to the point of production, as with *producer gas, *blast furnace gas and *refinery gases. It may be compressed and liquified in cylinders for use elsewhere, as with *propane and *butane. It may be distributed in a pipeline, like *natural gas, *town gas and *coke oven gas. The important characteristics of a fuel gas are *calorific value, *relative density and *flame speed.

WG

fuel oil. A high boiling point *petroleum fraction burned to heat ovens and furnaces, to heat water and to raise steam in an industrial boiler. The liquid is automized in a *burner to provide a cloud of fine droplets for ease of ignition and combustion. Viscous fuels require preheating to reduce their viscosity so that adequate atomization is achieved. The *flash point should be greater than 55°C.

WG

fuel substitution. In consumption or production, the modification of the composition of total energy use in terms of the quantities of the various fuels used, in response to changing *relative prices for the fuels and/or changing technologial opportunities. *See also* elasticity; substitution.

MC

fuelwood. *See* firewood.

full employment. An apparently simple idea which is, in fact, difficult to define either by reference to qualitative characteristics or numerical magnitudes. At one time it was widely thought, especially by *Keynesian economists, that there existed some proportion of the labour force unemployed which corresponded to full employment, this proportion being fairly constant over time. This view is now less widely held. Attention now focuses on the *natural rate of unemployment and its determinants. Objectives other than the achievement of a low actual unemployment rate, such as reducing *inflation, are correspondingly given more weight.

MC

fume. Particulate material with diameters smaller than one *micron. It results from the condensation of gaseous material into the liquid or solid state.

PH

furl. A term used with *aerogenerators when safety mechanisms operate on the turbine to safeguard the machine in wind speeds above the cut-out speed. The term arises from sail machines.

JWT

furnace. A chamber in which a combustion process takes place. It is normally lined with heat-resistant and insulating brickwork to prevent heat loss to enable an appropriate high temperature to be reached. It may be provided with a *grate on which a bed of solid fuel is burned. Alternatively liquid petroleum fuels, gases or pulverized solid fuels may be introduced through nozzles fitted into the walls of the furnace.

If steam generation is required, the walls of the furnace will be lined with metal tubing and water passed through them. If metallurgical processes are to be carried out, the furnace will be suitably shaped to contain molten metal, and the heating arrangements designed to achieve a particular reaction temperature rather than a maximum thermal efficiency.

WG

fusain. A soft black powdery material which is found in thin layers in coal. It retains a woody fibrous nature and resembles charcoal. A lump of coal tends to fracture at a band of fusain and thus this powdery material is generally responsible for the dust and dirt associated with coal handling.

WG

fusion. The process whereby two light *nuclei join to form a heavier one. For elements of *mass number less than about 50, the *binding energy per nucleon increases as mass number increases. The fusion of two light nuclei therefore produces a heavier nucleus of greater binding energy per nucleon, and the reaction is *exothermic.

The following nuclear fusion reactions, all of which involve the isotopes of hydrogen (*deuterium, ${}^{2}_{1}H$, and *tritium, ${}^{3}_{1}H$) and of *helium, are of importance:

$${}^{2}_{1}H + {}^{2}_{1}H \rightarrow {}^{3}_{1}H + {}^{1}_{1}H + 4.03\,MeV$$

$${}^{2}_{1}H + {}^{2}_{1}H \rightarrow {}^{3}_{2}He + {}^{1}_{0}n + 3.27\,MeV$$

$${}^{3}_{1}H + {}^{2}_{1}H \rightarrow {}^{4}_{2}He + {}^{1}_{0}n + 17.6\,MeV$$

$${}^{2}_{1}H + {}^{3}_{2}He \rightarrow {}^{4}_{2}He + {}^{1}_{1}H + 18.3\,MeV$$

The products ${}^{3}_{1}H$ and ${}^{3}_{2}He$ of two of these reactions are radioactive, and are the reactants in the other two reactions. If these four reactions are combined with the capture of two neutrons in hydrogen to form deuterium, i.e.

$$2\,{}^{1}_{0}n + 2\,{}^{1}_{1}H \rightarrow 2\,{}^{2}_{1}H + 4.2\,MeV$$

then the net effect is to convert deuterium to helium:

$$4\,{}^{2}_{1}H \rightarrow 2\,{}^{4}_{2}He + 47.7\,MeV.$$

The energy yield of this process is 5.7×10^{11} kJ/kg of deuterium, which is approximately ten million times greater than the energy released by the combustion of 1 kg of oil. Fusion reactions of the same type as those described above occur in the sun and provide the earth with a continuous supply of *solar energy.

An obstacle to fusion reactions is the coulomb repulsion force that exists between all atomic nuclei. Before fusion can take place, the reacting nuclei must have sufficient energy to overcome this coulomb force. The energy required corresponds to temperatures in excess of 10^7K. *See also* fusion reactor; nuclear energy.

DB

fusion reactor. A system in which a *fusion reaction can take place in a controlled fashion. The reactions of importance are those involving the *deuterium ($^2_1$H) and *tritium ($^3_1$H) isotopes of hydrogen and isotopes of *helium (*see* fusion). Tritium is radioactive, and another important reaction in a fusion reactor is neutron capture in *lithium-6 to produce it:

$$^6_3\text{Li} + ^1_0\text{n} \rightarrow ^4_2\text{He} + ^3_1\text{H}$$

Fusion reactions are *exothermic, the energy released being one to ten million times greater than that released by exothermic chemical reactions involving the same mass of hydrocarbon compounds.

The main problem in a fusion reactor is that the reacting nuclei must have a sufficiently high energy to overcome the Coulomb force of mutual repulsion before they can interact with each other. The energy required corresponds to temperatures of about 10^7 to 10^8 Kelvin, at which the deuterium and tritium form an ionized *plasma. Heating of the plasma in a fusion reactor can be achieved by passing a very large current through it by the discharge of a number of capacitors; this produces a few million amps for a fraction of a second. The magnetic field set up by this current compresses the plasma, this being the pinch effect, and produces a further increase in temperature.

The high-temperature plasma must be confined within the reaction vessel without coming into contact with its walls; this *containment is achieved by powerful externally applied magnetic fields. The shape of the reaction vessel and the magnetic field may be a *torus, a doughnut-shaped tube around which the magnetic field is continuous and from which the plasma cannot escape. Alternatively the reaction vessel may be a straight tube with the magnetic field stronger at the ends of the tube than at its centre, thus reducing leakage of the plasma from the ends. This method of containment is known as a magnetic mirror.

The cross-section of a possible fusion reactor is shown in the diagram. The innermost region contains the compressed high-temperature plasma in which the reaction takes place. The energy released by the reaction is transferred to liquid lithium coolant in a concentric region, and neutrons produced by the fusion process may be captured in the lithium to provide a supply of tritium for

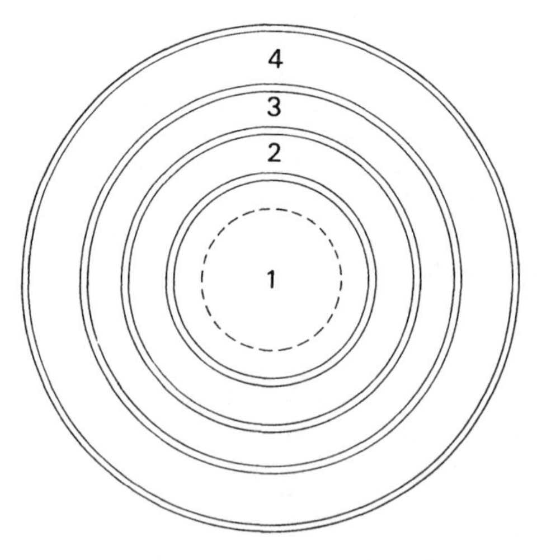

1. plasma 3. shield
2. coolant (lithium) 4. magnetic coil

the deuterium-tritium reaction. The coolant is surrounded by a thermal and radiation *shield, and the magnetic coil for plasma containment forms the outermost ring of the reactor.

It is essential that the energy released by the fusion reaction is considerably greater than the energy required to heat the plasma to its reaction temperature, and contain it. For a reactor operating in a pulsed fashion there will be a net output of energy if the product of the plasma density and the pulse duration exceeds a certain value, dependent on the plasma temperature. This is known as the Lawson criterion.

A possible alternative method for triggering a fusion reaction is to use high-energy lasers to bombard a target of deuterium and tritium in the form of a pellet. The heating and compression of the pellet is sufficient to cause the fusion reaction to proceed for a fraction of a second before the pellet explodes, and in this short time containment is achieved by inertia effects. One problem associated with this method is that of producing lasers with sufficient energy.

At the present time research into controlled fusion power is advancing slowly and without the certainty of eventual success. However, if fusion reactors can be designed to operate safely and satisfactory, then the vast resources of deuterium in the world's oceans will become an almost limitless source of energy.

DB

futility, point of. The point reached when the energy dissipated in acquiring, refining and delivering an energy supply equals the energy delivered. For example, if a *biomass energy system delivered one gigajoule of refined energy, but required the expenditure of one gigajoule of other energies to drive the process, then the point of futility would have been reached and the system would deliver zero net energy. It is conceivable that when fossil energy resources are acutely depleted, then the energy needed to win (say) oil may be as high as that in the oil so won. Under these circumstances one would only persist in the process if the product was valued for something besides its energy content.

MS

futures market [forward market]. A market in which transactions concern contracts to deliver or accept quantities at some future date at a currently agreed price. Such markets are especially evident in trading in *currency, *commodities and in oil. *See also* forward price; speculation.

MC

G

game theory. The analysis of rational behaviour in situations where the payoff to a course of action depends on the unknown courses of action to be followed by others. One of the first applications in economics was in the analysis of *oligopoly, where in considering, say, a price cut, one oligopolistic firm has to take account of the responses that such an action may elicit from his rivals, in determining his best strategy. Similarly, game theory can be used to analyse the behaviour of members of a *cartel, such as OPEC, where collusion is represented as a 'cooperative game'.

MC

gamma flux. A *flux of gamma-ray photons.

DB

gamma radiation. *Electromagnetic radiation of short wavelength and high energy. Its relationship to other components of the electromagnetic spectrum is shown in the diagram. Gamma radiation is emitted from atomic nuclei when undergoing changes in energy following processes such as radioactive decay or neutron capture. The emission of gamma radiation is the most common way for excited atomic nuclei to drop to their ground (or unexcited) energy level. Gamma radiation is highly penetrating and harmful to humans; consequently sources of gamma radiation must be shielded by materials such as lead, steel or concrete. The thickness of the shield depends on the energy and intensity of the radiation source. As an example, nuclear reactor cores, which are very intense sources of gamma radiation, require concrete shielding a few metres thick.

DB

gas. The least complex of the three recognized states of matter. Combustible gases, if available in commercial quantities, make excellent fuels since their

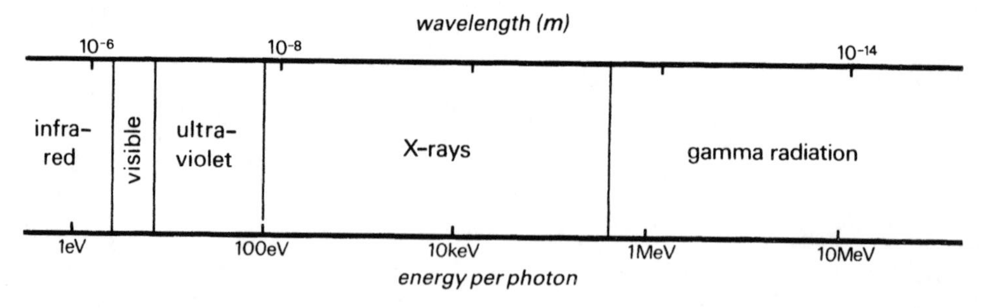

burning rate is easy to control and the combustion chamber is small. *See* coal gas; natural gas; producer gas.

WG

gas amplification. A process which occurs in a *gas-filled radiation detector whereby a single primary ion pair produced by ionizing radiation can, provided the voltage between the electrodes of the detector is high enough, produce a large number of secondary ion pairs. The amplification factor is typically 10^4 to 10^6, so that a single primary ionization event can produce an electron pulse which can be amplified and recorded.

DB

gas burner. *See* burner.

gas coal. A butuminous coal with a relatively high volatile matter content which when carbonized in a retort produces *coal gas and leaves a *coke residue. Since the major product is the gas, the quality of the coke formed is of less importance than with a coal used for the manufacture of metallurgical coke. When a very porous and chemically reactive coke forms, steam may be admitted to the hot retort and advantage taken of the water gas reaction to augment the supply of gas, albeit at a slightly lower calorific value.

WG

gas constant. *See* universal gas constant.

gas-cooled graphite-moderated reactor (GCGR). *See* nuclear reactor.

gaseous diffusion. *See* enrichment.

gas field A natural underground reservoir of *natural gas.

MS

gas-filled radiation detector. A type of *radiation detector in which incident radiation causes *ionization of the gas in the detector. A voltage applied between the electrodes of the detector induces the ions formed to migrate to these electrodes, causing a very small current to flow in an external circuit (*see* diagram). If the applied voltage is increased, the process of *gas amplification produces pulses of ions at the electrodes which can be amplified and recorded as voltage pulses in an external circuit.

DB

gas-gathering system. Where there exists a network of oil and gas fields, as in the North Sea or the Gulf of Mexico, it is economical for each producing rig to be joined to a *grid collecting the gas, which is then delivered to a single mainland point.

MS

gasification. A chemical process in which a solid or liquid fuel is converted into a material which is gaseous at *standard temperature and pressure. A large number of processes have been applied in the fuel industries ranging from the manufacture of *producer gas and *water gas from coke to the complete gasification of coal by reaction under pressure with steam and oxygen. The gasification of oils generally involves contact with a solid catalyst at elevated temperature and pressure in presence of hydrogen.

WG

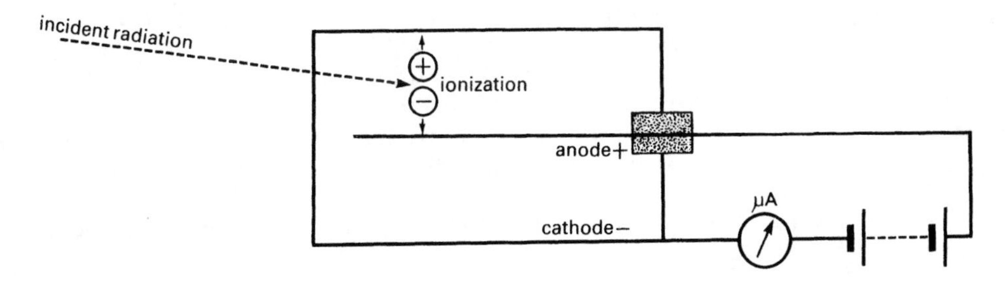

gasifier. A plant for the conversion of a solid or liquid fuel into a gaseous fuel. The simplest types of such plants are the *producer gas generator and the *water gas generator, which operate by blowing air and steam, respectively, through a fixed bed of red hot *coke. Such gasifiers have a limited application due to low *calorific values of the fuel gases made. The complete gasification of coal has for many years appeared to be an attractive process and should inevitably become important when reserves of *natural gas run low. The weight ratio of carbon to hydrogen in *bituminous coals ranges from about 14 to 17, while that for methane is only 3.0. Thus a process for the gasification of coal must be accompanied by the rejection of carbon as coke, or the addition of hydrogen or a substance with hydrogen in the molecule such as water applied as steam. If air is blown through the hot coal in the gasification process, the gas generated will contain a large proportion of the inert gas, nitrogen. Thus in modern coal gasification processes mixtures of steam and oxygen are blown through a bed of suitably sized coal particles held at a high temperature by the chemical reactions taking place. Practical difficulties encountered include the tendency of particles of many coals to fuse together when introduced into the hot gasifier, and thus reduce the amount of solid surface available for reaction with the gases, and the melting of the coal ash particles to form a molten slag if the gasifier temperature is too high.

Many processes have been proposed, but very few have progressed beyond the small-scale development stage, either due to mechanical problems or to adverse economics. Those which have been shown to be commercially viable include the following. In the *Lurgi, so called 'fixed-bed' process, coal is fed to a vertical cylindrical gasifier and moves downward through it as it is gasified. Oxygen and steam are introduced at the base of the unit which operates at a pressure of 20–30 bar, and ash is removed periodically through a rotating grate. The effluent from the gasifier top contains tars and other liquid products together with the fuel gas. The separation of these liquids from the gas complicates the process. In an improved version of the process, British Gas has operated at a higher temperature which allows the coal ash to be discharged as a liquid slag, and the tars and other liquid byproducts to be recycled to the gasifier so that they are converted to fuel gas. In the Koppers–Totzel entrained flow process, powdered coal is blown into an almost spherical gasifier in a mixture of steam and oxygen. Reaction between the gases and the solid particles takes place in about one second at over 1500°C and at atmospheric pressure. Most of the molten ash particles fall into a water-quenching bath, while the fuel gas is taken from the top of the gasifier, cooled and cleaned. Various modifications of this process have been developed by Texaco, Dow Chemical Co. and Shell. These in general relate to the manner of feeding the coal to the gasifier, operation under increased pressure and removal systems for the ash residue.

Fluidized-bed processes have been developed in Germany, in USA and in UK. The gasifiers are tall cylindrical vessels into which small coal particles are fed. The bed so formed is fluidized and gasified by a mixture of steam and oxygen. Care must be taken to keep the temperature below that of the fusion of the ash particles. In the Winkler process the bed is kept at temperatures on the range 800–1000°C. It is best suited to the gasification of lignites which do not form a coke and leave an ash with a high fusion point. In the UK, British Coal has developed a fluidized-bed process in a pressurized gasifier which uses coarsely crushed coal and operates at about 1000°C. Gasification of the carbon is incomplete. About 25% of it leaves the base of the gasifier with the ash in the form of a *char. This can be burned in a secondary fluidized bed fed with air for the generation of heat which is returned to the gasifier. Two similar types of fluidized gasification proccesses have been developed in the USA by the Institute of Gas Technology (U-Gas) and Westinghouse/KRW Energy Systems.

A process in which coal is gasified in a horizontal rotary kiln provided with special mechanical seals is being developed by Allis–Chalmers. An important factor in the selection of a gasifier is the ease with which efficient heat recovery is possible and the technique available for the removal of sulphur oxides from the fuel gas made. Thus for efficient operation the gasifier should be linked to heat exchangers and a gas turbine.

WG

gasohol. An American term originally denoting a mixture of 10% *ethanol, fermented from crops such as sugarcane, cassava and maize, and 90% unleaded gasoline (petrol), for use as a motor fuel. The word now has a wider meaning to cover any combination of ethanol/methanol (power alcohol) and petrol. The utilization of alcohols as constituents of motor fuels at the 10–15% level requires only minor and inexpensive modifications to road vehicles, though a larger fuel tank is needed to offset the reduced energy content per unit volume. However, energy consumption per unit distance travelled would be lower because higher compression ratios and simpler pollution controls could be effected. Also the octane rating will be increased over that of unleaded petrol. At alcohol levels of 20–40%, modifications to the carburettor and ignition system, together with a high-compression cylinder head, would be needed plus an alcohol-resistant tank and fuel delivery system and a heated inlet manifold.

In 1981 the Brazilian National Alcohol Programme was reported to be producing around 3 billion litres of ethanol fermented from sugar-cane and cassava, with a production target of over 8 billion litres by 1985, in order to substitute for most of the country's oil imports. The US gasohol programme centres around the fermentation of maize. At the beginning of 1980 there were 12 grain-alcohol plants in the USA with an overall annual capacity of 300 million litres. This is envisaged to rise to 155 plants producing 5.7 billion litres of ethanol in 1990 – a modest proportion of the projected unleaded gasoline demand. At present, gasohol in Brazil and the USA has to be subsidized to make it economically competitive with petrol at the filling station. Other countries such as Canada and West Germany are more interested in the gasification of wood followed by methanol synthesis as a transport fuel constituent, while Third World nations like India, Pakistan and Sudan are concentrating on molasses as the main substrate for power alcohol programmes based on ethanol. Yet other countries such as New Zealand are examining both the ethanol and methanol routes.

CL

gasoline [US: *motor spirit]. A highly volatile petroleum distillate fraction, boiling in a temperature range of 40–180 °C. Consisting mainly of hydrocarbons with five to eight carbon atoms in the molecule, it is likely to contain additives to ensure efficient combustion. *See also* antiknock; octane rating.

WG

gas scrubbing. The removal of some components of a gas mixture by intimate contact with an absorbing liquid or adsorbing solid. A simple form of gas scrubber is provided in a chemical plant by a tower loosely packed with irregular lumps of an inert solid down through which water is trickling.

WG

gas turbine. *See* turbine.

gas synthesis. A process whereby a gaseous fuel is created by chemical rearrangement of other compounds. Over 50 processes have been explored for this purpose, but few have been commercially developed. *See* gasification.

MS

Geiger–Muller counter. A *gas-filled radiation detector operated at high voltage in which the *gas amplification effect produces a large discharge pulse after each primary ionizing event.

DB

General Agreement on Tariffs and Trade (GATT). An international agreement about trade which came into operation in 1948. It serves as a centre for multilateral trade negotiations and sets out rules of conduct for members. These rules are based on three principles: multinational negotiations, non-discriminatory reductions and the dismantling of non-tariff barriers. Its aim is to liberalize world trade as far as possible, which it has done through a series of rounds of talks, the most recent two being the Kennedy and Tokyo Rounds. In the case of unfair competition GATT does permit the imposition of tariffs (*see* dumping). Difficulties have been encountered from three sources: the formation of regional groupings, e.g. the EEC; the problem of dealing with the interests of developing countries (*see* United Nations Conference on Trade and Development); exemptions of some commodities, especially agricultural commodities.

MC

general equilibrium analysis. An approach which considers simultaneously all the markets in an economy, allowing for *feedback effects between the individual markets. It is particularly concerned with the conditions under which there can exist simultaneously *equilibrium in all markets, and the determinants and properties of such an economy-wide set of equilibria. It avoids the *partial equilibrium analysis assumption that the links between individual markets are weak and can be ignored. This means that general equilibrium analysis is more complex and requires more informational inputs than partial equilibrium analysis. For this reason partial equilibrium analysis remains popular for applied studies. Thus, for example, in studying the energy market in an economy it is typical to ignore the effects which developments in that market will have on the rest of the economy, and the feedback from such effects to the energy market.

MC

generator. A machine for converting mechanical energy into electrical energy; in the case of alternating current, it is sometimes called an alternator. Electricity is generated, according to the principles of *electromagnetism, by changes in the magnetic flux surrounding a conductor; changes in the magnetic flux are created in the generator by the application of mechanical power from an *internal combustion engine, gas *turbine, steam turbine or similar.

TM

genetic effects. *See* biological effects of radiation.

genetic engineering. The controlled transfer of genetic information from one organism to another, invariably of a different species. Its more scientific definition of recombinant DNA (deoxyribonucleic acid) research is used to describe new biochemical techniques that allow fragments of DNA isolated from a plant, animal or microbe (the donor) to be inserted into another piece of DNA (the vector), and then finally transferred to a microbe (the host). In this way a micro-organism can acquire novel genetic properties. It can also gain an ability, unattainable through the more conventional means of selective breeding or mutation, to create a product, for instance, or utilize a substrate which it had been unable to before.

Examples pertinent to energy production might include improved *photosynthetic efficiency and hence higher *biomass productivity, better fermenting yeast strains tolerant to increased levels of *ethanol manufacture, and the transference of *nitrogen fixation genes from bacteria to cereal crops, allowing the latter to fix their own nitrogen and simultaneously reduce their need for energy-intensive chemical fertilizers. These are just three of many possible applications of genetic engineering to energy production and conservation. The feasibility of some are open to question, but certainly genetic engineering is opening up new horizons already which were almost unthinkable ten or even five years ago.

CL

geothermal energy. Heat obtained from the earth. Heat is flowing outwards from the interior of the earth due to cooling of the core (60%) and to radioactive decay (40%) of various naturally occurring species of atomic nuclei such as uranium. The average geothermal heat flow is about 0.01 W/m^2, which is negligible for energy supplies. However in certain regions of geothermal activity or thin earth mantle, this flow is greatly enhanced. Pipes sunk into the earth in these regions may tap aquifers of heated water and steam, usually heavily saline and brackish. Electricity generating stations may operate from heat engines driven directly or indirectly from such sources, and the heat can be used for domestic and industrial purposes. Examples are found in New Zealand, Italy and Iceland.

Geothermal heat in large masses of rock such as granite or sediments may be at useful temperatures. Water, passed through such 'hot rocks' in natural or explosively induced fissures, becomes heated and may be used, for instance, for district heating or preheating for conventional boilers.

JWT

giant kelp. A large seaweed, scientific name *Macrocystis pyrifera*, which grows prolifically off the Californian coast and elsewhere. It is the focal point of the ambitious US ocean farm project in which the kelp is to be cultivated in the Pacific Ocean over an area of about 80 km^2 at a depth of 15–25 m. The kelp will be grown on buoyancy control structures and submerged supporting lines. Fast growth would be achieved by pumping up cool nutrient-rich water from depths of 300–700 m. Mean *photosynthetic efficiencies of around 2% are expected, giving ash-free yields of organic matter in the range 75–125 t/ha.y. The biomass produced would be converted into *methane gas by *anaerobic digestion, with byproducts including fertilizers, salts and animal feeds. However the *net energy of the proposed system does not appear very attractive and it may well be that other projects involving kelp will have a more promising future.

CL

Gibbs free energy. *See* free energy.

Giffen good. A commodity for which, unusually, the quantity demanded increases as price rises. This occurs because the *income effect is positive and larger than the negative *substitution effect. A standard example of a Giffen good would be a low-quality food consumed by low-income households.

MC

giga-. A prefix denoting a multiple of one thousand million (US: billion), i.e. 10^9. For example, 1 gigawatt equals 10^9 watts.

PH

gilsonite. One of a variety of naturally occurring asphalts differing slightly in composition and properties. Gilsonite is a very pure form of brittle asphalt with a high melting point.

WG

Gini coefficient. A measure of economic *concentration or the degree of inequality.

MC

Girdler sulphide process. *See* deuterium.

glassification [vitrification]. A process proposed for the long-term disposal of *radioactive waste from nuclear power stations. *Fission products of long *half-life and *actinides are combined with a suitable glass compound (such as borosilicate glass) to produce a highly stable product thought to be capable of withstanding high temperature and leaching by water for thousands of years. *See also* Harvest process; radioactive waste management.

DH

Glauber salt. A chemical compound that can absorb and release thermal energy at environmental temperatures, and may thus be used as an *energy store. The process is a *latent heat process associated with a change in molecular structure. Such heat storage systems are particularly important for solar-energy heating systems whereby energy trapped on sunny days can be stored for night time and subsequent periods.

JWT

glazing. Those surfaces of a building envelope constructed in glass. Glazing is provided in the walls and roofs of buildings in order to allow penetration of daylight and to provide a view out of the building. The impact of the glazing on the thermal behaviour of the building is significant and can lead to excessive heat loss in winter and excessive heat gain in summer. Increasingly, glazing, particularly double and triple layers with intervening air gaps is being used as a passive solar collector which is wholly integrated into the building design.

Special types of glass are available for use in building construction: absorbing glass reduces the proportion of transmitted solar energy by storing heat energy; reflecting glass, which has a special film applied to its external surface, reduces the proportion of transmitted energy by reflecting an increased proportion of the incident solar energy.

TM

global radiation. (1) The total radiation from all sources and all directions onto a particular surface. Thus it includes direct sunshine, diffuse light from the sky and clouds, reflected light from buildings, etc.

(2) Radiation falling upon the entire earth's surface.

JWT

glycolysis [Embden–Meyerhof–Parnas (EMP) pathway]. The sequence of enzymatic reactions in which glucose is converted to pyruvic acid within living cells under *anaerobic conditions. The pyruvic acid is subsequently reduced either to lactic acid or to acetaldehyde and ethanol, depending on the particular organism's metabolism. The ethanol route is the principal one for ethanol production in the fermentation of sugars by yeast. The overall equation of glycolysis plus ethanol formation is:

$$\underset{\text{hexose sugar}}{C_6H_{12}O_6} \rightarrow \underset{\text{carbon dioxide}}{2CO_2} + \underset{\text{ethanol}}{2C_2H_5OH}$$

The free energy liberated by the above overall reaction in gram mole terms is 234.5 kJ, of which only 67 kJ is conserved as chemical energy within the two ethanol molecules produced. The remainder is dissipated as heat to give a thermodynamic efficiency of sugar conversion to ethanol via glycolysis of 28.5%.

CL

Gobar gas plant [biogas generator]. A simple apparatus for turning animal dung into *biogas plus nitrogen fertilizer. 'Gobar' is the Hindu word for cow dung. The family-sized plant was first introduced into Indian villages in 1954. There are now some 75,000 in India, each designed to produce around 2 m^3 (equivalent to 40 MJ) of gas each day from the dung input of about four cows. The Indian design is a variable volume/constant pressure reactor in which a mild-steel gas holder floats over a brick or concrete-lined pit sunk into the ground. Cheaper materials than steel are now being tested to lower the overall cost. Dried dung is also used in many developing countries as a domestic fuel and the impetus for the development of these plants came initially from concern over the loss of the nitrogen content of the dung when burnt. This meant loss of valuable nitrogen fertilizer. The problem is resolved by introducing the dung at daily intervals into the Gobar plant; the gas is liberated after a suitable *retention time, leaving a residual sludge which retains the original nitrogen, phosphorus and potassium content of the dung. This sludge can then be applied to the fields as a *biofertilizer.

Although the overall *anaerobic digestion process is marginally exothermic, the heat generated is insufficient to maintain the required fermentation temperature. Thus biogas plants work best in locations of high year-round ambient temperatures. Where the temperature does fall below the desired *mesophilic range of 30–35 °C, then reaction rates decrease, retention times increase and daily biogas yields drop. The influence of temperature may be judged from the fact that a temperature rise from 15 °C to 23 °C increases the rate of output by 80%.

Gobar gas plants are now operating throughout the developing world. A second major design, the so-called Chinese model, operates under conditions of constant volume/variable pressure and numbers 7 million on mainland China alone. The Chinese design is used primarily as a fertilizer production unit with the gas considered as a bonus; the input feed tends to be more variable than that for Indian plants, containing human faeces and vegetable matter in addition to animal wastes.

CL

good. In economics, a tangible output of some production process which commands a positive *price, e.g. wheat, shirts, coal, electricity. A 'bad' is something which has a negative price, i.e. costs would be incurred in order to have less of it. Examples are pollution, congestion, criminal activities, ill health. The term *commodity covers goods and intangible services, such as insurance and education.

MC

graphite. *See* carbon.

grate. A mechanical device consisting of horizontal or inclined parallel steel bars, or a perforated sheet of cast iron, used to support a bed of burning solid fuel. In a simple system the ash formed may be cleared manually from the grate to fall into an ash pit below by the use of one of a variety of tools. Rocker bars and similar devices may be also incorporated within the grate to clear the ash by an automatic mechanism. Various forms of moving grate are also available. *See also* stoker.

WG

gravitation. The universal tendency of every body to move towards every other body. The *force of attraction between two bodies (measured in *newtons) is proportional to the product of their masses and inversely proportional to the square of the distance between them. The constant of proportionality, the gravitational constant, is equal to 6.67×10^{-11} N m^2/kg^2.

In free fall this force between an object and the planet earth produces a downward acceleration of about 9.8 m/s^2. The exact value varies from this by about 1%, mainly due to latitude. The attractive force between an object and the earth, or any celestial body, is known as gravity.

TM, JWT

gravity. *See* gravitation.

gray (Symbol: Gy). The SI unit of *radiation dose. It is a measure of the energy of radiation of any type that is absorbed per unit mass of material, and is defined as

1 gray = 1 joule of absorbed energy/kg

It has replaced the earlier unit of dose, the *rad.

DB

Gray–King assay. A UK coal carbonization test for determining the caking and swelling properties of coal. A standard weight of powdered coal is heated in a horizontal silica tube placed in a standard furnace and at a prescribed rate of temperature rise to 600°C. The appearance of the coke residue is compared with a series of standard cokes labelled A to G, corresponding to a progressive increase in the extent of swelling in the transformation from coal to coke. This test in conjunction with the volatile-matter content forms the basis of coal classification used by the UK National Coal Board.

WG

green algae. Organisms dependent on light for energy and carbon dioxide as their principal carbon source. Species range in diversity from unicellular, through filamentous and colonial forms, to large multicellular structures. Their *photosynthetic efficiency within oxidation ponds at 30° latitude is around 2% on an annual basis, realizing 50 t/ha.y of ash-free dry weight biomass. Converting the algal product to *biogas can yield 500–600 GJ/ha.y as methane energy, but the energy inputs required to grow and harvest the algae and then run the digestion process currently render the overall system a *net energy sink. However, there is room for technological improvement, particularly in the energy-intensive harvesting procedure. Nevertheless green algae operating in concert with degradative aerobic bacteria within oxidation ponds can treat sewage, given favourable climatic conditions, at lower energy requirement than conventional treatment processes, and at the same time produce an algal byproduct for use as an animal feed. At the state of current technology, algae are better used in systems to conserve fossil fuel energy rather than as energy producers themselves.

CL

greenhouse effect. Glass can transmit about 80–92% of incident solar radiation at visible and near visible wavelengths, but does not transmit infrared radiation of long wavelength from heated objects. Thus a glass greenhouse allows solar radiation to enter, but does not transmit the radiation outwards from the heated interior. The solar energy is therefore trapped so producing increased temperatures. A similar effect, known as the greenhouse effect, occurs in the earth's atmosphere: carbon dioxide and water vapour allow solar radiation to pass onto and be absorbed by the earth's surface, but do not transmit back to space the long wavelength infrared radiation re-emitted from the earth's surface. Increasing concentration of carbon dioxide in the atmosphere can therefore be expected to increase the earth's temperature if no other effects occur.

JWT

grey body. A notional body in which the *emissivity of radiant heat energy does not vary with wavelength.

TM

grid. A distribution and/or collection network for energy supplies. Electricity grids distribute electrical power by means of high voltage networks of cables, either suspended from pylons or enclosed in insulation for burial in the ground or for submarine transmission. Gas may be distributed in pipes from sources at very great distance from consumers.

JWT

gross. In economics, an adjective applied to many economic variables and meaning measured before allowing for *depreciation. For example, gross *national product is the total of the economy's output of *final product before allowance is made for the replacement *investment necessary to maintain intact the stock of *capital. *See also* national income accounts.

MC

gross calorific value (GCV). *See* calorific value.

gross domestic product (GDP). A country's *domestic product, i.e. the value of the goods and services produced in it, measured *gross of depreciation. It differs from *gross national product by the amount of income received from abroad. *See also* national income accounts.

MC

gross energy requirement (GER). The sum of all the *primary energy sources, expressed in terms of the heat (enthalpy) that must be dissipated in order to deliver a good or service to a particular point in the market place. It is a convention devised by a workshop on energy analysis methodology, held under the auspices of the International Federation for Institutes of Advanced Study (IFIAS) in Sweden in 1974. The energies are counted in the following manner.

Fossil fuels: the gross calorific value of the primary energy sequestered in order to deliver all the necessary fuels to the production process, without regard to whether in the real process all energies are usefully or non-usefully combusted.

Nuclear fuels: the fissile energy of the fuels, irrespective of whether there is 100% burn-up, plus the energies required for reprocessing and disposing of wastes.

Renewable energies and human energies are not included in the sum. Thus as defined, the GER reflects the amount of depletion of the earth's inherited store of non-renewable energy in order to create and deliver a good or service. *See also* natural resources.

MS

gross heat of combustion. *See* calorific value.

gross national product (GNP). A country's *national product, the value of the goods and services produced in the country plus any income received from abroad, measured *gross of depreciation. *See also* national income accounts.

MC

ground level concentration. The highest concentration of a particular gaseous pollutant, measured at ground level, which results from the dispersion of the pollutant downwind from the chimney of an industrial plant. This concentration depends on the nature of the gas emitted, its temperature and velocity of emission, the height of the chimney, the wind velocity and the atmospheric temperature conditions. In the situation where there is a *temperature inversion, the waste gases from a chimney tend to collect in the region beneath the inversion layer with a resulting increase in ground level concentration.

PH

H

Haber process. A process for converting *methane (usually as *natural gas) to ammonia. Air, which contains nitrogen, is passed with methane over a catalyst at elevated temperature. Together with variants it is the major process in the world today for making nitrogen fertilizers.

MS

hafnium. A metallic element which because of its high neutron capture *cross-section may be used as the control material in nuclear reactors (*see* control systems).

Symbol: Hf; atomic number: 72; atomic weight: 178.5; density: 13,400 kg/m^3.

DB

half-life. A measure of the lifetime of a radioactive substance which is defined as the time required for the number of radioactive atoms of that particular nuclide in a sample to be reduced by one half. For example, if a sample of a particular radioactive nuclide contains initially N atoms, and the half-life is one day, then after one day the number of atoms of the nuclide will be $\frac{1}{2}$N, after two days it will be $\frac{1}{4}$N, etc. *See also* mean life; radioactivity.

DB

Halobacterium halobium. A salt-tolerant bacterium capable of *photosynthesis in the absence of chlorophyll by utilizing a purple pigment, *bacteriorhodopsin. This pigment absorbs *photons of light, releasing protons from within the enclosing membrane and setting up a small electric potential. If the pigment molecules could be integrated with a suitable artificial membrane on a large scale, then an electrical supply could be harnessed and utilized by man. The bacteriorhodopsin pigment provides a great advantage over the chloroplast pigments of higher plants and algae in that it is surprisingly stable, probably as a direct consequence of the extreme environmental conditions under which it normally has to survive.

CL

Hanford Engineering Development Laboratory. The site of the first US military plutonium-producing reactors, built in 1943, and still an important establishment for plutonium production.

Address: Richland, Washington 99352, USA.

DB

hard coal. Coal which is resistant to mechanical shock and abrasion and thus does not suffer much physical degradation when in a coal-handling plant. It has generally a dull appearance and consists largely of the coal constituent *durain.

WG

hard solar. A term that has come to imply the utilization of *solar energy through the application of sophisticated technology such as *solar towers, photovoltaic devices (*see* photovoltaic conversion) and *wave-power devices. *See also* soft energy.

MS

hardwoods. Deciduous tree species producing compact hard timber. Examples are the *alder, *eucalyptus, poplar, sycamore and willow, which are generally good coppicing trees, suitable for *short rotation forestry, and hence potentially excellent energy crops. They also have the advantage over *softwoods of a lower lignin content, allowing easier hydrolysis to fermentable sugars in the production of ethanol fuel. On the debit side hardwoods may contain three times more pentoses (5-carbon sugars) as do softwoods, the pentoses not being fermentable by the *Saccharomyces cerevisiae* yeast used for alcohol production, and even inhibiting hexose fermentation. These pentoses must therefore be removed prior to fermentation.

CL

harmonic decrement. *See* decrement.

Harvest process. A process being developed by the *United Kingdom Atomic Energy Authority for the disposal of high-level *radioactive waste. Liquid waste is mixed with silica and borax to form a slurry which is heated in a furnace at around 1000°C to form a glass. The glass is then cast into cylindrical stainless steel containers. Each cylinder would contain the waste arising from about 5.5 tonnes of spent fuel, reduced in volume to about 0.36 cubic metres of glass. The ultimate disposal of this vitrified waste is still being studied.

PH

H-coal process. *See* coal liquefaction.

head. The height of a water or other liquid supply above its point of use, or the pressure equal to that derived from a column of water of that height.

JWT

health physics. The study of the effects of ionizing radiation on human beings and the recommendation of safe limits of *radiation doses. *See also* biological protection.

DB

heat. Energy in transfer from one environment or system to another. Such transfer can only occur when there is a positive temperature gradient. Thus a heat flow has the effect of changing the heat content of one system at the expense of another. More precisely, it is the *internal energy of the system which is changing. Heat may be transferred by *conduction, *convection or *radiation. The term heat is often used imprecisely to signify energy. The units of heat are commonly the *joule, *calorie or *British thermal unit (Btu) or multiples thereof. The joule is the definitive SI unit. *See* conversion tables p. viii.

MS

heat capacity. (1) [specific heat; specific heat capacity] In thermodynamics, the energy required to raise unit mass of a substance by one degree of temperature. It is possible to carry out such an act at constant pressure or at constant volume, or with both varying. Heat capacities are listed in tables of properties of substances at constant pressure (C_p) and constant volume (C_V). For liquids and solids the difference between heat capacity at constant pressure and that at constant volume is trivial, but for gases it can be important. Where not otherwise specified, heat capacities are given at constant pressure. It can be shown that C_p and C_V are related by the formula

$$C_p - C_V = \left[\text{pressure} + \left(\frac{\delta U}{\delta V}\right)_V\right]\left(\frac{\delta V}{\delta T}\right)_p$$

where U is the *internal energy of the gas. For an ideal gas, this term reduces to R, the *universal gas constant. Heat capacity normally, but not invariably, increases with temperature.

Substances can vary greatly in the value of their specific heat capacities: at normal room temperature, water has a specific heat capacity of 4.187 joules per gram per kelvin (4.187 J/g.K); gold has a specific heat of 0.13 J/g.K.

(2) (of a building) *See* thermal capacity.

MS

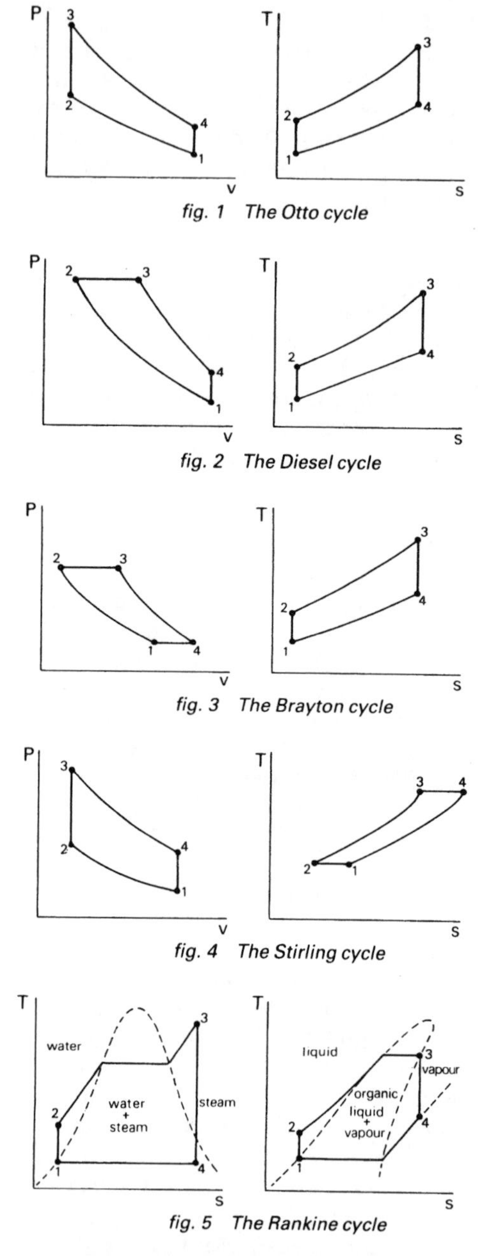

fig. 1 *The Otto cycle*

fig. 2 *The Diesel cycle*

fig. 3 *The Brayton cycle*

fig. 4 *The Stirling cycle*

fig. 5 *The Rankine cycle*

heat engine. A device for the conversion of heat energy into mechanical energy. The operation of a heat engine can be represented by a thermodynamic cycle in which a sequence of thermodynamic processes is performed on a fluid before it is returned to its initial state. During many of the processes in the sequence, one property of the fluid is held constant: these include constant temperature (*isothermal), constant pressure (*isobaric), constant volume (*isometric), constant *entropy (*isentropic), constant *enthalpy (*throttling) and zero heat flow (*adiabatic) processes. Any particular thermodynamic cycle can be uniquely represented by constructing two graphs of the fluid property coordinates, one with axes of pressure (p) and volume (V), the other with axes of temperature (T) and entropy (S). The cyclic integral represents, according to the *first law of thermodynamics, both the net work done and the net heat input. The thermal efficiency of the cycle (η) is conventionally defined as the ratio of net work output to the heat added at high temperature.

Heat engines can usefully be subdivided into two categories: internal combustion and external combustion engines.

Internal combustion engines: the Otto cycle is the basic thermodynamic power cycle for the spark-ignition internal combustion engine commonly found in petrol (gasoline) fuelled motor cars. Figure 1 shows the cycle as having four processes: in process 1–2 the air/fuel mixture is compressed isentropically; the spark ignition adds energy isometrically (process 2–3); the hot fluid then expands adiabatically (process 3–4); the heat is rejected isometrically (process 4–1). The diesel cycle (figure 2) is somewhat different, having compression as opposed to spark ignition. The four processes are, in sequence, isentropic, isobaric, adiabatic, isometric. *See also* diesel engine.

Internal combustion engines: the Otto cycle is the basic thermodynamic power cycle for the spark-ignition internal combustion engine commonly found in petrol (gasoline) fuelled motor cars. Figure 1 shows the cycle as having four processes: in process 1–2 the air/fuel mixture is compressed isentropically; the spark ignition adds energy isometrically (process 2–3); the hot fluid then expands adiabatically (process 3–4); the heat is rejected isometrically (process 4–1). The diesel cycle (figure 2) is somewhat different, having compression as opposed to spark ignition. The four processes are, in sequence, isentropic, isobaric, adiabatic, isometric. *See also* diesel engine. High-speed internal combustion engines may operate on combined (or mixed or duel) cycles in which both isometric and isobaric combustion of the fuel takes place.

External combustion engines: the Brayton cycle is the basic thermodynamic power cycle for the gas *turbine engine. Figure 3 shows the cycle as having four processes: in process 1–2 the fluid is compressed isentropically; heat is added isobarically (process 2–3); the hot gas expands isentropically (process 3–4); the heat is rejected isobarically (process 4–1). The Stirling cycle is receiving renewed attention as the basic thermodynamic cycle relevant to a low-pollution reciprocating car engine fuelled by hydrogen. Figure 4 shows the Stirling cycle; the four processes, in sequence, are isothermal, isometric, isothermal, isometric. The Brayton and Stirling cycles are essentially gas power cycles. The Rankine cycle on the other hand is a vapour power cycle. Water is the common working fluid in vapour power heat engines – steam reciprocating engines and steam turbines – but other fluids include potassium, sodium and ammonia. The Rankine cycle (shown in figure 5 for water and organic fluids) has the same sequence of processes as the Brayton cycle – isentropic compression, isobaric heat addition, isentropic expansion, isobaric heat rejection. TM

heat exchanger. A device to effect efficient transfer of heat from one fluid medium to another. The most common form of heat exchanger is that in which the two fluid streams pass through the shell of the heat exchanger in steady counter flow, the heat transfer taking place through a separating wall. TM

heat flux (Symbol: φ). A measure of the heat energy per unit time flowing normally to the direction the flux is propagating. The unit of measurement is kilowatts per square metre (kW/m^2). TM

heat gain/loss. The net gain or loss of heat energy to or from a building in relation to the external environment. Heat loss, in winter for instance, may be attributed primarily to conductive heat flow outwards through walls and roof and to ventilative *airchange; heat gain in a building may be made up of heat energy emitted by the building occupants, machinery and lighting, supplemented by solar energy falling on the building envelope and shining through the glazed areas. TM

heating system. *See* central heating; domestic heating.

heat island. An area of urban development which exists at a higher temperature than surrounding suburban and rural areas. The temperature difference (as much as 4 °C) is manifest mainly at night and in cold weather. It is due in part to the wind protection afforded by buildings and in part to the heat emission from the buildings. TM

heat limit (atmospheric). A concept embodying the idea that there may come a time when the rate of dissipation of energy on the earth's surface, through man's use of non-renewable energy sources, may be sufficiently high to create serious environmental consequences. It implies an eventual limit of the rate of energy dissipation that may be tolerated on earth. MS

heat loss. *See* heat gain/loss.

heat of combustion. *See* calorific value.

heat of formation. The heat absorbed or released when a chemical compound is formed from its elements. For example, when carbon dioxide (CO_2) is formed from carbon and oxygen, there is a considerable evolution of heat. In order to form *methane from its elements, carbon and hydrogen, there is a considerable absorption of heat. The fact that one may compute the heat of formation from *thermodynamic tables does not imply that such a chemical reaction is possible in a direct way, but it may be formed through the medium of intermediate reactions.

MS

heat of reaction. The heat gained or lost when two or more substances react chemically to form other substance(s). It is not a unique value for any chemical reaction, but varies with the temperature at which the reaction is carried out and the extent of the reaction. *See also* Le Chatelier's principle.

MS

heat pipe. A tube or pipe containing a small quantity of volatile liquid and perhaps a porous filling. Liquid vapourizes at the heated end and the vapour passes to the other, cooler end. Here condensation occurs with the release of heat, and the liquid returns directly or via the filling to the heated end again. The overall effect is that heat is passed rapidly, easily and at high intensity from the hot to the cold end of the pipe.

JWT

heat pump. Space heating plant operating on exactly the principles of refrigeration plant. Whereas refrigeration plant extracts heat from the controlled environment and discharges to the surrounding environment, heat pumps extract heat from the surrounding environment (the earth, a lake, a well, a river) and supply it to the controlled environment. The energy supplied to the pump may be mechanical (as in the *vapour compression chiller) or heat (as in the *absorption chiller). *See also* coefficient of performance.

TM

heat reclaim. *See* heat recovery.

heat recovery [heat reclaim]. The effective utilization of what otherwise would have been discarded or wasted heat energy. Systems for heat recovery can be employed on a large scale, e.g. the use of the waste heat from electricity generation for *district heating, and on a small scale, e.g. the use of a *thermal wheel to recover heat from the air being exhausted from a building.

TM

heat storage. *See* energy storage.

heat transfer. The phenomenon, occurring within a system or at the boundary of communicating systems, as a result of temperature difference. Heat is transferred by three modes: *conduction, *convection and *radiation, which may occur separately or in combination.

TM

heavy. As applied to petroleum fractions, the term signifies a high molecular weight substance with high boiling point, high viscosity, density greater than water and probably dark in colour.

WG

heavy distillate. A petroleum fraction with a boiling range between 350–

500°C. Depending on the molecular structure of the hydrocarbons present, it may be sold as *fuel oil or be further refined to produce lubricating oils and waxes.

WG

heavy fuel oil. A liquid hydrocarbon fuel from the higher-boiling fraction of crude *petroleum. It should have a kinematic viscosity of about 750 centistokes ($10^{-6} m^2/s$) at 38°C, a sulphur content less than 4.5% and a gross calorific value of about 42.5 MJ/kg. It is sometimes known as Bunker B. A still heavier grade is also available and is used as a fuel on large ships. It is known as Bunker C and has a viscosity of about 1500 centistokes at 38 °C. It should have a sulphur content less than 5% and a GCV of about 41.5 MJ/kg.

WG

heavy water. The commonly used name for *deuterium oxide, which is used as the *moderator in some types of *nuclear reactor, namely CANDU and the steam generating heavy water moderated reactor (SGHWR). The use of heavy water as the moderator in large thermal reactors has one major disadvantage: the large quantity required (typically a few hundred tonnes) and its very high cost. Precautions must be taken to prevent leakage, and this requirement has led to the adoption of the *calandria, a heat exchanger, as the core vessel for CANDU and SGHWR. In these reactors the heavy-water moderator is contained at low temperature and pressure, and leakage is thereby reduced.

DB

hedging. (1) Action by a trader who buys or sells forward in order to protect himself from variations in price. Suppose a purchaser of oil expects the price to rise in the future. He can then buy a forward contract for the quantity he wishes (*see* forward price). Since the price is agreed when the contract is made, there is no uncertainty. If the price does rise then the additional amount he paid out for the forward contract is likely to be less than what he would have paid out at the higher price.

(2) [hedging against inflation] Action in which an *asset is purchased whose rise in price is expected to be at least as great as the rate of inflation.

MC

heliodon. An apparatus which models how sunlight falls on buildings. A scale model of the building configuration is placed on a flat surface which can be rotated horizontally to represent latitude and vertically to correspond to the earth's rotation, in relation to a fixed lamp. Shadows cast by the lamp on the scale model are comparable to those cast by the sun on the buildings and their surroundings.

TM

helium. An inert gas which is used as the *coolant in high-temperature gas-cooled *nuclear reactors.

Symbol: He; atomic number: 2; atomic weight: 4.

DB

henry (Symbol: H). The SI unit of electric *inductance. A henry is defined as the inductance of a closed circuit in which an *electromotive force of one *volt is produced when the current in the circuit varies uniformly at a rate of one *ampere per second.

TM

heptane. A *flammable aliphatic liquid hydrocarbon (C_7H_{18}) obtained from crude petroleum. Its volatility brings it within the gasoline boiling range: n-heptane is used in the *octane rating test as a standard fuel with zero octane number. It finds use as a solvent. Its gross calorific value is 48.54 MJ/kg.

WG

Herfindahl index. An economic measure of *concentration.

MC

hertz (Symbol: Hz). The SI unit of frequency. One hertz is the frequency of a periodic phenomenon which has a period of one second. It is thus equivalent to one cycle per second.

TM

heterotrophic. Denoting or relating to organisms that require organic compounds as their main source of carbon. They can be divided into two groups. Photoheterotrophs are dependent on light for energy and use organic compounds rather than carbon dioxide as principal carbon source. They include some of the photosynthetic bacteria only. Chemoheterotrophic organisms require organic compounds both as energy and carbon sources. They include most of the nonphotosynthetic living world.

CL

hex. Short for uranium hexafluoride, UF_6. *See* uranium.

hexane. A *flammable aliphatic liquid hydrocarbon, C_6H_{14}, obtained from highly volatile fractions of crude petroleum. It finds use as a solvent. Its gross calorific value is 48.78 MJ/kg.

WG

hierarchy. A system classified in successively subordinate grades or levels. One can refer, for example, to hierarchies of control (management), needs (food before luxuries) or energy quality (electricity higher than wood).

PH

higher calorific value [higher heating value]. Alternative terms for gross calorific value. *See* calorific value.

high explosives [detonating explosives]. Explosives which require a detonator to set them off properly. Speed of burning is so rapid that a shock wave is created which precedes the combustion through the exploding mass. They thus cause a shattering effect rather than the bursting effect of the deflagrating explosive. A high explosive like nitroglycerin has both the fuel component and the oxygen necessary for its reaction contained in the one molecule.

WG

high-level waste (HLW). Highly radioactive material which is separated during the *reprocessing of spent (irradiated) nuclear fuel. It is composed of two basic fractions: a liquid HLW containing the highly radioactive *fission products and *actinides, and a *transuranium (TRU) waste containing plutonium and other actinides together with small quantities of fission products. Typically about 380 litres of concentrated waste is produced from reprocessing 1 tonne of spent fuel containing about 70 kg of fission products and *transuranic elements.

Management of high-level waste involves three steps: (a) solidification to immobilize the radionuclides; (b) *interim retrievable storage to allow for decay of shorter-lived nuclides; (c) permanent disposal.

The two most studied methods for the conversion of liquid HLW to a solid are firstly drying and calcination to form a dry powder or granules, and secondly *glassification to produce a borosilicate glass material. Other methods include the conversion of HLW to a ceramic product, as in the *synroc process.

Various concepts for the permanent (irretrievable) disposal of HLW are under study. They can be classified as disposal in geologic formations, on land, on the seabed or in polar ice sheets, and extraterrestrial disposal in space. *See also* radioactive waste management.

PH

high-octane fuel. Gasoline with a high value of octane numbers (*see* octane rating), generally 100 or more. *See also* antiknock.

WG

high-temperature carbonization. A *carbonization process in which coal is heated in a closed retort or oven to temperatures in the range 1000 to 1300°C for the production of coal gas and coke. *See also* low-temperature carbonization.

WG

high-temperature gas-cooled reactor (HTGR). *See* nuclear reactor.

homeostasis. The state of a system (body, building, etc.) when there is an energy balance between the system and its environment.

TM

homogeneous reactor. A *nuclear reactor in which the *fuel and *moderator are uniformly mixed together. Aqueous reactors, in which the uranium fuel in the form of a compound such as uranyl sulphate, UO_2SO_4, is dissolved in water or heavy water, are examples of homogeneous reactors.

DB

honeycomb collector. A solar-energy *collector that incorporates an open array of holes, A, between tube-like walls (the honeycomb) in the space between the cover and the *absorber, B (*see* diagram). The honeycomb structure lessens convection currents of air within the collector, so reducing heat loss, but does not seriously affect the incoming solar radiation.

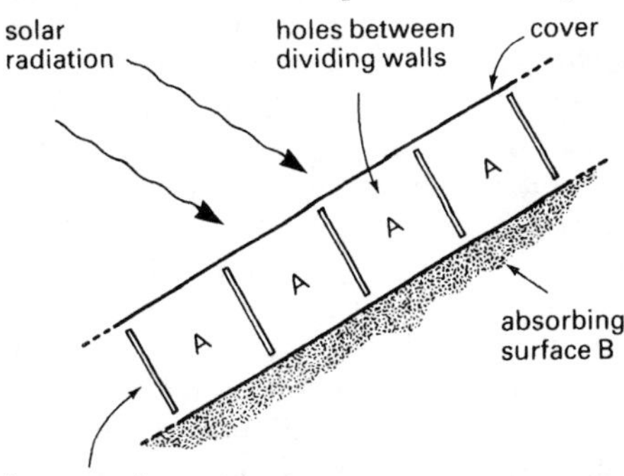

Solar collector with honeycomb structure beneath the cover

JWT

horsepower (Symbol: hp). A practical unit of mechanical power. The preferred SI unit is the watt (W). One horsepower is equal to 746 W.

TM

hot rocks. *See* geothermal energy.

Hottel – Whittier equation. An equation which gives the performance of a solar *collector. The heat (q) gained per unit time per unit area of *absorber equals the heat incident from the solar radiation less the heat losses. The equation is

$$q = \tau\alpha I - U(T_c - T_a)$$

where the solar intensity I enters through a cover of *transmittance τ onto a *collector surface of *absorptance α. The losses are from the collector surface at temperature T_c to the ambient environment at temperature T_a with a heat transfer coefficient U.

JWT

hot water supply. A system for the provision, within a building, of sufficient quantities of water at a temperature suitable for bathing, clothes washing, dish washing, etc. The water is normally heated in an electrical *immersion heater or in a *calorifier before being pumped, through flow and return pipework, to draw-off points around the building.

TM

hour angle. An angle used in solar *collector analysis to describe the daily motion of the sun relative to the collector. The hour angle ω is given by the equation

$$\omega = \frac{2\pi}{24}(12 - t_{zone}) - (\lambda - \lambda_{zone}) - C$$

where t_{zone} is local time in hours, λ is longitude, λ_{zone} is longitude defining the local time zone and C is a small correction factor associated with the earth's motion to the sun.

JWT

human energy. *See* metabolic energy.

humidity ratio. *See* relative humidity.

hydrate. The combination of a molecule of an inorganic chemical compound with one or more molecules of water. Hydrates are usually solid crystalline substances. When this water of crystallization is removed by heat, the compound is said to be anhydrous.

WG

hydrides. Chemical compounds of hydrogen, such as lithium hydride. Metal hydrides are particularly important as stores of hydrogen at relatively low pressures, since absorbed hydrogen can often be released with the application of a little heat. If hydrogen is used as a means of supplying energy, then the ability of storing the gas in this concentrated and mobile form is important. There are obvious applications for hydrogen-powered vehicles. *See also* energy storage.

JWT

hydrocarbons. Organic compounds composed only of carbon and hydrogen.

WG

hydrocracking. A petroleum refinery *catalytic cracking process, carried out in the presence of hydrogen at about 400 °C and at pressures in the range 80–135 atmospheres. The presence of hydrogen prevents the formation of *olefins. The product is thus entirely paraffinic and low in sulphur. The catalyst may be in the form of either a fixed bed or a fluidized bed. The process is useful in converting highly aromatic fractions into lighter paraffinic materials and is used for the production of *kerosene and *diesel fuel.

WG

hydrodesulphurization [dehydrosulphurization]. The removal of sulphur compounds from a petroleum fraction by treating it with hydrogen in the presence of a suitable catalyst. The sulphur comes off as the gas hydrogen sulphide and as volatile low molecular weight hydrocarbons containing sulphur (mercaptans). These gases may be burned to convert them to sulphur oxides for sulphuric acid manufacture. Alternatively, sulphur may be obtained by passing a mixture of the incompletely burned gases over an aluminium oxide catalyst (Claus Kiln process).

WG

hydroelectricity. Generation of electricity, or the electricity generated, from the flow of water, usually obtained from rainfall onto mountains and hills. The term is very occasionally used in the context of *wave power and *tidal range power generation. Water in streams and rivers may be piped directly to *turbines, or stored in reservoirs behind catchment *dams. The incoming flow in the penstock pipe propels the water at high pressure onto various forms of turbine, e.g. *Francis, *Kaplan or *Pelton turbines, connected to the electrical *generator. Conversion efficiencies are high (about 90%) and the equipment lasts for a long time (about 50 to 150 years). Once the high initial capital costs are repaid, hydroelectricity becomes cheap and of great industrial importance. One major handicap can arise from deposits of silt in catchment dams.

The total world potential of hydroelectric generation is about 1.2 million megawatts, of which about 0.3 million megawatts has now been harnessed. Much further major development is possible in many countries of the Third World. However if small-scale installations of less than 500 kW are considered, there are numerous applications in both the developed and developing countries. For domestic and small business use, installations as small as 1 kW capacity (microhydro) may be economic.

JWT

hydrofining. A petroleum refinery process in which petroleum fractions are treated with hydrogen over a catalyst at temperatures of 200–425°C and pressures in the range 4–55 atmospheres. *Hydrogenation reactions occur and sulphur, nitrogen and other non-hydrocarbon elements are removed, with corresponding improvements in odour, colour, stability and combustion characteristics of the material.

WG

hydroforming. The first *catalytic reforming process, developed in the USA in 1939. It was based on a molybdenum oxide aluminium oxide catalyst.

WG

hydrogen. (1) (chemical) A colourless odourless flammable gaseous element at normal temperatures and pressures. It is the simplest atom (comprising one proton and one electron) and is the lightest element. Hydrogen is found in nature largely combined with oxygen in the form of water or combined with carbon in living matter or in fossil fuels such as natural gas, petroleum and coal.

It may be prepared on the industrial scale by the reaction of water vapour with red-hot carbon or by the *steam reforming of natural gas. Alternatively an electric current may be used to split water into its oxygen and hydrogen constituents. It is also available as a by-product of many petroleum refining processes in which *cracking is involved.

Hydrogen is used as a raw material in the manufacture of ammonia, and as a fuel in welding due to the high *flame temperature achieved when burned with oxygen. It is required in the *hydrogenation of oils and fats and as a reducing agent which may be used to convert metallic oxides to the parent metal.

Hydrogen has great potential as an industrial fuel since the only product of combustion is water. However its low density creates storage problems. Storage as a gas in high-pressure vessels or as a liquid in refrigerated insulated vessels is extremely expensive.

(2) (nuclear) Hydrogen has several uses in nuclear energy. In the form of chemical compounds (water and heavy water) it is extensively used as the *moderator and *coolant for several different types of reactor. Its isotopes *deuterium and *tritium are important in the *fusion process.

Symbol: H; atomic number: 1; atomic weight: 1.008; gross calorific value: 141.8 MJ/kg.

WG

hydrogenase. An *enzyme which catalyses the combination of protons and electrons to form hydrogen gas. The enzyme may be extracted from certain *algae and bacteria and added to isolated plant chloroplasts; on illumination, biophotolysis (the biological splitting of water molecules in the presence of light) occurs in which hydrogen gas is evolved. The development of such a biological system for generating hydrogen has great potential as an energy supply in the future, particularly if hydrogen ever becomes a major fuel and energy carrier.

CL

hydrogenation. A chemical reaction through which the hydrogen content of a fuel is increased. The hydrogen atoms are taken up by double bonds in *hydrocarbon and other molecules. *See also* hydrotreating.

WG

hydrogen economy. A recent concept, arising from the perceived forthcoming shortfall in convenient portable fuels like gasoline, kerosene or natural gas, which it is argued could be solved by using *nuclear energy to form hydrogen in large quantities. This hydrogen could be distributed by a pipeline grid, as is done today with natural gas, and, by the development of hydrogen storage systems like *hydrides, could provide convenient portable forms of hydrogen for vehicle propulsion, etc. The hydrogen could also be liquified, and in such form could be used as a fuel for airplanes and ships. It was originally envisaged that the hydrogen would be made by nuclear heat, either through the route of electrical generation and then *electrolysis, or by direct water splitting using high-temperature sources such as the high-temperature reactor (HTR). Today, however, with the swift development of certain solar-based energy sources, it is clear that some of them, such as photovoltaic systems (*see* photovoltaic conversion), could also be most readily used through the creation of hydrogen.

The switchover to an economy based on hydrogen requires such major changes in operating procedures and devices for its use that such an economy is unlikely to come about in a short time. However there already exists, for the purpose of the chemical industry, a grid of hydrogen pipelines in Europe, the USSR and the USA.

MS

hydrogen storage. Hydrogen gas produced by an energy supply device and subsequently stored before use as a fuel for combustion or in engines. *See also* energy storage; energy density; hydrides.

JWT

hydrogen sulphide. A colourless toxic gas, H_2S, with a characteristic noxious odour associated with rotten eggs. Formed in biological degradation processes, it is found in natural gases and in the gases arising from petroleum refinery operations. It may be separated from the hydrocarbons and burned in air to form sulphur oxides for sulphuric acid manufacture.

WG

hydrometer. A device for measuring *relative density, consisting of a weighted glass bulb topped by a glass stem containing a calibrated scale. When placed in a liquid, it floats with the stem vertical and projecting from the surface of the liquid to an extent dependent on the liquid density. The relative density is read off on the calibrated scale at the point where the stem cuts the liquid surface.

WG

hydropower. Power generated from the movement of water. *See* hydroelectricity.

JWT

hydrotreating. The treatment of petroleum fractions with hydrogen in the presence of a *catalyst at elevated temperature and pressure. It removes sulphur and other catalyst poisons before the petroleum fractions are subjected to a *catalytic reforming process. *See also* hydrofining.

WG

hypothermia. That condition of the human body in which it is at a temperature significantly below normal. The very young and the elderly are particularly susceptible to illnesses resulting from hypothermia, including pneumonia and bronchitis.

TM

hysteresis. The lagging of an effect behind the cause of that effect. Most physical changes follow the same path, whichever direction they take, e.g. water changing to ice or ice to water. In systems where hysteresis occurs, they do not. The best-known example is magnetic hysteresis. The heat energy generated in taking a specimen of iron through a cycle of magnetization is known as hysteresis loss.

TM

I

IFIAS. *See* International Federation of Institutes of Advanced Study.

illuminance (Symbol: E). The quantity of light flux falling on unit area of a surface. The unit of illuminance is the *lux (lx).

TM

immersion heater. An electric resistance heater used for heating liquids. In its domestic form it is the heating element in hot water storage *calorifiers.

JWT

impedance (Symbol: Z). The total virtual resistance of an electric circuit to alternating current, arising from the resistance and reactance of the conductor. Impedance is measured in *ohms.

TM

imperfect competition. A term used in two ways, in both of which the essential point is that it refers to market situations in which selling firms face downward-sloping demand curves; thus they do not act as *price takers and do not operate where price equals *marginal cost. In one usage the term refers to any market structure which is not *perfect competition, thus covering *monopoly, *monopolistic competition, *ologipoly and *duopoly. In the second usage it refers to market structures lying between the extremes of perfect competition and monopoly. Energy industries are examples of imperfect competition in that they are typically either oligopolistic (oil) or monopolies (electricity and gas in the UK).

MC

import. Something utilized domestically, having been produced overseas. One country's import is some other country's *export. As with exports, imports can be classified as visible or invisible. *See also* balance of payments; invisible trade; visible trade.

MC

incandescence. The emission of light from a solid body due to thermal excitation of its molecules. The phenomenon is used in lighting technology: electric current is passed through a filament, typically made of tungsten, held within a gas-filled *lamp.

TM

incidence of taxation. The eventual real impact of taxes after taking account of all direct and indirect effects; 'incidence' denotes the effect the tax has on purchasing power (including its distribution) relative to the pre-tax situation. A sales tax is imposed on a retailer, who is able to pass this on in part or in whole to his customers. The impact is on the retailer but the incidence is on the customer and the retailer generally. Again, the real costs of the excise taxation of petroleum in the UK are borne little by the producers of petroleum and the proprietors of garages, but mainly by the purchasers of the petroleum. In any particular case the incidence of a sales tax will depend on the *elasticity of supply and of demand.

MC

incineration. (1) (chemical) Disposal by complete combustion. The term is generally used with reference to waste materials. Incinerators may be designed not only for the safe disposal of organic wastes but also for the generation of steam or the provision of hot gases to operate a gas turbine.

(2) (nuclear) A proposed method for the elimination of *actinides of very long *half-life which are produced in *nuclear reactors and are subsequently separated with the *fission products during reprocessing. The proposal is that these actinides should be irradiated in a reactor and be transformed by neutron capture into other actinides of shorter half-life, and consequently of less long-term hazard.

WG, DB

inclined surface. A surface which slopes with respect to the horizontal. The angle of inclination or slope is particularly important for solar *collectors.

JWT

income. The value over some period of the flow of goods and services which an individual or nation could consume without a reduction in *wealth or *capital. This is sometimes referred to as *real income to indicate that changes in the money value of wealth as a result of *inflation are to be ignored. For an individual, income is usually regarded as a flow of money receipts, but this ignores imputed income arising from unrealized *capital gains. With rising oil prices, for example, the owner of an oil well has income even if in some period he chooses not to extract and sell any oil. *See also* national income accounts.

MC

income effect. A change in the price of a commodity, with other prices and income constant, changes an individual's *real income since there is a change in the total quantity of commodities which can be bought. The effect of this on the consumption of the commodity is the income effect, which may be negative (consumption rises as price falls) or positive (consumption falls as price falls); in the latter case the commodity is an *inferior good. For most commodities the income effect is negative. If a positive income effect is larger than the always negative *substitution effect, the commodity is a *Giffen good. Typically, fuels are neither inferior goods nor Giffen goods.

MC

index numbers. Numbers which compare the general level of a group of distinct but related variables in two or more situations. They can refer to quantities, prices, values, consumption, or to less precise magnitudes like intelligence, etc. Comparisons can be between time periods, places or categories. In simple terms, they are indicators of the change in a set of distinct but related items taken as a whole. For example, an index of energy consumption denotes the change in the various fuels, which are the components of energy consumption, taken together over, say, a period of time. Differences in measurement can arise because of different weights that can be attached to the individual components of the set. *See also* Laspeyres index; Paasche index; price index.

MC

indifference curve. A graphical device used in the representation of what economists typically assume about individuals' preferences. It is the locus of combinations of consumption levels for commodities which yield equal levels of satisfaction, as indicated by the consumer's *utility function. In the diagram,

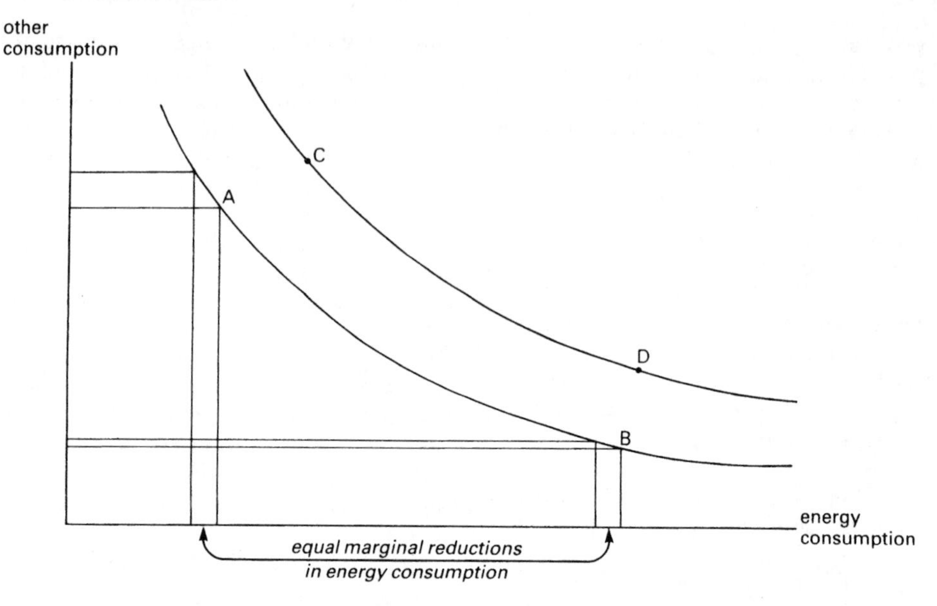

the point A yields the same utility as point B so that the consumer is indifferent between them, whereas points C and D are preferred to points A and B. As the amount of energy consumed increases, so the slope of the indifference curve decreases, reflecting that as more energy is consumed so less of 'other consumption' is required to maintain satisfaction constant in the face of a given *marginal reduction in the consumption of energy. The precise shape and position of the indifference curves for a particular consumer are determined by his preferences and hence his utility function. The general form shown in the diagram is that assumed for a typical consumer of typical commodities.

MC

indirect energy. Energy not used directly in a production process, but used in the preparation of intermediate inputs to the process (*see* intermediate product). *See also* direct energy.

MS

indirect taxation. Taxes levied on goods and services, e.g. beer and tobacco taxes, value-added tax. The *incidence of such taxation depends upon the degree to which sellers can shift the burden onto their customers.

MC

induced draught. An increase in the flow rate of products of combustion from a fuel-burning appliance, caused by a fan placed between the combustion-chamber exit and the base of the chimney.

WG

inductance (Symbol: M). The property of an electric circuit by which a change of current in the circuit produces a change in the magnetic flux surrounding the circuit. Inductance, which is measured in *henrys (H), is utilized in the design of *electric motors.

TM

industry. A group of firms producing the same or a closely similar product. Thus the coal industry and the oil industry are among the energy industries. Where there are many firms producing the same product and selling to many customers, there is approximately *perfect competition. Where there is just one firm on the selling side, *monopoly obtains. A small number of firms in an industry is *oligopoly, a special case of which is *duopoly with just two firms. *See also* concentration.

MC

inelastic scattering. A neutron-induced nuclear reaction in which the *compound nucleus formed by the absorption of the neutron emits a neutron and a gamma-ray photon. Thus the kinetic energy of the neutron and nucleus is not conserved in this reaction (hence the name inelastic), but is reduced by the amount of the energy of the gamma photon. Inelastic scattering occurs mainly with intermediate and heavy elements and high-energy neutrons. *See also* elastic scattering; scattering.

DB

inert gas. In the context of fuels, a gas or mixture of gases which will not support combustion. This comprises a wider category than the simple so-called chemically inert gases such as helium, neon and argon. A tank containing air and small proportions of flammable vapour may be rendered safe from explosion by pumping in inert gases such as nitrogen, carbon dioxide or even cooled flue gas containing little oxygen.

WG

inferior good. A commodity for which, with price constant, consumption falls as incomes rise. It is also a commodity for which the *income effect is positive. Typically fuels are not inferior goods, though coal may be in some circumstances with higher real incomes enabling households to switch to more convenient fuels.

MC

infiltration. The ingress to a building, through its envelope, of air. The extent of infiltration, influenced by wind pressure on the building facade and gaps in the building fabric, will affect the *airchange rate, and consequently the energy consumption of the building.

TM

inflation. A sustained rise in the general level of prices, including the price of inputs to production. While *relative prices may change during a period of inflation, this is a separate matter from inflation as such. Inflation can also be defined as a fall in the value of money, since with prices expressed in money terms, the quantity of goods which can be exchanged for a unit of currency falls with inflation. The rate of inflation is the proportionate increase in the general price level per unit of time. It is important to correct for inflation when considering relative price movements and their implications. Thus the *real price of oil has risen since 1973-74 by much less than its *absolute price, since all other prices have risen as well. An *energy theory of value argues that real energy price inflation is linked to *energy requirement for energy.

MC

infrared radiation. A type of *electromagnetic radiation with wavelength longer than that of red light (0.7×10^{-6}m) but shorter than that of microwave radiation (10^{-3}m). About 40% of solar energy is in the infrared region of the *electromagnetic spectrum, at wavelengths between 0.7×10^{-6}m and about 2×10^{-6}m; this is called the near infrared region. Materials at environmental and moderate temperatures (up to about 300°C) emit infrared radiation at the longer wavelength of about 10×10^{-6}m, i.e. in the far infrared region.

JWT

inherited energy resources. Those energy resources known to exist and still to be found within the earth. The term 'inherited' simply reinforces the fact that mankind inherits these resources; though we can search for them, we cannot increase their total but must eventually use them up.

MS

initial boiling point (IBP). The temperature at which a petroleum fraction starts to distil in a standard distillation test. *See also* boiling range.

WG

inorganic. Denoting or relating to chemical compounds which do not contain carbon atoms combined in structures associated with organic chemistry, i.e. in long chains and in rings. The term thus tends to relate to chemical compounds not involved in living processes. However such divisions have become increasingly difficult to make since the discovery of organo-metallic compounds, in which inorganic elements such as metals form part of organic structures containing carbon, hydrogen and oxygen atoms. *See also* organic.

WG

input–output analysis. The study of the inter-relationships between the sectors of an economy, and their implications for the economy's ability to meet various patterns of demand for *final product, given available *resources. It is a practical implementation of the approach required in *general equilibrium analysis, made possible by the compilation of input–output tables which record in monetary units the transactions between the various sectors. Such compilation is expensive so that for any economy input–output tables exist only for a few widely separated years.

In the analysis of input–output data, the economy is treated as a set of inter-related sectors, or industries; each of these produces output using as *inputs both primary inputs (*capital services, *labour services and *imports) and *intermediate products (the outputs of other industries). Each industry is assumed to have a linear *production function with fixed coefficients, and hence constant *returns to scale. With such restrictive assumptions, a single input–output table can be used to fix the numerical values of the coefficients in the production functions. This is advantageous because of the infrequent availability of data for input–output tables. However, where for a single economy there are available two or more input–output tables referring to different times, it is always seen that the production coefficients do change over time. Hence, results derived from input–output analysis, while useful, must be treated with caution. Given numerical values for the production-coefficients, it is possible to compute the output level required in every industry if any given pattern of demand for final products is to be met; this computed output level for an industry will include the amount required to meet the needs of all other industries for its output as input.

Input–output analysis has been widely used in economic planning and policy analysis. It has also been used to study the way economies use energy. It identifies the way the outputs of the fuel industries enter the production processes of other industries directly and indirectly. The monetary data appearing in the published input–output tables can be (approximately) converted into data, in energy units, using the appropriate prices for the various fuels. With this data the total amount of energy, or the amounts of the various fuels, required per unit of the output of any industry can be computed.

This measurement of the *indirect energy as well as the *direct energy inputs to production can reveal important aspects of the pattern of energy use. For example, input–output methods reveal the energy used in making the steel inputs to car production, as well as the energy input direct to car-assembly plants. The indirect energy input to car production is a large proportion of the total energy required for car production; considering only the direct energy inputs would seriously understate the energy-use implications of an expansion of the output of cars. Information on indirect energy use, and its distribution across the various fuels, is also useful in assessing the impact on the *relative prices of all commodities, consequent upon changes in fuel prices. *See also* energy analysis.

MC

input–output analysis (energy). Adapted from economic *input–output analysis to show the *direct energy used at each element of the matrix of an input–output table. The sum of all the intersectoral direct energies for a given industry adds up to the *gross energy requirement of the output of that industry, which is therefore often given in the form of *energy intensity, the energy used per unit value of output. However for certain industries with a single main product, like the steel industry, it can lead to values in terms of energy per unit mass (*GER). For the energy industries themselves, the value arrived at is the *energy requirement for energy.

Such input–output tables, though exceptionally useful for making swift *energy analyses, suffer from a number of inaccuracies, whose magnitude is not easy to quantify. For example, the original input–output tables from which they are deduced note intersectoral transfers in money terms, yet each industry may pay a different price for energy. There are problems with internal transfers, leading to double-counting. Input–output tables treat capital goods as being at final demand whereas *energy analysis does not. Finally, all energies are treated as of equal quality.

The greatest disadvantage from an analytical point of view is that the values derived are industry averages, and thus cannot account for energy economies of scale or the energy implications of degrees of intensiveness, as in agriculture, where wide differences occur even for a given crop.

MS

input–output table. *See* input–output anaysis.

inputs. In economics, the material things and the services used in production processes. Inputs are classified into *factors of production (also known as primary inputs) and *intermediate products.

MC

insensible heat. Heat absorbed into a material without change in temperature of the material. It is associated with a change of phase in the structure of the material, e.g. melting. *See also* latent heat; sensible heat.

JWT

insolation. The flux of radiant solar energy falling on a surface. The total solar insolation on a given surface is composed of a direct and a diffuse component, as well as short-wavelength radiation reflected from other terrestrial surfaces. The direct insolation on a surface normal to the sun's rays depends on the time of year, the time of day and the latitude of the surface, as well as the atmospheric conditions. Maximum insolation at the earth's surface is about 1 kW/m^2. *See also* solar constant; solar energy.

TM

Institut Economique et Juridique de l'Energie (IEJE). A major economics research group supported by the Centre Nationale de la Recherche Scientifique at the University of Grenoble, France.

Address: Université de Grenoble, St. Martin d'Hyères. BP 47X, Centre de Tri, Grenoble, France

MS

Institute for Energy Analysis. An institute operated by Oak Ridge Associated Universities. It assesses energy policy, energy research and development options and analyses alternative energy supply and demand projections from technical, economic and social perspectives.

Address: PO Box 117, Oak Ridge, Tenn 37830, USA.

PH

Institute of Energy. A UK professional institution devoted to professional representation in the field of energy and in maintaining professional standards. It was formerly the Institute of Fuel.

Address: 18 Devonshire St, London W1.

MS

Institut Français de l'Energie. A society which arranges for the teaching and recording of information on energy topics.

Address: 3 Rue Henri-Heine, 75016, Paris.

WG

Institute of Petroleum. A publishing society which promotes and coordinates the study of petroleum and its allied products.

Address: 61 New Cavendish Street, London W1M 8AR.

WG

Institution of Fire Engineers. A publishing society which promotes and seeks to improve the science of fire engineering and technology.

Address: 148 New Walk, Leicester, England.

WG

Institution of Gas Engineers. A society which promotes research and education in the production, distribution and utilization of gas and the byproducts of its production.

Address: 17 Grosvenor Crescent, London SW1X 7ES.

WG

Institution of Nuclear Engineers (INucE). An institution which acts as an organizing body for professional engineers engaged in all aspects of nuclear technology in the UK.

Address: 1 Penerley Road, London SE6 2LQ.

DB

insulation. Any substance which offers substantial resistance to the flow either of heat energy or electrical energy; also the effect produced by using such substances. Thermal insulation materials are generally employed to help preserve the temperature within an enclosure at a level above or below the temperature outside the enclosure; hence the high levels of insulation in, for example, a thermos flask, a domestic refrigerator and a well-designed house. Electrical insulation is extensively used to maintain a potential difference between the parts of an electric circuit and earth; hence the high level of insulation on electric power lines, cables and appliances.

TM

integrated environmental design. An approach to the design of buildings which recognizes and attempts to take simultaneous account of the thermal, acoustic, lighting and aesthetic considerations, and their interaction. The approach has been advocated primarily in the interests of energy conservation and has led to an increased interest in methods for *heat recovery.

TM

intensity, solar. The energy flow per unit area per unit time. Solar radiation intensity, for example, has a maximum value of about 1 kW/m^2 at the earth's surface.

JWT

intercropping. The planting of different crops in alternate rows to increase the effective area available for *photosynthesis, and thus enhance primary production per unit area (*see* primary productivity). This technique can be used to good effect in mixed *energy farms. Cereals such as *maize can be grown intercropping with *legumes such as the soyabean. Some of the nitrogen fixed by the legume is made available to the maize crop, thus reducing the need for applying energy-intensive chemical fertilizers to secure good yields. In practice the tall-growing maize is planted first and the second crop, in this case soya bean, of shorter stature, is planted later in intervening rows. This also increases the annual effective *canopy cover giving high *biomass yields per unit area.

CL

interest rate. The money paid on borrowed money. For example a 10% interest rate per annum would mean that for each £100 borrowed, the borrower would have to pay back, in addition to the amount (or part of the amount) borrowed, £10 at the end of each complete year on each £100 of the loan still outstanding. The actual interest rate depends on such factors as the duration of the loan, the risks and the type of money being borrowed (e.g. domestic or *eurocurrency).

MC

interim retrievable storage. A concept in *radioactive waste management in which high-level radioactive waste is placed in an interim storage facility prior to its transportation to a final disposal site. This procedure allows time for the development of new improved technology for permanent disposal.

PH

intermediate product. A commodity used in the production of other commodities, as opposed to a commodity used as a *final product. Many commodities are both intermediate and final products. For example, coal, gas, electricity and refined oil are all consumed directly by households and are also used in the production of other commodities. In an advanced economy, the value of trade in intermediate products typically exceeds the value of trade in final products. The method of *input-output analysis considers both types of trade, and its implications for, for example, the total output required of the oil industry to meet a specified demand pattern for all final products.

MC

intermediate technology. Technological and commercial activities of intermediate economic scale, as originally expounded by the late E. F. Schumacher. In his influential book *Small is Beautiful*, Schumacher outlined his philosophy and quantified the scale of intermediate technology according to the numbers of workers involved in a productive unit and the capital cost of buildings and equipment. As originally conceived, Schumacher's ideas were applied to developing countries but they have now gained considerable influence in industrialized countries. The concepts of intermediate technology are closely related to ideas of self-sufficiency, the use of local resources, *renewable energy and a range of alternative economic proposals that challenge the large-scale centralized policies of traditional industrialization. *See also* appropriate technology.

JWT

Intermediate Technology Development Group (ITDG). A registered charitable group in the UK for the encouragement and development of *intermediate technology as expounded by the late E. F. Schumacher. It is involved in technical, research and commercial ventures in many countries, especially in the Third World, and publishes the journal *Appropriate Technology*. Many programmes involve small-scale *renewable energy developments.

Address: 9 King Street, London WC2.

JWT

intermittent heating. The provision of an adequate thermal environment within a building without running the heating system continuously. Typically the heating plant is closed down during periods of non-occupancy; it is necessary to have the plant restart in advance of building reoccupation if comfort levels are to be ensured.

TM

internal combustion engine. An engine in which the heat produced by the combustion of a fuel gas or vapour in air in a closed cylinder is converted into mechanical power by a piston, connecting rod and crankshaft. In a car engine the mixture of gasoline vapour and air is ignited by means of an electric spark. In the *diesel engine the compression of the air in the cylinder raises its temperature so that when the fuel is injected it ignites spontaneously. *See also* heat engine.

WG

internal conversion. A process whereby an excited atomic *nucleus transfers its excess energy to an orbital electron, which is then ejected from the atom.

DB

internal energy (Symbol: U). An important concept in thermodynamics, used to define formally the *first law of thermodynamics. It states that when work W is done by a system which is absorbing heat Q, the change in the internal energy is given by

$$\Delta U = Q - W$$

The change in internal energy is dictated by the difference between the state of the system before and after the change, and not in any way by the path or manner of the change.

MS

internal rate of return. A proportional measure of the earnings arising from the outlay on some project which takes account of the time pattern of the earnings and outlay. It is the *discount rate which makes the project's net *present value equal to zero. With C as the initial capital expenditure at time 0, V_t as the time stream of net cash flow arising from the project, and with a project lifetime of T years, the internal rate of return is the value of r in the equation

$$C = \sum_{t=0}^{T} \frac{V_t}{(1+r)^t}$$

The decision rule is to go ahead with the project only if the internal rate of return exceeds some target rate, usually the relevant *interest rate. This is an alternative approach to the calculation of the project's net present value using a given discount rate. In practice the net present value method is usually preferred as it shows more clearly the size of the payoff to the project, and avoids the problems of possible multiple solution values for r in the above equation. *See also* cost–benefit analysis; rate of return.

MC

International Atomic Energy Agency (IAEA). An autonomous intergovernmental organization and member of the UN system. It is directed by a Board of Governors, which is composed of representatives from 34 member states, and a General Conference of the entire membership of 110 states. Its main objectives are to 'seek to accelerate and enlarge the contribution of atomic energy to peace, health and prosperity throughout the world, and to ensure, so far as it is able, that assistance provided by it or at its request or under its supervision or control is not used in such a way as to further any military purpose.'

The IAEA organizes meetings, publishes books, establishes safety standards for all types of nuclear activity, prepares feasibility and market studies, applies safeguards to nuclear materials, finances research, arranges loan of equipment, and operates an international nuclear safeguards inspectorate.

Address: PO Box 100, A-1400 Vienna, Austria.

PH

International Bank of Reconstruction and Development (IBRD) [World Bank]. An international organization set up at the Bretton Woods Conference in 1944 to assist the economic development of its member countries and to raise the standards of living of the peoples of the world. It is an agency of the United Nations. Initially set up to assist in recovery after World War II, it now largely guarantees investment loans to developing countries and provides them with technical assistance. Its loans are made on special projects, are long term and are made to governments or private firms with the government as guarantor. Its members must also be members of the *International Monetary Fund.

Address: 1818 M Street NW, Washington DC 20433, USA.

MC

International Commission of Radiological Protection (ICRP). An independent non-governmental body of experts, established, in 1928, to recommend the maximum *radiation doses to which the human population can be safely exposed. The recommendations of the ICRP, which are published in scientific reports, are accepted by both national and international bodies responsible for radiation protection. These recommendations are expressions of opinion based on deductions from the interpretation of scientific data.

Address: Clifton Avenue, Belmont, Surrey, England.

PH

International Energy Agency (IEA). An autonomous body within the *Organization for Economic Cooperation and Development (OECD), and formed in 1974; some OECD members (France, Australia) did not join. The major agreed functions of the IEA are developing secure supplies of energy, promoting energy conservation, and operating a scheme for sharing oil supplies in an emergency. Particular emphasis has been given to the objective of reducing the dependence of members on imported oil.

Address: 2 Rue André Pascal, 75775 Paris.

MC

International Federation of Institutes of Advanced Study (IFIAS). Founded in 1972 in the wake of the first UN Environmental Conference in Stockholm to 'further transmational and transdisciplinary research addressed to contemporary global problems'. It convened an international workshop to consider an appropriate set of conventions for *energy analysis, now known as the IFIAS conventions.

Address: c/o Institute for Environmental Studies, University of Toronto, Ontario, Canada or
Onze Lieve Vrouwelein 21, 6211 HE Maastrict, Netherlands.

MS

International Institute for Applied Systems Analysis (IIASA). An institute formed in 1972 in order to apply systems analysis to global problems, including energy. It is supported by 18 countries, and is the only institute of its kind providing facilities for scientists from both communist and non-communist countries to work together.

Address: 2361, Laxenburg, Austria.

MS

International Monetary Fund (IMF). A fund established in 1945 and a specialized agency of the United Nations since 1947. Its aims are to promote international monetary cooperation, to facilitate the balanced growth of international trade, to promote exchange-rate stability, to assist in the multilateral system of payments in respect of international indebtedness, to give confidence to members, and to shorten the duration and lessen the degree of disequilibrium in members' balance of payments. Each member is assigned a quota based on its national income, trade and international reserves; 25% of the quota is payable in gold or US dollars and the balance in its own currency. The quotas are reviewed every five years. Voting rights in the Fund, along with the allocation of *Special Drawing Rights (SDRs), are related to the size of the quota. Since 1972 the Fund's standard unit of account has been the SDR. The Fund was a central institution of the *Bretton Woods System (sometimes called the IMF System). Although the system broke down finally in 1973, the Fund remains the fundamental institution in world payments, and operates a special facility designed to assist members in meeting the cost of oil and oil-related imports in the wake of the 1973–74 and subsequent oil-price rises.

Address: 700 19th St NW, Washington DC 20431, USA.

MC

International Nuclear Fuel Cycle Evaluation (INFCE). A programme set up as a result of a proposal made by President Carter at a seven-nation summit conference in May 1977. The organizing committee of INFCE met in Washington in October 1977 when it agreed to a two-year programme of work (report published in 1979). Fifty-six countries and five international agencies participate in INFCE, including some who have not signed the *Nuclear Non-proliferation Treaty. The major concern of the committee is how to make each part of the *nuclear fuel cycle more resistant to nuclear proliferation. PH

International Solar Energy Society (ISES). An international organization to encourage the research, development and commerce of *solar energy, including ocean thermal energy, wind power, photochemistry, photovoltaics and solar thermal energy. It publishes the journal *Solar Energy*. National branches exist in many countries or continental centres.

Addresses: (World HQ) PO Box 52, Parkville, Victoria, Australia.
(UK) 19 Albemarle Street, London W1.
(USA) PO Box 1416, Killeen, Texas 76541.

JWT

inventory. A stock held by a firm consisting of finished output, of work in progress, or of inputs of raw materials and/or *intermediate products. Inventories are held to meet expected or unexpected fluctuations in the rate of output sales, production throughput or input arrivals, without the need for a change in the rate of production. Inventories are a form of *capital, and the distinction between inventories and buildings, machines, etc., is sometimes made by referring to the latter group as fixed capital. Inventories are important in the energy industries, especially now that large oil stocks are being held against the possibility of interrupted supplies from producing countries. The electricity-supply industry holds large stocks of coal in winter so as to be able to meet weather-induced variations in demand.

MC

investment. The flow which increases or maintains intact a *capital stock. Investment for the purpose of offsetting *depreciation of the existing capital stock is known as replacement investment. Gross investment is replacement investment plus investment to increase the size of the capital stock, which is known as net investment. Investment, for the economy as a whole, is that part of *final product output which is not consumed. *See also* national income accounts. MC

invisible trade. Trade in services rather than goods. These usually include payments and receipts from shipping, insurance, banking, tourism, aviation interest on overseas investment, remitted profits from home firms established abroad, and government expenditure abroad. The invisible balance is the difference between receipts and payments from invisibles. *See also* balance of payments.

MC

iodine. An element which is of importance in nuclear enegery as several of its isotopes are produced as *fission products in a reactor. The isotope iodine-131, whose *half-life is 8 days, is hazardous should fission products be released to the atmosphere. This did happen as a result of the reactor accidents at Windscale, UK, in 1957, at *Three Mile Island, USA, in 1979 and at Chernobyl, in 1986. The isotope iodine-129 has a half-life of 17×10^6 years and presents a long-term hazard in *radioactive waste management.

Symbol: I; atomic number: 53; atomic weight: 126.9.

DB

ion. An atom with less or more electrons than normal, thus possessing, respectively, either a positive or a negative charge.

TM

ionization. A process whereby, as a result of interactions between charged particles (such as *alpha or *beta particles) and atoms, an *electron is ejected from the atom. As a result an ion pair is formed; the negative ion is the ejected electron and the positive ion is the atom with the missing electron.

DB

ionization chamber. A type of *gas-filled radiation detector, operated at low voltage, whose output current is a measure of the intensity of the radiation being detected.

DB

ionizing radiation. Radiation capable of causing *ionization. Examples include gamma radiation, X-rays, ultraviolet radiation or a stream of high-energy electrons, alpha particles, or protons.

DB

irradiance (Symbol: *I*). The flux of radiant energy falling on unit area of surface at the angle of incidence of the radiation. Irradiance is measured in watts per square metre (W/m^2).

TM

irradiated fuel [spent fuel]. Nuclear *fuel which, having been in an operating nuclear reactor for some time, has undergone changes in is composition such as a depletion of its *fissile content and a build up of *fission products. *See also* radioactive waste management.

DB

isentropic process. A state change of a fluid during which its *entropy remains constant.

TM

isobaric process. A state change of a fluid during which its pressure remains constant.

TM

iso-butane. *See* butane.

isolux. A surface on which lie points of equal illumination.

TM

isomerization. The rearrangement of the structure of a molecule without changing its molecular weight. It occurs in many *cracking and *reforming processes carried out in a petroleum refinery, but its principal application is in the conversion of the straight-chain hydrocarbon *butane into the branched-chain isomer, isobutane. The latter is subjected to an alkylation process to produce gasoline. Higher molecular weight paraffins may be isomerized to increase their octane number. The process consists of passing the feedstock mixed with hydrogen through a fixed bed of catalyst at a temperature between 100 and 200°C.

WG

isomers. Chemical compounds which have the same molecular formulae but different structural formulae. For example, n-butane and iso-butane are isomers:

$$CH_3—CH_2—CH_2—CH_3$$
n-butane

$$(CH_3)_2—CH—CH_3$$
iso-butane

WG

isometric process. A state change of a fluid during which its volume remains constant.

TM

iso-octane. *See* octane.

isoquant. In the economic analysis of production, the locus, on a two-dimensional graph, of input combinations which yield a constant level of output. The only input combinations relevant are those corresponding to technical *efficiency, so that a set of isoquants represents a *production function. In the diagram, I_1 corresponds to a higher output level than does I_2, and each output level can be produced by any of the combinations of energy and other inputs lying along I_1 and I_2. The input combination actually adopted by a firm depends on the relative prices of the inputs. The line AB represents the possible combinations of inputs which the firm could purchase with a given sum of money when the relative prices are given by the slope of AB – all money spent on energy gets an amount OB, all spent on other inputs gets OA. The firm's maximum output is I_1, produced using OE_1 of energy and OO_1 of other inputs. If the price of energy increases, such that with the given money sum available only OB′ is the maximum purchasable energy input, the relative prices of the inputs are given by the slope of AB′. Output falls and the input combination used moves away from energy.

The extent of these effects depends upon the shape of the isoquants in the relevant region, and thus on the nature of the production function. The implications of higher energy prices depend on the substitutability of other inputs for energy. Where the possibilities are limited, there will be only a small change in the input mix, with a large change in output and/or its price (according to the extent to which the firm can raise price). Where input substitution is relatively easy, the higher energy price will pro-

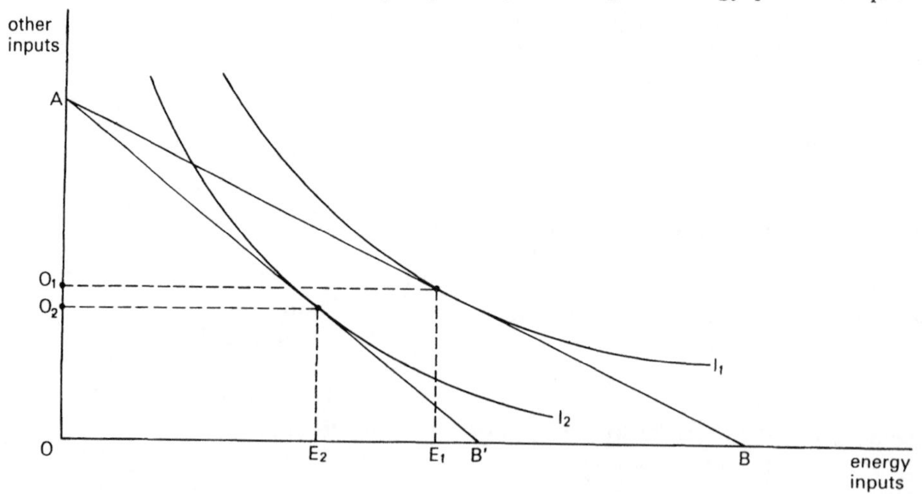

duce a large change in the input mix, with a small change in output and/or its price. The substitution possibilities in production are measured by the *elasticity of input *substitution, defined as the proportionate change in the optimal input ratio divided by the associated proportionate change in the relative prices of the inputs. General statements about the size of the impact of higher energy prices on output levels and prices are potentially misleading, as the value for the elasticity of input substitution for energy and other inputs varies considerably between different production activities, between countries, and over time.

MC

isotherm. A surface on which lie points of equal *temperature.

TM

isothermal process. A state change of fluid during which its *temperature remains constant.

TM

isotopes. Elements that are chemically identical but physically different. The number of *protons in the *nucleus defines the atomic number, and hence its chemical properties. However, the number of neutrons in the nucleus may vary from one atom to another, giving rise to the different isotopes of the element which may have different physical properties. Isotopes are identified by stating their mass number, *A*. For example, carbon-12 is the most common naturally occuring isotope of carbon (atomic number, $Z=6$); its symbol is $^{12}_{6}C$. The other naturally occurring isotope of carbon is carbon-13, symbol $^{13}_{6}C$. These two isotopes of carbon have six and seven neutrons in their nuclei respectively.

DB

isotope separation. The separation of different *isotopes of an element. The two most important examples in nuclear engineering are the separation of *deuterium from hydrogen (or *heavy water from light water), and the separation of uranium-235 from uranium-238 to produce enriched uranium (*see* enrichment).

DB

Ixtoc-1 blowout. The catastrophic escape of oil which occurred on 3 June 1979 when an offshore exploratory oil well, Ixtoc-1, in the Gulf of Campeche, 80 km off the Mexican Yucatan coast, blew out and caught fire. The initial flow rate was estimated at 10,000–30,000 barrels of crude oil per day. By 12 June an oil slick 180 km by 80 km had formed, moving in a northwesterly direction across the Gulf of Mexico. By August the oil had reached the Mexican and Texan beaches in the form of tar balls and mats and *chocolate mousse.

The well spewed oil and gas for 295 days before it was finally sealed in March 1980, having released an estimated 3 million barrels into the Gulf. This makes it one of the worst oil spills ever experienced. The Mexican national oil monopoly (PEMEX) who owned the well spent around $132 million on controlling the well and containing the environmental damage, and lost $87 million in oil revenues. The US Coast Guard spent $8.5 million on clean-up operations.

The environmental consequences of this spill included serious pollution to Mexican and Texan beaches, which affected the tourist industry and caused disruption of the shrimp and oyster fisheries in the Gulf of Mexico. It also posed a serious threat to various endangered species, such as the brown pelican and Kemp's Ridley sea turtle.

PH

J

J-curve effect. An empirically observed time lag between the price changes of imports and exports occurring from a *devaluation (or *depreciation) and the response on the quantity of imports and exports. As a result there is an initial worsening of the balance of trade (*see* balance of payments) before it gets better. Plotting the balance of trade against time, as in the diagram, traces out a J-curve. It may be two or more years before losses are recouped.

MC

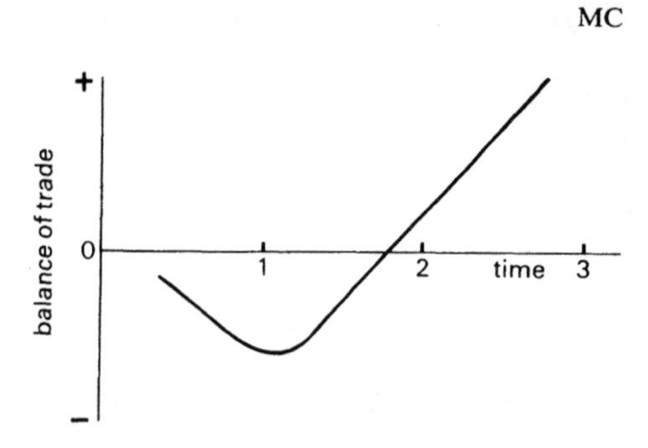

jet fuel. Two types of petroleum fractions are used as fuels for the gas turbines of jet aircraft. These are *kerosene with a boiling range of 150–290°C and *wide-cut gasoline, boiling range 30–280°C. The former is considered somewhat safer in use since it has a higher *flash point and is less liable to be ignited accidentally if spillage takes place. It is thus preferred for civil aircraft. Since only about 5% of crude petroleum is available as *straight-run kerosene, military aircraft tend to be fuelled with wide-cut gasoline.

WG

Joint European Torus (JET). The machine being developed at Culham, England by several European countries to carry out research into controlled *fusion reactions.

DB

joint production. A situation where a production process yields more than one output. The joint products may, but do not necessarily, arise in fixed proportions. A standard textbook example is wool and mutton. In the energy area examples are: oil and gas from an oil well; heavy and light products from an oil refinery; electricity and plutonium from nuclear power plants; electricity and atmospheric pollution from a coal-fired electricity plant. Joint production gives rise to severe analytical and practical problems. It is difficult to allocate costs to the separate products: changes in the demand for one product, if met, will lead to changes in the supply of the product with which it is joint. *See also* partitioning.

MC

joule (Symbol: J). The basic unit of energy within the *SI system. It is defined as the energy conveyed by one watt of power for one second, and is based upon standards of electromotive force and resistance maintained in various national standardization laboratories.

Prior to 1948 there existed an 'international' joule and an 'absolute' joule, which differed by 0.0165%. Thermophysical data prior to that date may have been made with reference to the international joule, which is now discarded. This difference, however, explains why some conversion tables list a *calorie (for example) as 4.187 joules, whereas in absolute joule terms it is 4.184. *See also* conversion tables.

MS

K

Kaplan turbine. A machine to convert the movement of a fluid, usually water, to rotary mechanical motion, e.g. for electricity generation. The Kaplan turbine is shaped somewhat like a conventional airplane propeller with *blades of adjustable pitch. Such turbines are common in large-scale hydroelectric installations.

JWT

K-capture. A radioactive decay process in which one of the electrons in the innermost or K-shell of an atom is captured by the nucleus of the atom and combines with a *proton to form a *neutron. An example is the decay of beryllium-7 to form stable lithium-7:

$$^{7}_{4}\mathrm{Be} + {}^{0}_{-1}\mathrm{e} \rightarrow {}^{7}_{3}\mathrm{Li}$$

DB

kelp. *See* giant kelp.

kelvin (Symbol: K). The unit of temperature on the *absolute temperature scale and the basic unit of temperature in the *SI system. A temperature of 273.16 K is equivalent to zero on the *centigrade (Celsius) temperature scale. Temperature intervals are equal on the absolute and centigrade scales, i.e.

$$1\,\mathrm{K} = 1\,°\mathrm{C}$$

JWT

kerogen. The complex organic matter present in *oil shales and carbonaceous shales. It is insoluble in all common solvents but on distillation at high temperatures it yields oil, gas and other compounds. It is formed by the biochemical conversion of plant and animal remains in carbonaceous rock and is the most common form of organic carbon on earth. It varies in composition, consisting of 77–83% carbon, 5–10% hydrogen and about 2% nitrogen together with some sulphur and oxygen.

PH

kerosene. A petroleum fraction with a boiling range of about 150–300 °C. Formerly used extensively in paraffin lamps, it now finds its principal outlets as a fuel in domestic stoves and jet aircraft. Domestic kerosene, or paraffin as it is commonly known in the UK, must have a flash point in excess of 230 °C. It is available in two grades; the higher or premium grade contains negligible amounts of sulphur and may be burned in lamps and flueless heaters. The regular grade is used in flued central heating appliances.

Aviation turbine kerosene burned in aircraft jet engines is thought to be a safer fuel for that purpose than gasoline due to its lower volatility. Owing however to the limited availability of kerosene it is reserved largely for civil jet aircraft. Military jet aircraft also utilize aviation turbine petrol which is a wide distillation cut, boiling between about 100 °C and 290 °C. It is thus a mixture of kerosene and some of the higher boiling components of petrol.

WG

Keynesian. Relating to or denoting a theory (or person) arguing that full employment can only be achieved by governments carrying out *fiscal and *monetary policy to create the necessary level of economic activity. The main instrument of policy, however, is fiscal policy. *See also* monetarist.

MC

kilo- (Symbol: k). A prefix indicating a multiple of one thousand units. For example one kilogram is one thousand grams.

JWT

kilowatt-hour (Symbol: kWh). The transfer of one thousand *watts of energy for one hour. Since a watt is a *joule per second, the kilowatt-hour is equal to 3.6 megajoules (MJ). In times past, electricity was measured, recorded and paid for at the domestic and industrial level in kilowatt-hours (the so-called 'unit' of electricity) while heat was measured in units such as the *calorie, *British thermal unit or joule. With the advent of the *SI system of units there is now no semantic distinction between a kilowatt-hour of electricity and one of heat. This leads to some confusion, for while one kilowatt-hour of electricity will yield one kilowatt-hour of heat, the reverse is not the case. For this reason the kilowatt-hour in some texts carries a subscript, kWh_e when the numbers refer to electricity and kWh_t when the numbers refer to heat.

MS

kinematic viscosity. A measure of the resistance to gravity flow of a liquid. Since the pressure head causing flow is proportional to the density of the liquid, kinematic viscosity is the ratio of dynamic *viscosity coefficient to liquid density. It is measured by determining the time for gravity flow of a known volume of the liquid through a standard orifice. For comparison purposes it is thus convenient to record kinematic viscosity as a time in seconds.

WG

kinetic energy. Energy of motion. A mass m with velocity v has kinetic energy $\frac{1}{2}mv^2$. *See also* potential energy.

JWT

Kirchoff's laws. Relationships established by Gustav Kirchoff concerning, on the one hand, the flow of electric current in a simple circuit and, on the other hand, the *emissivity and *absorptivity of radiative heat energy.

Kirchoff's first law states that the total current flowing towards a node in an electric circuit is equal to the total current flowing away from that node; his second law states that in a closed circuit the algebraic sum of the products of the current and the resistance of each part of the circuit is equal to the resultant electromotive force in the circuit. A further law established by Kirchoff states that, in the case of radiation exchanges to and from a *grey body, absorptivity is equal to emissivity at any given temperature.

TM

knock [knocking; pinking]. The tapping noise produced in a gasoline engine when the air/fuel mixture ignites prematurely, due to the action of heat and pressure, before the spark is passed at the spark plug. The word is also used as a verb: to emit such a noise. *See also* octane rating.

WG

Koppers–Totzek. *See* gasification.

Krebs cycle [tricarboxylic acid (TCA) cycle]. A key biochemical cycle for yielding energy in living organisms. It occurs in the *mitochondria, the sites of cellular respiration. An acetyl (CH_3CO) group is oxidized and phosphorylated through intermediate acids, like citric and oxaloacetic acids, to carbon dioxide and water. At the same time chemical energy is converted into a form suitable for storage within the cell as energy-rich phosphate bonds. The cycle also has an important biosynthetic role since several of its intermediate components give rise to the synthesis of amino acids (the building blocks of proteins) and other cell constituents.

CL

krypton. An inert gas which is of importance in nuclear engineering because the isotope krypton-85, *half-life 10.8 years, is produced as a fission product in nuclear reactors. Being an inert gas it cannot be processed or stored as a solid or liquid compound, and it is thus normally discharged to the atmosphere during fuel reprocessing. It is possible that this discharge of krypton-85 to the atmosphere may become hazardous in the future.

Symbol: Kr; atomic number: 36; atomic weight: 83.8.

DB

***k*-value.** *See* conduction.

L

labour. The human effort input to production: one of the *factors of production distinguished in economic analysis. Due to variations in levels of skill and application it is difficult to find a useful standard of measurement for labour input.

MC

la Hague. A nuclear fuel *reprocessing plant at Cap de la Hague near Cherbourg on the Normandy coast of France.

PH

laminar flow. The motion of fluids, gas or liquid, in a regular streamlined manner without *turbulence or eddies.

JWT

lamp [light]. A device used as a source of *light. Lamps are of two main types: those exhibiting *fluorescence and those exhibiting *incandescence. Fluorescent (or discharge) lamps typically contain mercury or sodium vapour, which transmits electric current as a discharge accompanied by radiation; the radiation excites a fluorescent coating on the inner surface of the lamp tube, resulting in a near-white shadowless light. Incandescent lamps contain a filament, typically tungsten, surrounded by an inert gas; electricity passing through the filament brings it to white heat, providing radiation at all wavelengths within the visible range. Fluorescent lamps are approximately four times more efficient than incandescent lamps.

TM

land. The use of this term in economics has changed and now means essentially the same as in everyday usage. Originally land was one of the *factors of production, and the term referred to all *natural resources. Nowadays economics explicitly distinguishes between types of natural resource according to their main characteristics. Land economics is now concerned primarily with spatial relationships in economic activity, and with the analysis of locational decisions. However, devices for capturing solar energy are area-dependent, so that consideration of such devices as major sources of energy have to be thought of in terms of land area utilized. *See also* photosyntheis; photovoltaic conversion.

MS

land geologic disposal. *See* radioactive waste management.

langley. A unit of energy used for solar radiation measurement and equal to 1 calorie/cm^2, i.e. 4.184×10^4 joule/m^2.

JWT

La Rance. An estuary in Brittany, France, with the world's first and most important tidal power plants. They have a maximum electrical output power of 240 MW.

JWT

laser. Acronym for *l*ight *a*mplification by *s*timulated *e*mission of *r*adiation. A device producing an intense narrow beam of light or of ultraviolet or infrared radiation. The radiation is monochromatic, i.e. of near enough a single wavelength, and is coherent, i.e. the waves move in step with each other. *See also* enrichment; laser fusion reactor. PH

laser fusion reactor. A proposed type of *fusion reactor in which a pellet of deuterium and tritium in the form of suitable compounds such as lithium deuteride is contained in a reaction vessel and bombarded by pulsed laser beams. The intense energy of the lasers creates a high enough temperature for fusion and causes a compression of the pellet, while inertia effects hold it together for long enough for the fusion reaction to proceed for an instant. This method of containing the reaction without the use of magnetic fields is called inertial *containment.

DB

laser isotope separation. *See* enrichment.

Laspeyres index. An *index number using a base year as a means of weighting the elements making up the index. For example, let p^0 and p^t represent base-period and current-period prices respectively, and similarly for quantities q^0 and q^t. Then a Laspeyre *price index is given by

$$P_L = \sum_{i=1}^{n} w_i p_i^t = \frac{\Sigma\, p_i^t q_i^0}{\Sigma\, p_i^0 q_i^0}$$

where $w_i = \dfrac{q_i^0}{\Sigma\, p_i^0 q_i^0}$.

A Laspeyre quantity index is given by

$$Q_L = \sum_{i=1}^{n} w_i^* q_i^t = \frac{\Sigma\, p_i^0 q_i^t}{\Sigma\, p_i^0 q_i^0}$$

where $w_i^* = \dfrac{p_i^0}{\Sigma\, p_i^0 q_i^0}$.

The Laspeyre method for the construction of index numbers is that most widely used in official statistics, such as the various cost of living indices compiled by governments. The base year needs to be changed fairly frequently if the weights are not to become out of date. Thus a Laspeyre index for the price of energy which used, as weights, fuel quantities in 1950 would be seriously misleading by 1975, due to the large changes in the relative quantities of fuels used. *See also* Paasche index.

MC

latent heat. Heat which flows to or from a material without a change in temperature. It arises from changes in the phase or structure of the material, e.g. melting.

JWT

latent period. *See* biological effects of radiation.

latitude. Angular position on the earth's surface relative to the poles and the equator. It is the angle between a position's vertical direction to the earth's centre and the plane of the equator (*see* diagram). During the year the sun is

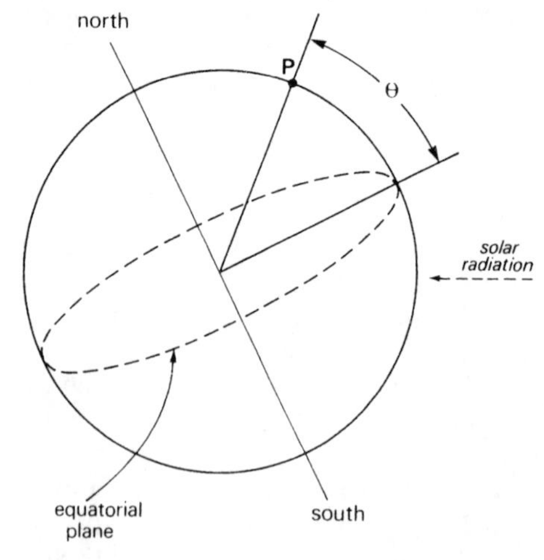

position **P** has latitude θ

overhead in the tropics between the northern and southern latitudes of 23.45° (23° 27′) and the equator, which is at latitude zero. The earth's axis of rotation is inclined to solar radiation so there are marked seasonal variations of solar energy with change in latitude.

JWT

Law. *See* thermodyamics.

Lawrence Livermore Laboratory. A unit of the US Department of Energy which is operated by the University of California and carries out research in nuclear science and the use of energy. Non-military aspects include fusion research, including laser-induced fusion, and solar, wind and battery storage systems.

Address: PO Box 808, Livermore, California 94550, USA.

DB

lead. A soft heavy bluish-white metallic element. It occurs in a number of minerals from which it may be separated by roasting in a furnace with carbon. It has a low melting point (327°C). Because of its high density, lead is used in nuclear medicine and industry as a shielding material for sources of high-energy radiation, such as gamma rays and X-rays. A considerable tonnage is used in the manufacture of the electrode plates in storage batteries or accumulators.

Lead is a toxic element and although corrosion-resistant to a wide variety of chemicals, it is sufficiently soluble in mildly acid drinking water to become a health hazard. Lead compounds are also poisonous so that the fumes from processes involving lead must be avoided. *Tetraethyl lead used as a gasoline additive causes lead to be present in motor car exhaust fumes, again considered to be a health hazard.

Symbol: Pb; atomic number: 82; atomic weight: 203.97; density: 11,340 kg/m^3.

WG

lead dioxide candle. A traditional device for measuring the levels of sulphur dioxide in the atmosphere.

PH

lead response. The extent to which the octane number of a gasoline may be raised by the addition of a known amount of *tetraethyl lead. *See also* lead susceptibility.

WG

leads and lags. The advanced payment or delay in payment of international transactions. If an oil importer, for example, expects a *devaluation then he will speed up (lead) his payment; overseas debtors will delay (lag) their payments. The result is a short-term capital outflow and a worsening *balance of payments position. The converse occurs for an expected *revaluation. Such movements can aggravate a crisis situation.

MC

lead susceptibility. The extent to which the effect of *tetraethyl lead on the octane number of a gasoline is reduced by the presence of organic sulphur compounds in the gasoline. *See also* lead response.

WG

lead-time. The time it takes between a decision to provide a new facility, such as a power station, and the completion of the facility. Such periods are very often measured in years. It has been known for a nuclear generation plant to take seven or more years.

MS

lean-burn internal combustion engine. Recent developments in the design of spark-initiated internal combustion engines have made it possible to construct gasoline engines which operate successfully on lower fuel/air ratios than formerly. This makes for more efficient combustion and less pollution by toxic exhaust fumes.

lean gas. A fuel gas of low *calorific value containing negligible proportions of hydrocarbons. It may thus be preheated without its composition being affected by *thermal cracking. *See also* blast furnace gas.

WG

lean mixture. A mixture of fuel gas to vapour with more air than that required for complete combustion.

Le Chatelier's principle. A principle stating that any reacting chemical system will respond to a change of environment, such as temperature or pressure, in such a way as to reduce the effect of the environmental change. For example, in the reaction between *ethene and hydrogen to form *ethane, two molecules become one; an increase in pressure therefore aids the reaction to ethane. The principle is also found to have its equivalent in the social sphere. *See also* feedback.

MS

legumes. Plants capable of incorporating atmospheric nitrogen into chemicals which they themselves can use. This is achieved via a *symbiosis with *Rhizobium* bacteria in their roots. Some legumes are high-protein food crops, others provide forage, fruits, luxury timber or fast-growing firewood, while many produce a combination of these. Within an energy context, the important contribution is that nitrogen in a form suitable for plant nutrition is produced at no external energy cost. This reduces the requirement for energy-intensive synthetic fertilizer. A leguminous crop can add up to 500 kg of nitrogen to the soil per hectare in a year. World-wide, cultivated legumes provide more nitrogen to the soil than do all synthetic fertilizers. As suppliers of wood energy, **Leucaena leucocephala* and **Sesbania grandiflora* are two of the fastest growing trees known. Both are legumes, making them excellent species for *silviculture energy plantations.

CL

Lenz's law. *See* Fleming's rule.

Leucaena leucocephala. A versatile shrub and tree crop originating in Central America. It probably offers the widest range of uses of any tropical *legume for it can provide forage, *fertilizer, *firewood and timber. Its dense wood has a *calorific value of 17.6–19.3 GJ/t. Individual trees have produced among the highest ever recorded annual tree yields. Although native to southern Mexico it has been successfully introduced as a firewood and timber source into Africa and Asia, where the 'Hawaiian Giant' variety is a prolific producer. Average annual growth can amount to 30–40 m^3/ha, equivalent to 650 GJ/ha of wood energy. As the stumps readily coppice, the *Leucaena* tree has the potential to be a truly renewable fuel source. Furthermore it has the ability to thrive on steep slopes, in marginal soils and in locations with extended dry seasons. This latter versatility makes it ideal for the reforestation of many regions of developing countries where shortages of firewood are a cause for great concern, both for the well-being of the local inhabitants and for their environment.

CL

leukaemia. Any of a group of diseases which are characterized by excessive production of white blood cells. Exposure to *ionizing radiation and certain chemicals increases the likelihood of developing the disease.

PH

lichen. A plant composed of a fungus and an alga in very close *symbiosis. It is found as crusty patches or bushy growth on trees, rocks, etc. Lichens are very sensitive to air pollution, particularly to *sulphur oxides, and very few grow near to heavily industrialized areas. Different species of lichen have different sensitivities to air pollution so that the number of species growing in an area, and their relative rates of growth, can be used as a crude indicator of air pollution episodes during the preceding year.

PH

lifetime. The useful life of an industrial plant, manufactured device or other capital investment. In cost–benefit analysis or in assessing the relative merits of one type of investment (e.g., nuclear power reactor) versus another type, the anticipated lifetime has a considerable bearing on the perceived economics of the investment proposal.

When considering new technologies it is almost impossible to arrive at reliable predictions of lifetime. It is the major determinant in computing the rate of depreciation of an investment. MS

lift. The force acting on an object in a moving fluid, either liquid or gas, perpendicular to the relative direction of motion. On an airplane wing the lift force is upwards. On a wind turbine blade lift moves the blade sideways to produce rotation. *See also* drag. JWT

lift-type machines. Devices moved by *lift forces in a moving fluid, usually wind turbines with blades powered to move perpendicular to the relative fluid motion. *See also* drag-type machines. JWT

light. (1) *Electromagnetic radiation to which the human eye is sensitive. It forms a very narrow region of the electromagnetic spectrum, the wavelengths of light falling approximately between 0.4 micrometres (violet light) and 0.7 micrometres (red light).

(2) *See* lamp. VI

light distillate. A petroleum fraction with a boiling range between 30°C and 250°C. It contains hydrocarbons in the gasoline, naphtha and kerosene range and thus is probably the most valuable intermediate product of the petroleum refinery. WG

light fuel oil. A liquid fuel with a *kinematic viscosity of about 50 centistokes at 38°C. The sulphur content should be less than 3.5% and the gross *calorific value about 43.5 MJ/kg. WG

light water. The name applied to ordinary water, H_2O, to distinguish it from *heavy water, D_2O, containing deuterium. DB

lignite. A material which is at an intermediate stage of *coalification. Lignites vary in character from those resembling a bituminous wood to others almost indistinguishable from *coal. As mined, they can contain as much as 50% moisture. The range of chemical composition of the dry ash-free material is considerable but on average would be 65% carbon, 5% hydrogen and 30% oxygen. The volatile-matter content is seldom less than 50%. The *calorific value on a dry ash-free basis is about 23 MJ/kg. Lignites have their principal outlet in large-scale steam generation. WG

limit pricing. A policy which restricts prices of existing firms so as to prevent new entrants into the industry. This can be accomplished by setting a low price such that no new entrant can cover their large initial costs. Thus, for example, an important factor in OPEC pricing policy for its oil is the price at which synthetic substitutes become commercially viable. If OPEC were to raise prices too rapidly it may simply destroy its existing market power. MC

Limits to Growth. The title of the first report prepared for the *Club of Rome's Project on the Predicament of Mankind. It was written by an international research team under the direction of Dennis L. Meadows with financial support from the Volkswagen Foundation. Using techniques of systems analysis developed by Jay W. Forrester, the team attempted to simulate the interaction of the five major variables in the world *ecosystem: population, agricultural production, natural resources, industrial production and pollution. Their global computer model investigated the long-term consequences of changing these variables in various ways.

The report was widely publicized as it was a pioneering attempt to model the future of the world. Though the model was heavily criticized for its lack of social feedback, it nevertheless aroused a new interest in long-term global problems and in the limits of the earth's physical resources. PH

linear accelerator. A particle *accelerator in the form of a long straight tube or tunnel along which an electric field is set up to accelerate electrons or other charged particles to very high energies. A small magnetic field is used to keep the particles travelling straight down the tube.

DB

linear programming. A practical technique for finding the arrangement of activities which maximizes, or minimizes, some criterion defined on the activities, subject to the operative constraints. For example, it can be used to find the most profitable set of outputs which can be produced from a given type of crude oil input to a given refinery with given output prices. The technique can deal only with situations where activities can be expressed in the form of linear equalities or inequalities, and where the criterion is also linear. If x_1 and x_2 are inputs and y is the output, the technique applies only if their relationship is of the form

$$y \leqslant ax_1 + bx_2,$$

for example. Techniques which can handle non-linear relationships, such as

$$y \leqslant ax_1 + bx_2 + cx_1x_2 + dx_1^2,$$

do exist but are much more difficult to use in practical situations.

MC

liquefied natural gas (LNG). Natural gas which has been liquefied by compression and cooling for ease of transport or storage.

WG

liquefied petroleum gas (LPG). Petroleum hydrocarbons which, although gases at normal ambient temperatures, have been liquefied by the application of pressure of the order of a few atmospheres. Hydrocarbons containing three or four carbon atoms in the molecule come within this range. *See also* bottled gas; butane; propane.

WG

liquidity. A property of an *asset denoting the ease with which it can be converted into money. Assets range from cash, the most liquid asset, through 'near money' and long term *securities to durable goods. Thus, for example, securities representing part ownership of an international oil company are a more liquid asset than a particular oil deposit, given the relative ease with which each can be sold for money. Liquid assets are those which can be easily and quickly converted into money. International liquidity denotes currency and other assets, such as gold or *special drawing rights, which are internationally acceptable. A liquidity ratio denotes the ratio between liquid assets and total liabilities. Banks are normally required to keep some specified cash or liquidity ratio.

MC

liquid metal-cooled fast breeder reactor (LMFBR). *See* nuclear reactor.

liquid metals. Metals or alloys which have relatively low melting points and which have uses as liquids. In nuclear engineering, *sodium (melting point: 97.8°C) and sodium-potassium alloys (melting points from −11°C upwards) can be usd as *coolants for *nuclear reactors. At present fast breeder reactors use liquid sodium coolant because of its excellent heat-transfer properties and high boiling point (883°C) at atmospheric pressure. Sodium becomes activated during its passage through the reactor, forming radioactive sodium-24, and a secondary loop of non-radioactive sodium is needed between the primary loop and the steam generators.

DB

lithium. The lightest metallic element. An important aspect of lithium is that under irradiation by neutrons its isotope lithium-6 forms *tritium, the radioactive isotope of hydrogen:

$${}^{6}_{3}\mathrm{Li} + {}^{1}_{0}\mathrm{n} \rightarrow {}^{3}_{1}\mathrm{H} + {}^{4}_{2}\mathrm{He}$$

Tritium, which does not occur naturally, is the isotope which reacts with *deuterium in one of the principal *fusion reactions. Lithium can therefore be regarded as one of the raw materials for the production of energy from fusion.

Symbol: Li; atomic number 3; atomic weight: 6.94; density: 530 kg/m^3. DB

load. In electrical and electronic engineering, the acceptance of power from some system providing it. If the accepting system is drawing as much power as the delivering system, the latter is said to be on full load. The ratio of average load to *peak load over a period is called the load factor.

Base load refers to the perpetual load on the system, and is thus approximately the lowest load in a given time period. Base load electrical provision, for example, is furnished most efficiently by generators that run best at *steady state, such as nuclear reactors, or large coal-fired installations. MS

load factor. *See* load.

load forecasting. *Forecasting the *load (usually in the context of electrical demand) for some years ahead. This is important given the long *lead-times for the provision of additional facilities, and the economic disadvantages of under provision and over provision. MS

load management. The exercise of control over the demands made on a mechanical or electrical power system. In the interests of efficiency and economy, an even distribution of load on the system over time may be desirable. *See also* off-peak. TM

load-on-top system. *See* ballast water.

London Dumping Convention. Short for the Convention on the Prevention of Marine Pollution by Dumping of Wastes and Other Matter, an international agreement concluded in London in 1972. It was an initiative of IMCO (International Governmental Maritime Consultative Organization), a United Nations agency. It applies to the seas of the whole world and has aims similar to those of the *Oslo Dumping Convention. Substances which may not be dumped into the seas include mercury, cadmium, persistent plastics and other non-degradable synthetic materials, carcinogens, highly radioactive materials, chemical/biological warfare materials, and oil. PH

long-flame coal. A coal which contains a high proportion of *volatile matter (e.g. 35% by weight on a *dry basis), burns readily with luminous flames, and has little or no ability to form a coke. This type of coal was preferred for the open domestic grate, and for coal-fired smelting furnaces where heat radiation and a reducing atmosphere were required. WG

long run. A period of time long enough to alter the input of all *factors of production, but not long enough to alter the present technology. It allows firms time to alter their input–output mix to the most *efficient possible. For example, for the electricity supply industry the long run is determined by the length of time it takes to bring on line a new generating station. This is longer than the long run for, say, peasant farming. *See also* short run. MC

long-wave radiation. Radiation in the far *infrared region of the *electromagnetic spectrum. It is emitted by surfaces at environmental and atmospheric temperature (about 10°C) and has a wavelength range of roughly 10–20 micrometres. It contrasts with *short-wave radiation of

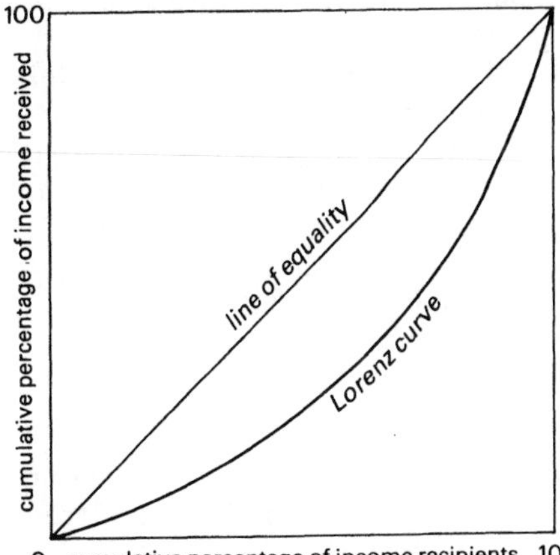

the solar infrared spectrum. Long-wave radiation is not transmitted by glass. JWT

Lorenz curve. A curve showing how the size distribution of something, such as wealth, income, firms, or population, departs from one of 'equality', or whether one distribution is more or less unequal than another. On one axis is plotted the cumulative percentage of the variable being considered (e.g. the percentage of income recipients with a particular size of income or less); on the other axis is plotted the percentage of the cumulative frequency distribution of a variable (e.g. the percentage of cumulative incomes of all recipients with a particular income size or less). The diagonal line represents 'equality' in the sense that 10, 20, 30, etc. per cent of income recipients receive 10, 20, 30, etc. per cent of all income. The extent by which Lorenz curves deviate from the line of equality shows the degree of inequality. A measure of this degree of inequality is the *Gini coefficient. MC

Los Alamos Scientific Laboratory. A unit of the US Department of Energy operated by the University of California. It carries out research in nuclear energy and plasma physics.

Address: PO Box 1663, Los Alamos, New Mexico 87545, USA. DB

loss of coolant accident (LOCA). A postulated accident in *nuclear reactors of potential seriousness. It is particularly relevant to pressurized and boiling water reactors, in which a rupture of a pressure vessel or pipe could lead to depressurization of the reactor, loss of coolant and flashing of water to steam. These events could in turn lead to overheating and meltdown of the *core unless effective alternative cooling systems are provided. *See* Three Mile Island. DB

lower calorific value [lower heating value]. Alternative terms for net calorific value. *See* calorific value.

low-grade fuel. A material of low *calorific value containing large proportions of non-combustible substances such as mineral matter, water or inert gases. WG

low-grade heat. A source of heat at a temperature not much above *ambient temperature. MS

low head. The *head of a water supply which is considered nearly minimal for *hydropower generation. In practice, for small turbines low head corresponds to a water height of about 1m.

low-level waste. That solid and liquid waste of low *radioactivity, produced in nuclear power stations, reprocessing plants for *irradiated fuel and other plant handling radioactive wastes. DB

low-temperature carbonization. A process in which small-sized coal is heated to 600–700°C in a retort for the production of a reactive coke, which will burn readily in the domestic grate, and tarry liquids, which may be employed as raw materials for chemical manufacture. *See also* carbonization; high-temperature carbonization. WG

lumen (Symbol: lm). The SI unit of luminous flux, i.e. of the rate of flow of radiant energy which is seen to be emitted by a light source or received by a surface. TM

luminance (Symbol: L). A quantitative expression or brightness, measured in *candela per square metre (cd/m^2). TM

luminous flux. *See* lumen.

lump sum taxation. Taxes which are fixed in size and do not vary with the amount or size of the item being taxed. A lump sum tax levied on a firm raises its fixed costs but leaves marginal costs unaffected.

MC

Lurgi process. A process developed before 1939 for the *gasification of coal using a mixture of oxygen and steam under pressures of 20–50 bar. The gas generator is a strong cylindrical steel shell lined with refractory material and fitted with a cooling water jacket. Oxygen and steam are injected into a bed of finely divided coal, and ash is removed by means of a revolving grate. The gas leaving the generator is cooled and sprayed with an alkaline solution to remove *hydrogen sulphide and *carbon dioxide. The washed gas has a gross *calorific value of about 14.8 MJ/m^3; it thus has to be enriched with hydrocarbons to enable it to be blended with coal gas with a gross calorific value of 18.6MJ/m^3.

If the gas is passed through a catalytic converter, advantage may be taken of the water gas equilibrium to convert most of the carbon monoxide present to carbon dioxide, which may be subsequently removed, and to convert water vapour to hydrogen. Passage through a second catalyst under pressure will then promote the formation of methane and the creation of a synthetic natural gas. Thus this combination of processes converts coal to natural gas.

WG

lux (Symbol: lx). The SI unit of *illuminance, defined as the illuminance produced by light from a source of one *candela falling directly on a surface at a distance of one metre.

TM

M

macroclimate. *See* climate.

macroeconomics. The study of economic aggregates and the relationships between them. The targets of macroeconomic policy are the level and rate of change of *national income (i.e. *economic growth), the level of unemployment and the rate of *inflation. Macroeconomics uses *models which look at the behaviour of groups of economic agents and abstract from relationships between individual economic agents. Whereas *microeconomics studies, for example, the way individuals allocate their total expenditure over different commodities as relative prices change, macroeconomics studies how total consumption expenditure by all individuals together varies as national income varies, and ignores the distribution of the total across commodities. In macroeconomics the questions about energy are how its price and availability affects economic growth, unemployment and inflation, and how economic growth affects the demand for energy. *See also* multiplier; national income accounts; numeraire; quantity theory.

MC

Madison process. A process developed by the US Forest Products Laboratory to hydrolyse woody substances using dilute sulphuric acid. The sugars formed are then fermented and distilled to produce power *ethanol as a transport fuel. This process is an improvement on the German *Scholler process in that the hydrolysis time is reduced to 3 hours and the yield of fermentable sugars is greater, though variable. Demonstration plants were initially built in 1946 and 1952 and although the technology is now in an advanced stage of development, the process is not widely practised even in the USA. Around 14 plants are reported to be operating in the USSR and the only other country apparently using the process today is Japan.

CL

magnetic mirror. *See* containment.

magnetism. The phenomenon exhibited by our planet and by certain naturally occurring or suitably treated materials containing iron. An object which exhibits magnetism, such as a bar of iron, has what is known as a north pole and a south pole situated at its opposite ends; each pole exercises an attractive force on unmagnetized ferrous objects. In the case of two objects exhibiting magnetism, i.e. two magnets, their opposite poles exercise an attractive force while their similar poles exercise a repulsive force.

Magnetism can also be produced by the passage of an electric current and is then known as *electromagnetism. This is the basis of the conversion of electrical energy to mechanical energy and vice versa. TM

Magnox. An alloy of magnesium with small quantities (less than 1%) of aluminium and beryllium. It has been developed as the *cladding material for the fuel in the first generation of British gas-cooled *nuclear reactors. DB

Magnox reactor. *See* nuclear reactor.

maize [Indian corn]. The third most cultivated grain crop in the world; scientific name *Zea mays.* Only 10% of production enters world trade, of which 60% comes from the USA. It is a *4-carbon plant, capable of high yields under favourable conditions: 30 t/ha.y have been reported in the subtropics, which is an excellent figure for a cereal crop, while short-term productivity of 400 kg/ha.day are not unknown. The latter value is equivalent to a *photosynthetic efficiency of 3.4%, approaching the maximum practical limit.

Maize starch is a good substrate for grain alcohol production and has been used extensively in the manufacture of *gasohol in the USA, where the average corn yield is over 5.7t/ha. Wheat, sugarbeet, and molasses from *sugar-cane are also used in gasohol production, but in less quantity. During the 1930s almost all US ethanol was fermented from grain and molasses and some cars were run on 100% crop-fermented alcohol. In 1962 grain fermentations accounted for 22% of US industrial ethanol – mostly from grain surpluses or damaged crops in the Midwest. The present gasohol programme in the USA originated in the state of Nebraska in 1971. At the beginning of 1980 there were 12 grain-alcohol plants in the USA with an overall annual capacity of 300 million litres. The 1979 US maize grain production was 197 million tonnes (half the world total) of which 110 million tonnes were used as animal feed. Maize starch is also used as a staple human food and the feedstock for the production of high-fructose syrup. CL

manioc. *See* cassava.

marginal. In economics, additional and arising as the result of a small change. For example, in the analysis of output level decisions, marginal cost is the addition to total production costs for a unit increase in the level of output, and marginal revenue is the addition to total sales receipts arising from an extra unit of output. Typically marginal cost rises and marginal revenue falls as the level of output increases. To maximize profit (the excess of total revenue over total cost), a firm would operate at the output level for which marginal cost is equal to marginal revenue. If at a given output level marginal cost is less than marginal revenue, profit can be increased by increasing output by one unit; if marginal cost exceeds marginal revenue, profit can be increased by cutting the output level by one unit. The use of marginal analysis is one of the main distinctive characteristics of *neoclassical economics. *See also* diminishing marginal returns. MC

marginal cost pricing. A situation where a private firm or public corporation sets its price equal to *marginal cost. Where costs are *social costs, marginal cost pricing is necessary for *efficiency in the allocation of resources, and it has been advocated as the rule which publicly owned or controlled energy supply *monopoly industries, such as electricity, should be made to follow. However, there are difficulties of principle and practice involved. As a matter of principle, some would argue that the rule ignores the *second best problem arising from the fact that not all prices elsewhere in the economy are set equal to marginal social cost. In practice it is not easy to identify social marginal cost as it is relevant to the goal of efficiency in allocation. There is a distinction between *short run and *long run marginal cost, with the former refering to output variation with fixed plant and the latter to output variation where plant size, and characteristics, can be varied. Different prices would be set according to whether equalization was to short- or long-run marginal cost, except in special circumstances. The principle which appears to command most support among economists is that price should be set equal to short-run marginal cost, with the size of the energy supply industry being adjusted until this is equal to long-run marginal cost.

MC

marginal energy requirement. The additional energy, *direct and *indirect, which is required to create one more unit of output in a process. *See also* marginal cost pricing.

MS

marker crude. A crude oil of specific quality which is used as a reference against which other crudes are priced. In the case of the *Organization of Petroleum Exporting Countries (OPEC) it is Arabian light crude of 34° *AP1 gravity. A denser (higher AP1 gravity) crude has a higher value as it yields on refining a greater proportion of high value products like *gasoline. A less dense oil, in contrast, attracts a lower price.

MS

market. *See* market equilibrium; market structure.

market equilibrium. A market is where buyers and sellers are brought into contact. A market equilibrium denotes a situation where, at some specified price, what is being demanded is matched exactly with what is being supplied; there is thus a balance between intended purchases and intended sales. The price at which demand equals supply is called the equilibrium price. *See also* partial equilibrium analysis.

MC

market failure. The inability of competitive markets to solve the *allocation problem by bringing about a pattern of efficient economic activity. Sources of market failure include *externality, *imperfect competition and *public goods. The existence of market failure constitutes a case for government intervention to seek *efficiency in allocation. The possible forms of intervention range from legislation to establish private property rights where they did not formerly exist, to taking a *monopoly into public ownership. In practice there are several sources of market failure so that intervention has to face the *second best problem. *See also* external cost.

MC

market structure. The characteristics of a market, including the number of firms, the degree of product differentiation and the ease of entry and exit from the market. Typical market structures are *perfect competition, *imperfect competition, *oligopoly and *monopoly. Energy markets usually involve a small number of sellers, being either examples of oligopoly (oil) or monopoly (electricity in European countries).

MC

marsh gas. Methane produced by natural microbiological processes in the shallow waters and mud of a swamp or marsh.

WG

mass balance. A balance of the materials entering and leaving a defined space or system. For example, the mass of crude oil entering a fractionation column in an oil refinery equals the sum of the mass of the various fuels and other products leaving the column. *See also* energy balance.

PH

mass defect. The difference between the mass of an atom (or an atomic nucleus) and the sum of the individual masses of its constituent particles. *See also* binding energy.

DB

mass–energy equation. *See* Einstein's equation.

mass number (Symbol: A). The number

of neutrons and protons in the *nucleus of an atom. Elements can have the same mass number but different *atomic numbers. The latter defines the chemical characteristics of an element.

DB

materials balance principle. As a consequence of the conservation of mass, the economic system returns to the natural environment, as waste, approximately the same mass of material as it extracts from it as inputs to production. The approximation arises from the lag introduced into the materials cycle by the production of *durable commodities: its magnitude depends on the period of measurement. The principle implies that insofar as economic growth involves increased extraction of natural resources, including energy resources, it will also involve increased pollution to an amount determined by the composition of the wastes returned in relation to the assimilative capacities of the natural environment.

MC

materials unaccounted for (MUF). The difference between theoretical and actual stocks of valuable material. At nuclear fuel reprocessing plant there are procedures for accounting for stocks of uranium and plutonium. One procedure relies on keeping a physical inventory of these materials; the other relies on a 'book-keeping' inventory. The difference between these two inventories is the MUF. This difference reflects the great difficulty of knowing precisely the amounts of plutonium and uranium in *irradiated fuel elements and subsequent reprocessed fuel.

DB

maximum permissible body burden (MPBB). *See* radiological protection standards.

maximum permissible concentration (MPC). The maximum level or concentration of a radioactive substance in the environment, e.g. in water or air, that can be regarded as having no significant hazardous effect on human health. *See also* radiological protection standards.

DB

maximum permissible dose (MPD). The amount of ionizing radiation to which humans may be exposed without the risk of a significant health hazard. The MPD for workers in the nuclear industry, who are subject to continuous monitoring and health checks, is greater by a factor of ten than the MPD for the general public. The maximum permissible doses specified by the *International Commission on Radiological Protection (ICRP) are: for workers in the nuclear industry 50 millisieverts/y and for the general public 5 millisieverts/y. *See also* radiological protection standards.

DB

maximum permissible dose (MPD). The amount of ionizing radiation to which humans may be exposed without the risk of a significant health hazard. The MPD for workers in the nuclear industry, who are subject to continuous monitoring and health checks, is greater by a factor of ten than the MPD for the general public. *See also* radiological protection standards.

DB

maximum permissible level (MPL). The level of radiation or radiation dose rate which, assuming humans are exposed to it for 40 hours per week, 50 weeks per year, will not produce an accumulated dose greater than the *maximum permissible dose. The currently accepted value of the MPL for occupationally exposed workers is 0.025 millisieverts per hour, giving a maximum yearly dose of 0.05 sieverts. *See also* radiological protection standards.

DB

maximum sustainable yield. The largest amount which can be taken from a renewable natural-resource stock per period of time, without reducing the stock of the resource. With a fish stock, for example, it is the catch which corresponds to the natural growth of the fish stock at the stock size for which natural growth is greatest. Biologists regard it as self-evident that renewable resources should be managed so as to maintain stock size at that corresponding to maximum sustainable yield. Economists do not, noting that larger catches now at the expense of smaller future catches can be justified by the existence of a positive rate of *time preference. *Biomass energy sources give rise to the same management possibility and the same issues in inter-temporal choice. MC

mean life (nuclear). The average lifetime of all the atoms in a radioactive substance. The mean life is 1.44 times the *half-life. *See also* radioactivity. DB

mechanical equivalent of heat. The heat generated by a unit of work. When heat and work are both quoted in SI units, i.e. in joules, no conversion factor is needed. *See also* conversion tables. MS

medium fuel oil. A liquid fuel with a *kinematic viscosity of about 220 centistokes at 38°C. The sulphur content should be less than 4.0% and the gross *calorific value about 43.0 MJ/kg. WG

mega- (Symbol: M). A prefix indicating a multiple of one million units. For example, a megawatt (MW) is a million (10^6) watts, a megajoule (MJ) is a million joules. One megajoule is equal to 0.278 kilowatt-hours, 948 British thermal units, and 0.239×10^6 calories. JWT

mercury. A naturally occurring liquid metallic element which is widely dispersed in the form of mercury compounds throughout the earth's crust, the atmosphere and hydrosphere. Since the metal and its compounds are toxic to all living organisms, the ways in which it is released into the environment by man and natural processes is of considerable interest. In the context of the energy industry, mercury is released into the atmosphere through the combustion of fossil fuels since fossil fuels contain on average around one part per million of mercury. On a world-wide basis it has been estimated that the combustion of coal releases at least 3000 tonnes of mercury per year. This figure far exceeds the amount released into the aquatic environment by the largest industrial user of mercury, the chlor-alkali industry.

Symbol: Hg; atomic number: 80; atomic weight: 200.59.

PH

mesophilic. Denoting or relating to organisms that have optimum growth rates between 30–45°C, but cannot grow below about 10°C and above around 47°C. Some indeed cannot grow at temperatures below 15°C or above 35°C. Microbial *fermentations for energy production are generally mesophilic. For example, fermentations taking place in most *biogas generators operate best at 30–35°C. Should the temperature be much less than 30°C the rate of biogas evolution falls significantly unless the digester is heated, so lowering the *net energy of the system.

CL

metabolic energy. Energy required to maintain the biochemical reactions of living organisms. Metabolic energy is dissipated as heat and work. The metabolic energy requirement for humans is usually 100–150 watts, but in deprived areas with near starvation metabolic energy may be as low as 40 watts. The energy is apparent predominantly as body heat, with a relatively small amount going to mechanical work. *See also* work (human).

JWT

metal hydride. *See* hydride.

metallurgical coke. Coke produced as a fuel and a source of *carbon monoxide as a reducing agent in the blast furnace or cupola. The physical characteristics of the coke are as important as its chemical reactivity since it has to support the weight of the charge in the furnace and permit the passage of gases through it. Important characteristics include size, strength and porosity. Ash and sulphur contents must also be low.

WG

meteorology. The scientific study, recording and prediction of *climate. Meteorological data are logged at stations set up throughout the world by the meteorological authorities in each country. The completeness of the climatic records and the form in which they are stored varies from station to station.

TM

methane. An odourless flammable gaseous aliphatic hydrocarbon, CH_4, which has the simplest structure and the smallest molecule of any organic compound. It is the chief component of *natural gas. Methane is generated by *anaerobic microbial degradation processes and can thus be obtained from activated sewage sludge. Various schemes have been proposed for producing it locally from agricultural wastes. By far the greatest proportion of methane is burned in industrial and domestic applicances to generate heat. However it is also a valuable source of chemicals. When mixed with steam and passed over a catalyst, a mixture of carbon monoxide and hydrogen (synthesis gas) is formed which may be further processed to obtain alcohol and higher molecular weight alkanes. Its gross calorific value is 55.67 MJ/kg.

WG

methanogenic bacteria. Bacteria which ferment organic substrates, chiefly acetate, to *methane under *anaerobic conditions. The acetate is made available following the bacterial hydrolysis of (mainly) *carbohydrate input to the *biogas plant. The methanogenic bacteria themselves are difficult to classify because of their reluctance to grow in pure culture, though some have now been isolated. *Methanobacterium*, *Methanosarcina*, *Methanococcus* and *Methanospirillum* species are claimed to be the principal organisms involved – responsible for the overall reaction:

$$\underset{\text{carbohydrate}}{(C_6H_{10}O_5)n} + \underset{\text{water}}{nH_2O} \rightarrow \underset{\text{carbon dioxide}}{3nCO_2} + \underset{\text{methane}}{3nCH_4}$$

The specific methanogenic reaction is

$$\underset{\text{acetate}}{CH_3COOH} \rightarrow CH_4 + CO_2$$

The rate-limiting reaction of the entire process is usually considered to be the methanogenic step itself, but might well be more closely associated with the final mass transfer of dissolved methane and carbon dioxide to the gaseous phase. Additionally the metabolism of the methanogenic bacteria, following *Le Chatelier's principle, actually inhibits the product gases formed. Methanogenic bacteria also occur in the gut of ruminant animals where they produce methane mainly from hydrogen.

CL

methanol. The systematic name for methyl alcohol, CH_3OH. It is a colourless volatile flammable and toxic liquid. It is prepared industrially by the *steam reforming of hydrocarbons to produce a mixture of carbon monoxide and hydrogen (synthesis gas); this is passed over a copper catalyst at 50–100 atmospheres pressure and a temperature of 250 °C to produce methanol. Although it has a potential value as a motor fuel, it has never been used as such on a commercial scale. The greatest outlet for methanol is as a chemical raw material in the manufacture of a wide variety of products, such as polymers, resins, cleaning materials, paint removers, herbicides and fungicides. Its gross calorific value is 21 MJ/kg.

WG

methyl alcohol. *See* methanol.

metric ton. *See* tonne.

microclimate. *See* climate.

microeconomics. The branch of economics dealing with the choices made by individual units, i.e. persons, firms or industries. It is particularly concerned with *relative prices, the way the decisions of individual economic agents respond to changes in them, and whether the totality of such decisions will, in given circumstances, produce a desirable pattern of economic activity. In microeconomics the questions about energy are: how do fuel prices reflect the relative scarcity of fuels; how do responses to those relative prices promote the appropriate patterns of fuel use; can the determination of the way energy in general is used (and also particular fuels) be left to competitive markets? *See also* macroeconomics.

MC

microenvironment. The local, as opposed to global, surrounding of a specified system. *See* environment.

microhydro. *See* hydroelectricity.

micron. A micrometre, i.e. one millionth of a metre, 10^{-6} m. The wavelength of infrared radiation is often given in microns.

JWT

middle distillate. A petroleum fraction with a boiling range between 200 and 350°C. It contains hydrocarbons in the kerosene, gas oil and light fuel oil range. It is a source of diesel fuel and is used as a cracking feedstock.

WG

Midwest Fuel Recovery Plant. A commercial nuclear fuel reprocessing plant constructed near Morris, Illinois. It never became operational as a result of unforeseen engineering problems and is now used as a store for spent fuel elements.

PH

mine. An underground working of a deposit of coal or other mineral which the workers (miners) reach on foot or by rail car down a sloping roadway or tunnel from the surface.

WG

mineral matter. In mining, those minerals which are closely associated with in-ground coal and remain associated after cleaning and crushing; they thus contribute to the ash residue when the coal is burned. The weight of mineral matter which leads to a particular weight of ash may be calculated, but only if the chemical composition of the ash is known and assumptions are made as to the types of mineral present. When this is done it becomes possible to calculate the composition of a coal to a *mineral matter-free basis rather than an *ash-free basis.

WG

mineral matter-free basis. A basis for reporting the analysis of a coal after the *mineral-matter content has been subtracted from the total. Since the exact content of mineral matter is difficult to determine directly, the ash content of the sample is determined; by making assumptions about the chemical nature of the minerals originally present, the weight of mineral matter is calculated from the weight of ash.

WG

mitochondria. The sites of energy-yielding metabolism within living cells via the reaction sequence of the *Krebs cycle. They are also the location of some biosynthetic pathways. Mitochondria are not found in bacteria. In plants they also perform *dark respiration, much of which is inefficient as an energy supplier to the cell, and could be eliminated to increase net *biomass production. *Genetic engineering techniques can block the wasteful biochemical pathway so that instead of carbohydrates being needlessly oxidized to carbon dioxide and water, they remain to add to the overall biomass quantity available for use as an energy source by man.

CL

mixed base crude. A crude petroleum containing both bitumen and paraffin wax.

WG

mixed cropping. The interposed planting of two or more crops. Mixed cropping and *intercropping are not quite synonymous since the latter implies the symmetrical cultivation of two crops in alternate rows, while mixed cropping refers to more of an asymmetrical arrangement. Greater productivity can be achieved by cultivating two plant species out of phase with each other's growth patterns so as to prolong the period of light interception by the increased leaf area. The leaf canopy can also be increased via mixed cropping so that upper leaves, adapted to high levels of sunlight, and lower leaves, adapted to lesser levels, can together utilize these differences to give greater *biomass yields per unit of land area. Leaves on tall plant species may preferentially absorb the primary photosynthetic region of the spectrum, while those below are able to make use of light of 'inferior' quality equally well. A plantation containing two or more crop species is also less susceptible to infection and destruction by external parasites and predators than is a monoculture.

CL

model. (1) A physical replica on a smaller scale of an actual or proposed structure.

(2) An abstraction of a real world system in simplified terms. Since we can never know what is the real world, the modelling process acts as follows:

real world → perceived world → model

That is, it takes a perception of the real world as seen through the mind of the modeller or from the consensus of those who study a given system (e.g., economists, engineers, planners), and expresses it in simplified terms. The decision process requires that the decision maker relates the factors involved and emerges with a conculsion. This is a mental model. An explicit model (economic, engineering, etc.) sets down the important perceived interactions of a system as a set of quantitatively expressed relationships. The model is then tested against real data to establish whether it sufficiently accurately reflects reality. Dynamic models seek to quantify the evolution of a system through time, and hence can be used for forecasting. There is a danger that the basis of such models may change with time leading to false forecasts. *See* energy analysis; macroeconomics.

MS

moderator. The component of the *core of a thermal *nuclear reactor whose function is to slow down neutrons from fission energy to thermal energy by successive *scattering collisions. The best materials are those of low *mass number (or compounds containing elements of low mass number), high scattering *cross-section and low capture cross-section. Three materials in common use as moderators are water (H_2O), *heavy water (D_2O) and *carbon (graphite).

DB

moisture content. The total moisture content of solid fuels is made up of free moisture and inherent moisture. The former is removed by exposing the sample to the air of a room at ambient temperature until it reaches a constant weight. Inherent or equilibrium moisture may be removed by drying the coal in an oven to constant weight at 100°C. As the sample may change weight due to oxidation of the coal, the oven drying is normally carried out in an atmosphere of nitrogen or under vacuum. *See also* air-dried coal; dry coal. WG

mole. The amount of a substance containing the same number of elementary particles as there are atoms in 12 grams of carbon-12. The particles may be atoms, molecules, ions, etc., and a mole will contain 6.022×10^{23} (Avogadro's number) of them. One mole of a compound is equivalent to its gram molecular weight. It has been shown experimentally that one mole of all gases at 273K and one atmosphere pressure occupies 22.41 litres. WG

molecular formula. A shorthand description of a chemical compound which shows the numbers and grouping of the various types of atoms in the molecule, but not their spatial or chemical relationship. Examples are H_2O (water), C_2H_6 (ethane) and U_3O_8 (uranium oxide). MS

molecular weight. The sum of the *atomic weights of all the atoms in a molecule. WG

monetarist. An adherent to the school of thought known as monetarism, the main tenet of which is that it is supply-side disturbances in the monetary sector which are the principal source of instability in the economy. Thus the observed changes in *nominal *national income are to be traced back to prior changes in the *money supply. As opposed to *Keynesian economists, monetarists argue for minimal government intervention, and for *monetary policy being more effective, as a cure for *inflation especially, than *fiscal policy. MC

monetary policy. The use by government of monetary instruments, such as the *money supply or *interest rates, to influence the way the economy behaves, in pursuit of the objectives of *macroeconomic policy. MC

money. Anything that is widely accepted as a medium of exchange. Usually it serves two additional functions: a measure of value and a *unit of account. Unlike *barter it acts as an intermediary between the exchange of goods and so avoids the problem of the double coincidence of wants (each wanting what the other has). Any commodity can act as money and in the past these have included sea shells, cattle and cigarettes. Precious metals, such as gold and silver, are the standard. We now have coins and notes which have little intrinsic worth but act as money. *See* numeraire; money supply. MC

money illusion. A situation where behaviour responds to *nominal changes in economic variables which do not properly reflect real changes. For example, with *inflation at 10% per annum and the nominal, or *absolute, price of oil rising at 10% per annum, the price of oil relative to prices in general is not changing, i.e. the relative price of oil is constant on average. If, in this situation, an individual or firm acts upon the belief that oil is becoming more expensive, by, say, reducing oil use by substituting something else for oil, that individual or firm is subject to money illusion. MC

money supply. The amount of money in an economy. There are a variety of definitions, of which the most common are as follows. M_0 is notes and coins, M_1 is M_0 plus current (in the USA 'sight') account deposits, M_2 is M_1 plus time account (or 'deposit account') deposits with commercial banks, M_3 is M_2 plus time account deposits with other financial institutions. There is disagreement over which of these defines that which is the most appropriate indicator of the stance of *monetary policy, and that which commands the most support, and the most attention from government, changes over time. In part this reflects changing conditions in financial and monetary institutions and habits. Thus, whereas it was once the case that few retail institutions would accept a cheque drawn on a commercial bank, most will now accept one drawn on an account with a variety of financial institutions or a credit card. MC

monochromatic. Of one colour. The term is applied to all forms of *electromagnetic radiation having the same wavelength or frequency, whether within or outside the visible spectrum. (The wavelength of light determines its colour.) Monochromatic *transmittance, *absorptance, *reflectance, etc., refers to these properties at one specified wavelength or frequency only. JWT

monopolistic competition. A market structure where there are a large number of firms in an industry, each producing and selling similar but not identical products. Hence each firm faces a downward sloping demand curve and has some *monopoly power. Garages selling petrol illustrate the case of monopolistic competition. There are a small number of different brands of petrol, but each garage's 'product' is also differentiated by the precise location of the garage, ancillary services offered, etc. As a result garages sell petrol at similar but not identical prices, and do not act as *price takers. MC

monopoly. An industry consisting of only one producer supplying a homogeneous product to many consumers. In this case the firm's demand curve is also the industry's demand curve and it is downward sloping, unlike *perfect competition when it is horizontal. This results in a divergence between marginal revenue and average revenue and so price (equalling average revenue) diverges from marginal cost (*see* marginal cost pricing); this gives rise to monopoly profits. Monopolies occur in both the public and private sector but governments tend to oppose private monopolies because they do not price at marginal cost and can restrict output. They can, however, supply more than under competition if *economies of scale can be reaped. The electricity supply industry is a monopoly supplier of electricity which has achieved economies of scale. In most countries the electricity supply industry is either publicly owned or its operations are publicly controlled. MC

monopoly power. A situation where a firm has some control over the market price, in the sense that it can raise the price without losing all (or nearly all) of its customers. Energy-supply industries are characterized by consisting of firms with monopoly power. *See also* monopoly; imperfect competition. MC

monopsony. A sole buyer of a product or *factor of production which is supplied by many firms or from many sources. In the UK the Central Electricity Generating Board is in a position of monopsony with respect to the manufacturers of large electricity-generating sets. MC

motor. *See* electric motor; internal combustion engine.

motor spirit. A volatile liquid fuel that may be vaporized when mixed with air in a carburettor and burned in a spark-initiated *internal combustion engine. The vast majority of such engines provide the motive power for automobiles. *See* gasoline. WG

moving bed. A solid/gas reaction system in which the solid particles move downwards while the gas passes upward between them. *See also* catalytic cracking; fixed bed; fluidization. WG

multiplication factor (Symbol: k). The number which expresses whether or not a *chain reaction is possible in a *nuclear reactor. If F is the fraction of *fission-produced neutrons which cause further fission in the reactor, and ν is the average number of neutrons produced per fission in the reactor, then

$$k = \nu F.$$

If $k = 1$, then the reactor is critical and can sustain a chain reaction.

If $k < 1$, then the reactor is subcritical and cannot sustain a chain reaction.

If $k > 1$, then the reactor is supercritical and can sustain a diverging chain reaction.

See also criticality.

DB

multiplier. An essential feature of *Keynesian economic analysis. It refers to the ratio between the size of an *exogenous increase (or decrease) in the level of economic activity to the size of the consequent increase (or decrease) in the level of *national income. For example, if government expenditure on educational services is increased, then more people are employed in the education sector. These additional employees receive income, a large part of which they spend on goods and services produced in the private sector of the economy. This new expenditure generates additional employment and income in the private sector; part of this income is itself spent on goods and services, generating further additional employment and income. Thus there is a chain of induced increases in employment, income and spending. Given that the propensity of individuals to spend out of additional income is less than unity, successive increases in this chain get smaller and smaller, and the process is self-limiting. There will be a new level of national income which is larger than the original level by an amount which exceeds the size of the initial increase in government expenditure. The ratio between the size of the increase in national income and the increased government expenditure is the government expenditure multiplier, the size of which depends on the propensity to spend out of additional income. This simple example illustrates the Keynesian view of how a government can use fiscal policy to manage the level of national income and employment.

MC

mutagen. A substance or agent that causes a *mutation.

PH

mutation. The process whereby inheritable changes in the genetic material (*DNA) arise.

PH

N

nacelle. The framework and casing of the gearing, generator and associated equipment of a wind-powered machine. A horizontal-axis machine usually has the nacelle at the top of a tower, but a vertical-axis machine can have the nacelle at ground level.

JWT

nano- (Symbol: n). A prefix indicating a submultiple of one thousand millionth (10^{-9}) of a unit. For example, one nanometer is 10^{-9} metre.

JWT

naphtha. A petroleum fraction separated from crude petroleum by distillation and having a boiling range of about 70–200 °C. It is used to some extent in the preparation of gasoline and as a hydrocarbon solvent. Since it may have too low an octane number for direct use as a gasoline and be too volatile for burning in paraffin (kerosene) lamps and stoves it finds a large outlet as a petrochemical feedstock. Steam reforming of naphtha produces a mixture of carbon monoxide and hydrogen suitable for synthesizing natural gas.

WG

naphthalene. A crystalline solid aromatic hydrocarbon, $C_{10}H_8$, in which the carbon atoms in the molecule are arranged in two *fused rings. It is a major component of tars from the *high-temperature carbonization of coal and has a characteristic odour associated with mothballs. It finds application in the chemical industry in dye and resin manufacture.

WG

naphthenes [cycloparaffins]. *Alkanes in which the carbon atoms are arranged in closed rings, i.e. they are alicyclic hydrocarbons. The have the general formula C_nH_{2n}. Cyclopentane (C_5H_{10}) and cyclohexane (C_6H_{12}) both occur in petroleum. Similar hydrocarbons with two or more six-membered rings occur in higher boiling fractions, such as diesel fuel.

WG

napier grass [elephant grass]. A *4-carbon plant, *Pennisetum*, which has provided some of the highest recorded yields of any species: 85 t/ha.y of dry weight *biomass growth has been registered in Central America. This corresponds to a mean annual *photosynthetic efficiency of 2.4%, with short-term efficiencies as high as 4.2%. The grass is ideal for planting in tropical *energy farms and has been chosen in Australia as being one of the five most desirable high-yielding energy crops which can be harvested throughout the year.

CL

National Association of Water Power Users. A UK organization for the encouragement of small-scale *hydro-power.

Address: PO Box 27, Exchange Chambers, 10b Highgate, Kendal, Cumbria, England.

JWT

national debt. The outstanding borrowing of the government, on which interest has to be paid to the holders of the *securities issued against the borrowing. A large part of national debt is matched by securities which are tradeable, and trade in such financial *assets accounts for a large proportion of the activity on the *stock exchange.

MC

national expenditure. The total value of the purchases of *final product giving rise to productive activity in the economy in some period. It may be measured *gross or *net. *See also* national income accounts.

MC

national income. The total of the payments to *factors of production arising from their employment in the generation of *national product in an economy in some period. It may be measured either *gross or *net. In the latter case it is the value of the goods and services which the economy could use up without reducing its *capital stock; it is therefore the proper measure for purposes of *welfare comparisons over time or across countries. *See also* national income accounts.

MC

national income accounts. The principal source of data for quantitative *macroeconomics in terms of description, analysis and forecasting. The essential idea is the adoption of a set of accounting conventions in terms of which the level of economic activity can be measured in *value terms (i.e. everything appears as price times quantity) in three ways: *national income, *national product and *national expenditure; these are by definition equal to one another, each being a different way of measuring the level of economic activity.

On the expenditure side the principal aggregates are defined as follows:

	expenditure on *consumption
plus	expenditure on *investment
plus	expenditure by the government
plus	expenditure on *inventory accumulation
plus	expenditure by foreigners on the nation's *exports
less	expenditure by residents on *imports
equals	*Gross National Expenditure at market prices*
less	*indirect taxation
plus	*subsidies
equals	*Gross National Expenditure at factor cost*
less	the total of *depreciation for all firms
equals	*Net National Expenditure*

On the other side of the accounts:

	wage and salary payments
plus	profits
plus	rent
plus	interest payments
equals	total payments to *factors of production
equals	*Gross Domestic Income*
equals	*Gross Domestic Product at factor cost*
plus	income received from abroad
equals	*Gross National Income*
equals	*Gross National Product at factor cost*
less	the total of depreciation for all firms
equals	*Net National Income*
equals	*Net National Product at factor cost*

Note that since all payments to factors of production everywhere in the economy are counted, national income is just the value of sales of *final product, i.e. national product. Note also that for the purposes of using national expenditure or product as an indicator of the level of economic activity, in relation to the demand, say, for energy, it is the factor cost measure which is appropriate. If the government in pursuit of *fiscal policy raises *indirect taxation, national product measured at market prices increases, but there is no increase in production and no increase in the demand for energy.

MC

national product. The value of the output of *final product in the economy in some period. It can be measured *gross or *net. Movements in the size of national product indicate variations in the level of economic activity, with implications for energy demand. *See also* national income accounts.

MC

National Radiological Protection Board (NRPB). A group created by the UK Radiological Protection Act of 1970. Its role is to advance knowledge and provide information and advice about *radiological protection. The staff of the Board carry out research into many aspects of radiological protection and organize training courses. The Board also advises the government on the acceptability of ICRP recommendations concerning *radiological protection standards.

Address: Chilton, Didcot, Oxfordshire, England.

PH

natural draught. The rate of flow of air into a combustion chamber and of products of combustion up through the chimney caused by the difference in density between the hot gases in the chimney and the cold air outside. It is thus dependent on the temperature of the hot waste gases and the height of the chimney. The rate of flow may be reduced by interposing a moveable shutter, called a damper, between the combustion chamber and chimney.

WG

natural gas. A mixture of gaseous aliphatic hydrocarbons found underground in porous rock structures, from which it may be obtained by suitably placed boreholes. The major component is *methane (80–98% by volume) with small proportions of gaseous hydrocarbons with higher molecular weight and small amounts of nitrogen, carbon dioxide and helium. Hydrogen sulphide may also be present giving the gas an offensive smell; such natural gas is said to be sour.

When underground, the natural gas may be associated with coal seams or reserves of crude petroleum; it is said to be wet if it contains small amounts of condensable hydrocarbons when brought to the surface. Such materials when separated from the 'dry' gas are known as *natural gas liquids. The gas may be transported as such by pipeline or it may be liquified by pressure and cooling and transported in specially designed insulated tankers.

When burned as a commercial fuel, natural gas has a lower *flame speed and larger *flame size than *coal gas, but a *calorific value more than twice as great. However, per unit volume, it requires more than twice the air for complete combustion as coal gas. The gross calorific value of natural gas will vary with its composition but is normally in the range 47–57 MJ/kg.

WG

natural gas liquids. Higher molecular weight hydrocarbons in *natural gas which are liquid at moderate conditions of temperature and pressure. They include propane, butane and pentane. Hydrocarbons in natural gas with an even higher molecular weight are liquid at atmospheric pressure and ambient temperature and are termed natural gasoline. *See also* casing head gas.

WG

natural gasoline. *See* natural gas liquids.

natural radioactivity. *Radioactivity

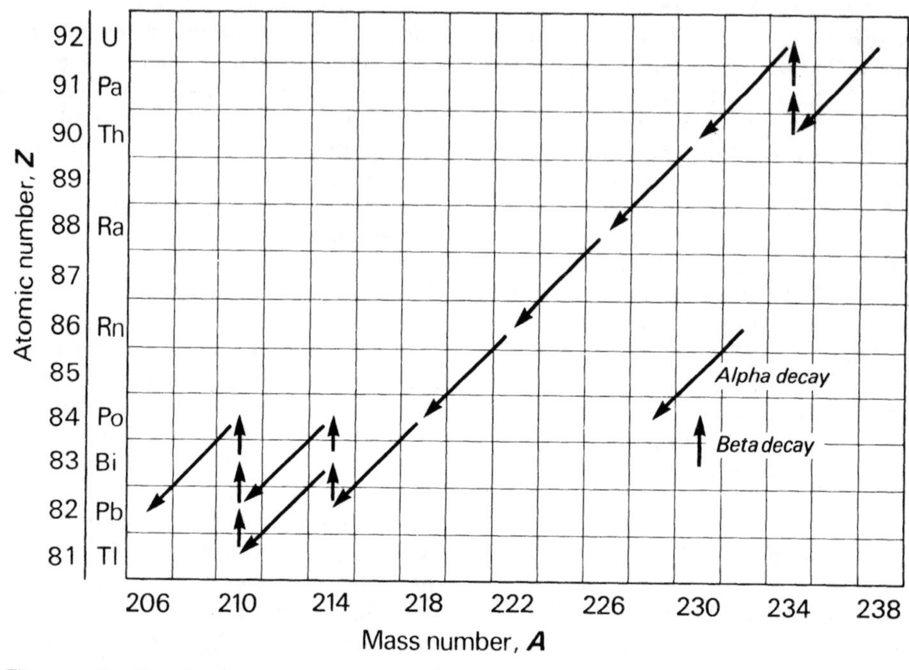

The natural radioactive decay chain of uranium-238

which emanates from naturally occurring radioactive substances and their *daughter products. The parents of these natural *radioactive decay chains have very long *half-lives. For example, the half-lives of uranium-238, uranium-235 and thorium-232 are respectively 4.5×10^9 years, 7.1×10^8 years and 1.4×10^{10} years. The decay chains of these three naturally occurring radioactive nuclides contain many daughter products and lead eventually to stable isotopes of lead.

DB

natural rate of unemployment. The level of unemployment at which the labour market is in *equilibrium. It does not correspond to zero unemployment since there will always be some workers changing jobs, and there will always be some unemployment associated with the changing structure of the economy. It is the level below which *fiscal and *monetary policy cannot be used to push unemployment except at the cost of *inflation. The concept is an essential feature of the *monetarist approach to the problems of *macroeconomics, and is somewhat controversial.

MC

natural resources. The material phenomena of nature available for use as inputs to economic activity, either by direct consumption (e.g. wilderness recreation) or by use in production (e.g. oil deposits, forests, fish stocks), or by serving as the receptacle for wastes arising in the activities of consumption and production (e.g. watercourses receiving domestic and industrial effluent). The principal distinction in economics is between renewable resources, which reproduce biologically so that perpetual use is possible (fish stocks, timber stands), and non-renewable resources, which do not so reproduce so that use must imply eventual exhaustion (oil deposits, mineral deposits generally). Renewable resources may be used in such a way that exhaustion, or extinction, does occur: this is likely to happen where a renewable resource is a *common property resource.

The question of the size of the *reserves of a non-renewable resource gives rise to some confusion: the commercially exploitable reserve is not constant over time. While use acts to reduce such reserves, it is typically associated with either actually rising prices, or the expectation of such. This works to increase the size of such reserves by making formerly non-commercial deposits commercially viable, and by encouraging exploration for new deposits. These effects are especially apparent in the oil industry, for example. While currently in advanced economies energy use is based principally on non-renewable resources (oil, gas, coal, uranium), it is argued that rising energy prices will make energy-delivery systems based on renewable resources (wind, wave, solar energy) commercially viable. *See also* land; materials balance principle; maximum sustainable yield.

MC

natural resource taxation. Where a natural resource, such as an oilfield, is not a *common property resource, ownership can give rise to large economic *rents accruing to the owner. Historically governments have been concerned to tax such rents, for reasons of equity as well as for the purpose of raising revenue. In principle it is possible to tax rent in such a way as to leave the owner's *depletion programme unaffected by the tax. In practice it is difficult to devise such tax systems. For example, a tax levied on the value of an oilfield will speed up the rate at which it is depleted, while a system of capital gains taxation would tend to slow down depletion. This does not mean that neither type of taxation is ever appropriate, as in certain circumstances a government may wish to influence the depletion rate in a particular direction. Thus in setting up the taxation arrangements affecting North Sea oilfields, the UK government had as one of its objectives 'the attainment as early as practicable of net self-sufficiency in oil' by the UK for *balance of payments and national security reasons. It is argued that in pursuit of this objective, the taxation system adopted is not entirely successful in effectively taxing the rent accruing to the oil companies.

MC

natural uranium. *See* uranium.

negative feedback. A stabilizing interaction between cause and effect. In the design of safe systems it is necessary to seek such a relationship. For example, the design of a fast fission reactor (*see* nuclear reactor) can be such that in the event of a rapid rise in temperature of the core, the neutron flux decreases; this tends to diminish temperature and so acts as a stabilizing effect. Without negative feedback any energy system has a tendency to explode. *See also* feedback.

MS

negentropy [exergy]. That portion of the energy of a fuel which is able to do *work. The name derives from *entropy, which is a measure of the fuel energy unable to do work. *See also* available energy; free energy.

MS

neoclassical economics. Economics based on marginalist concepts which were developed in the late 19th century (*see* marginal). It was classical in the sense of being concerned about resource allocation but was new (neo) in the sense that the analytical approach was *microeconomic and used the concepts of utility and profit maximization to derive *demand and *supply functions. These were then combined to determine prices.

MC

neptunium. A *transuranic element, some of whose *isotopes are produced in nuclear reactors. For example, neptunium-239 is formed as a result of neutron capture in uranium-238 and the subsequent decay of uranium-239 by *beta particle emission. Neptunium-239 decays by beta emission to form plutonium-239. Another isotope, neptunium-237, whose *half-life is 2.1×10^6 years, is an important isotope in the management of *irradiated nuclear fuel.

Symbol: Np; atomic number: 93.

DB

net. An adjective meaning that some offsetting effect has been allowed for. In economics the most common use is to indicate measurement after allowing for *depreciation, as in net *national income or net *investment. With respect to *indirect taxation, the price net of tax is the price before the addition of the tax. Net *present value is the present value of the stream of *benefit (or revenue in private sector *project appraisal) less *cost. Net *profit is profit remaining after meeting depreciation charges and tax liability. *See* net energy.

MC

net calorific value. *See* calorific value.

net energy. The amount of energy available from a system after the deduction of the energy expended in creating the necessary capital equipment for exploitation and refining and the energy required for operating the system itself. It is an important concept since it is the net energy that is actually available to do work within the economy. *See also* energy requirement for energy.

MS

net photosynthesis. *See* dark respiration.

net radiation. The resultant energy radiation transfer when two surfaces radiate back and forth to each other.

JWT

neutrality of taxation. A property of a tax instrument such that its imposition leaves the previously existing allocation of resources and pattern of economic activity unaffected. Neutrality of taxation would be attractive to a government which considered the existing allocation of resources satisfactory, but which needed to raise revenue. Tax instruments possessing the property and being politically and institutionally feasible are hard to devise. Thus, while in principle it should be possible to devise taxes on the *rents arising in natural resource extraction which do not affect the *depletion programme, existing taxes on, for example, oilfield ownership and operation do appear to influence decisions about oil extraction rates.

MC

neutrino (Symbol: ν). A particle of zero mass and charge, emitted in radioactive decay by *beta particle and *positron emission. The neutrino carries some of the energy released in the process.

DB

neutron (Symbol: ${}^1_0n$, n). One of the constituent particles of the atomic *nucleus, possessing zero charge and a mass very slightly more than that of the *proton, i.e. approximately one *atomic mass unit. When bound in atomic nuclei, neutrons are stable. Free neutrons however are unstable and decay with a *half-life of 12.8 minutes to form a proton, an *electron and a *neutrino:

$${}^1_0n \rightarrow {}^1_1p + {}^{\;0}_{-1}e + \nu$$

This is equivalent to *beta particle emission. The radioactive decay of neutrons is of little importance as free neutrons almost inevitably undergo some other process, such as *absorption, within a fraction of a second of being produced.

Neutrons can be produced by nuclear reactions, of which one example is the interaction between *alpha particles and light nuclides such as *beryllium-9:

$$^{4}_{2}He + ^{9}_{4}Be \rightarrow ^{1}_{0}n + ^{12}_{6}C$$

Thus beryllium mixed with a long half-life alpha emitter, such as radium-226 or plutonium-239, acts as a source of neutrons. It was this type of reaction that led to the discovery of the neutron in 1932.

A key characteristic of neutrons is that, being uncharged and so undergoing no Coulomb force of repulsion as they approach atomic nuclei, they can interact readily with a nucleus, either undergoing *scattering or absorption. The essential features of these reactions are that in scattering the neutron 'bounces off' the nucleus with reduced energy, and in absorption the neutron is absorbed into the nucleus and a *compound nucleus is formed at an excited state. The compound nucleus in turn decays instantaneously and may emit *gamma radiation, or a particle such as a neutron, proton or alpha particle. In the case of the heaviest elements, the excited compound nucleus may split, i.e. undergo *fission, and produce more neutrons. This in turn could lead to a *chain reaction.

DB

newton (Symbol: N). The SI unit for *force. A newton is that force which, when applied to a mass of 1 kg, gives it an acceleration of one metre per second per second (1 m/s^2).

TM

niobium. A metallic element which has a low neutron capture *cross-section and retains its strength at high temperatures. It was used as the *cladding material for the fuel in the Dounreay Fast Reactor.

Symbol: Nb; atomic number: 41; atomic weight: 92.9; density: 8570 kg/m$_3$.

DB

nitrogen. A colourless odourless non-flammable gaseous element at normal temperatures and pressures. It forms about 79% by volume of air, from which it may be separated by cooling to form a liquid and then distilling in a *fractionating column.

Although it has a very low chemical reactivity, nitrogen is an essential structural element in the chemical compounds involved in living processes. The gas, being non-combustible, may be used as a protective barrier to prevent fire and explosion in tanks containing flammable vapours. Atmospheric nitrogen is employed in chemical processes making *ammonia and nitrates. The latter, together with other organic nitrogen compounds, are used in the manufacture of explosives.

*Isotopes of nitrogen are produced in a water-cooled *nuclear reactor when the naturally occurring isotopes of the oxygen in the *coolant interact with neutrons:

$$^{16}_{8}O + ^{1}_{0}n \rightarrow ^{16}_{7}N + ^{1}_{1}H$$

$$^{17}_{8}O + ^{1}_{0}n \rightarrow ^{17}_{7}N + ^{1}_{1}H$$

These isotopes of nitrogen are both radioactive, with short *half-lives, and the radioactivity may be carried over from the core of the reactor to the heat exchangers and, in the case of direct-cycle boiling-water reactors, the turbines.

Nitrogen cannot be used as the coolant for a nuclear reactor because neutron capture in nitrogen-14, which is the predominant naturally occurring isotope, produces radioactive carbon-14.

Symbol: N; atomic number: 7; atomic weight: 14.001.

WG, DB

nitrogenase. An *enzyme present in a small number of bacterial and *blue-green algal species and capable of *nitrogen fixation. It mediates the initial reaction:

$$\underset{\text{nitrogen}}{N_2} + \underset{\text{hydrogen}}{3H_2} \rightleftharpoons \underset{\text{ammonia}}{2NH_3}$$

The enzyme consists of two proteins containing both iron and sulphur and acquires its reducing properties from other electron-carrying proteins called ferredoxins. The power for reduction is applied by the energy-carrying molecules of *adenosine triphosphate (ATP), which drive the whole process. In the absence of nitrogen, nitrogenase can decompose water by reducing protons (H^+) to hydrogen gas in the process of *biophotolysis. Biophotolysis also probably occurs during biological nitrogen fixation. Nitrogenase is irreversibly destroyed by oxygen, and so nitrogen fixation requires an *anaerobic environment.

CL

nitrogen fixation. The process in which atmospheric nigrogen is turned into a chemical form usable by plants. Nigrogen fixation can also be achieved by various chemical processes, such as the *Haber process. Biologically, a small number of bacterial species and *blue–green algae contain the enzyme *nitrogenase, which catalytically converts nitrogen gas to ammonia. The ammonia in turn may be converted into forms of nitrogen which can be assimilated by plants. The use of organisms capable of biological nitrogen fixation in agriculture reduces the requirement for the application of energy-intensive chemical fertilizers.

CL

nitrogen oxides. Compounds of nitrogen and oxygen which are formed in small amounts by the oxidation of atmospheric nitrogen at high temperatures. They occur in the gaseous products from most combustion processes, e.g. the internal combustion engine, welding operations, flame cutting and high-temperature furnaces. The mixture of oxides, which is given the general formula NO_x, is injurious to health. Oxides of nitrogen are involved in the production of an irritant smog from automobile exhaust emissions in such cities as Los Angeles and Tokyo.

WG

nominal. Denoting or relating to measurement in current money prices, or *absolute prices, as opposed to measurement using a *constant price series. Thus, if the general price level increases by 10% and the price of oil by 15%, the nominal increase in the price of oil is 15%; the real increase in the price of oil is just 5%.

MC

non-caking coal [non-coking coal]. A coal which when carbonized does not fuse to form a coke but leaves only a non-coherent powdery residue after the volatile matter has distilled off.

WG

non-commercial energy. Energy sources, usually *renewable energy sources like firewood, which are used locally and not traded, or are not traded through a recognized market where the amounts traded are recorded for statistical purposes.

MS

non-renewable. Not replenished or not capable of being replenished. Non-renewable energy includes coal, oil, uranium and natural gas. In some cases such energies are being renewed but over time scales so large, e.g. five million years, that they may be considered effectively non-renewable. Nuclear energy supplies are strictly non-renewable, although, supplies may extend for so long, e.g. water used in nuclear *fusion reactors, that they are sometimes considered virtually infinite. *See also* renewable energy.

JWT,MS

North of Scotland Hydro-electric Board. A UK government-appointed board responsible for generation and distribution of electricity in the Highlands and Islands area of Scotland. It had its origin in the extensive development of *hydropower in the 1950s. Some 90% of the Board's generation comes from *renewable energy.

Address: Rothesay Terrace, Edinburgh 9.

MS

Nostoc. A *blue–green alga capable of *nitrogen fixation.

CL

NTP. Normal temperature and pressure; more precisely 1 atmosphere pressure (=1.0133 *bar) and O°C.

nuclear energy. The energy which has its origin in the binding forces in the atomic *nucleus, and is released by *nuclear reactions. Nuclear energy can be distinguished from chemical energy as the latter has its origin in chemical reactions, in which the atomic nucleus remains unchanged. The immense amount of energy that can be released by a nuclear reaction is due to the very strong short-range forces which bind together the *protons and *neutrons in an atomic nucleus. The *binding energy of a nucleus is a measure of the energy required to break it into its constituent particles, or alternatively the energy released when it is formed from its constituent particles. These processes involve the conversion of mass to energy in accordance with *Einstein's equation, $E = mc^2$.

The nuclear reactions which are of most importance are energy-releasing (exothermic) reactions in which some mass is destroyed and the binding energy of the products of the reaction is greater than the binding energy of the original constituents.

The binding energy curve shows two types of nuclear reaction which proceed in the direction of increased binding energy and are therefore exothermic: the *fusion of very light elements to form heavier ones, and the *fission of the heaviest of elements to form others of intermediate mass. These two processes are sources of very intense energy, and are basic to the operation of *nuclear reactors and *fusion reactors. In both

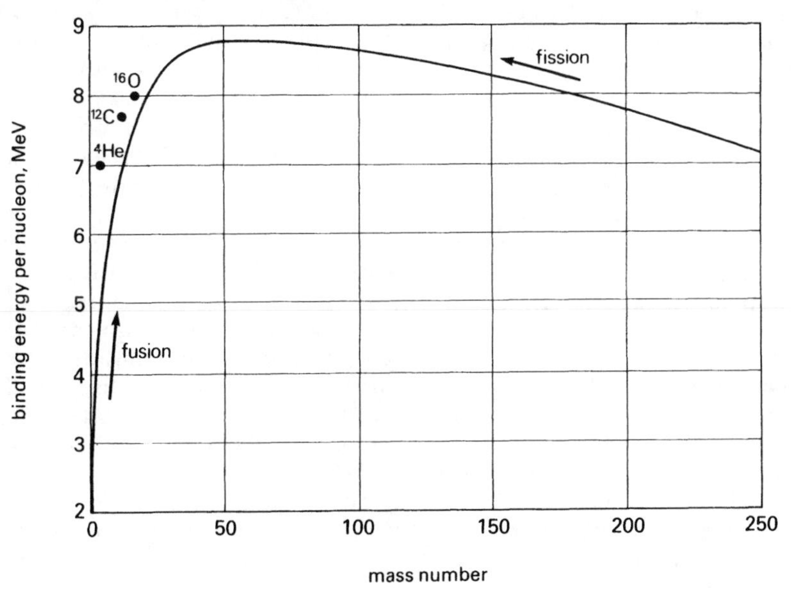

Curve of binding energy per nucleon versus mass number

fission and fusion, the energy release is of the order of one to ten million times greater than in chemical reactions, as the following figures indicate:

> The fission of 1 kg of uranium-235 releases 8.2×10^{10} kJ.
> The fusion of 1 kg of deuterium to produce helium-4 releases 5.7×10^{11} kJ.
> The combustion of 1 kg of oil releases 4.5×10^{4} kJ.

Another exothermic nuclear reaction which occurs spontaneously is radioactive decay (*see* radioactivity). In this process an unstable *nuclide changes to form a stable one. Since these processes are *exothermic, they can be used as sources of energy. For example, plutonium-238 is used as the energy source in a heart pacemaker, but the amount of energy released is very small.

DB

Nuclear Energy Agency (NEA). An agency set up in February 1958 by the Council of the Organization for European Economic Cooperation, later to become the *Organization for Economic Cooperation and Development (OECD). Its main purpose is to promote international cooperation within the OECD area for the development and application of nuclear power for peaceful purposes, through international research and development projects and exchange of scientific and technical experience and information.

Address: 38 Boulevard Suchet, 75016 Paris.

PH

nuclear energy levels. All atomic nuclei have energy levels above their ground (or zero) state of energy which correspond to the displacement of one or more *nucleons within the nucleus. When a nucleus is excited, as would be the case, for example, when a neutron is absorbed to form a *compound nucleus, it exists momentarily at an excited energy level above its ground state. As the excited nucleus decays to its ground state of energy, it emits *gamma radiation or a particle such as a neutron.

DB

nuclear fuel. *See* fuel.

nuclear fuel cycle. The sequence of processes which nuclear fuel undergoes, from the *uranium mine to fuel fabrication plant, *nuclear reactor and finally to the *reprocessing plant. The cycle also includes the transportation of the fuel between each process and the disposal of the *radioactive waste produced by each process.

PH

Nuclear Installations Inspectorate (NII). A branch of the Health and Safety Executive, responsible for the safety assessment and inspection of nuclear power plant in the UK. Nuclear power plant in the UK cannot be built, commissioned, operated or returned to service after overhaul without an NII licence, and is always open to inspection.

DB

Nuclear Non-Proliferation Treaty (NPT). A treaty which came into force on 5 March 1970 after it had been ratified by the three depositary states – the US, USSR and UK and 40 other states. More than 90 states have now ratified the treaty. Non-signatories include the two nuclear-weapon states, China and France, and other states such as India, South Africa, and Israel which are believed to have nuclear-weapons capabilities. The major objective of the treaty was to ban the acquisition of nuclear weapons by all non-nuclear states, and to prohibit the dissemination of such weapons by the nuclear states, which in 1970 were the US, USSR, UK, China and France.

The major provisions of the treaty which must be followed by all ratifying states can be summarized as follows.

Article I prohibits the transfer of nuclear weapons to any state, and forbids nuclear states to assist non-nuclear states to acquire nuclear weapons or devices.

Article II prohibits non-nuclear states to manufacture or otherwise acquire nuclear weapons.

Article III puts an obligation on non-nuclear states to accept nuclear safeguards, as laid down in a special agreement with the *International Atomic Energy Agency, on their peaceful nuclear activities in order to ensure that there is no diversion to the manufacture of nuclear explosives.

Article IV affirms that all states ratifying the treaty have the right to undertake research, production and exploitation of nuclear power for peaceful purposes, and that all states in a position to do so have an obligation to assist other less-developed countries.

Article V obligates nuclear states to make available nuclear explosives for peaceful purposes to non-nuclear states under international observation and procedures.

PH

nuclear park. A large nuclear site comprising several reactors together with their associated fuel fabrication and reprocessing plants. *See also* energy park.

PH

nuclear reaction. Any reaction involving the *nucleus of an atom. In *nuclear reactors the most important reactions are those caused by neutrons and include *scattering, capture (*see* absorption) and *fission. In *fusion reactors the important reactions involve deuterium, tritium and lithium.

DB

nuclear reactor. A system in which a *fission *chain reaction takes place under controlled conditions, giving a continuous output of energy. Several different and not mutually exclusive classifications of nuclear reactor exist.

Thermal reactor: one in which the fission chain reaction is maintained by low-energy or thermal neutrons (in the energy range from zero up to about 0.1 electron-volt [eV]).

Fast reactor: one in which the fission chain reaction is maintained by high-energy or fast neutrons (in the energy range from 10 keV to 2 MeV).

Power reactor: one whose primary function is to provide the energy source for the generation of electricity.

Breeder reactor: one whose primary function is to produce more *fissile fuel than it consumes; most breeder reactors

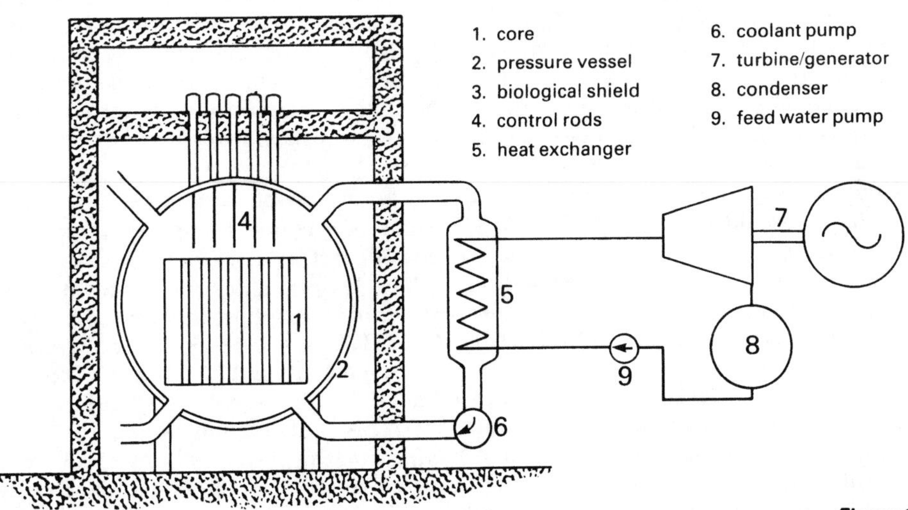

Layout of a typical nuclear power reactor and power plant

Figure 1

also operate as power reactors.

Research reactor: one whose primary function is to provide a source of neutrons for irradiation and research purposes in laboratories and universities; energy production is of secondary importance, and some research reactors operate at very low powers of only a few kilowatts.

The layout of a typical power reactor and power plant is shown in Fig. 1. The essential components of a reactor are the *fuel in which fission takes place, and (in the case of thermal reactors) the *moderator in which fission neutrons are slowed down to the low (thermal) energy at which they cause further fission. These two components make up the *core (numbered 1 in the diagram) of a thermal reactor, the fuel being distributed throughout the core in the form of a large number of fuel elements or assemblies in a regular array in the moderator. The *coolant, which is required for all reactors except very low power research reactors, is circulated through the core and transports the energy released by fission to *heat exchangers (5) in which (in power reactors) steam is generated for a conventional thermodynamic power cycle (7–9). The core is usually enclosed in a *pressure vessel (2) whose function is to allow the coolant to be circulated under high pressure to improve the thermal efficiency of the plant, and reduce the power required to drive the coolant pump (7). The pressure vessel, and in some reactors the heat exchangers also, are surrounded by a *biological shield (3), a concrete structure designed to prevent the escape of radiation from the core. In some types of reactor a prestressed concrete *containment vessel serves both as a biological shield and pressure vessel.

Fast reactors differ from thermal reactors in that the core, having no moderator, is a compact assembly of fuel and coolant. The fuel must contain a fairly high fraction, greater than about a quarter, of fissile material, and the coolant should not act as a moderator.

The fuel is enclosed in *cladding, usually metal tubes, whose purpose is to prevent the escape of the radioactive fission products from the fuel into the coolant, and provide structural support for the fuel and extended surfaces for heat transfer. The reactor is controlled by materials of high neutron capture *cross-section in the form of control rods (4) which can be driven into and out of the core.

Many different designs of nuclear reactor have been evolved since the first one was built in the USA in 1942. The vast majority of existing reactors are thermal, the principal reason for this being that whereas fast reactors require fuel with 25% or more of fissile material, thermal reactors can achieve *criticality with natural or slightly enriched *uranium (2–3% of uranium-235). Thermal reactors are also considered to be safer and easier to control than fast reactors. One disadvantage of thermal reactors is that existing types, fuelled with natural or enriched uranium, are not capable of acting as breeder reactors. On the other hand fast reactors can be designed to operate as breeders, hence the interest in fast breeder reactors for extending the life of uranium resources.

Nuclear reactors may also be classified according to the materials used for their main components. The table overleaf lists the most important materials in use and their principal applications.

The following descriptions refer to some of the most important types of power reactors in use at present:

The *Magnox reactor*, also known as the gas-cooled graphite-moderated reactor (GCGR), is a type of thermal reactor developed in the UK from the original Calder Hall power reactors built by the UKAEA and commissioned in 1956. Nine nuclear power stations of this type have been built in the UK for electricity generation, with two more overseas as

Fuel	*Application*
Natural uranium metal	British Magnox reactors
	French gas-cooled reactors
Natural uranium dioxide (UO_2)	CANDU reactors
Enriched uranium dioxide (UO_2)	Advanced gas-cooled reactors
	Pressurized-water reactors
	Boiling-water reactors
	Steam-generating heavy-water-moderated reactor
Enriched uranium carbide (UC_2)	High-temperature gas-cooled reactors
Plutonium oxide (PuO_2)	Fast breeder reactors
Moderator	
Graphite	Gas-cooled reactors
Water	Pressurized-water reactors
	Boiling-water reactors
Heavy water	CANDU reactors
	Steam-generating heavy-water-moderated reactor
Coolant	
Carbon dioxide	British Magnox reactors
	French gas-cooled reactors
	Advanced gas-cooled reactors
Helium	High-temperature gas-cooled reactors
Water	Pressurized-water reactors
	Boiling-water reactors
	Steam-generating heavy-water-moderated reactor
Heavy water	CANDU reactors
Liquid sodium	Fast breeder reactors
Cladding	
Aluminium	Low-power research reactors
Magnox (magnesium alloy)	British Magnox reactors
Magnesium-zirconium alloy	French gas-cooled reactors
Zircaloy (Zirconium alloy)	Pressurized-water reactors
	Boiling-water reactors
	CANDU reactors
	Steam-generating heavy-water reactor
Stainless steel	Advanced gas-cooled reactors
	Fast breeder reactors

export orders. (Several gas-cooled graphite-moderated reactors of similar design have been built in France.) A typical layout, that of the Wylfa power station (the last of the UK series) is shown in Fig. 2.

The fuel for these reactors is natural uranium metal rods enclosed in finned *Magnox cladding tubes. The core (1) is a cylindrical graphite structure with vertical channels for the fuel. The coolant is carbon dioxide, passing upwards through the core and downwards through the heat exchangers (2) in which steam is generated for the thermodynamic cycle (7). The gas is pumped back to the base of the core by gas circulators (4). The core is enclosed in a steel pressure vessel (prestressed concrete (3) in the case of the last two power stations of the UK series). These reactors are designed for *on-load refuelling, and the *charge-discharge machine (6) is located above the core and can be con-

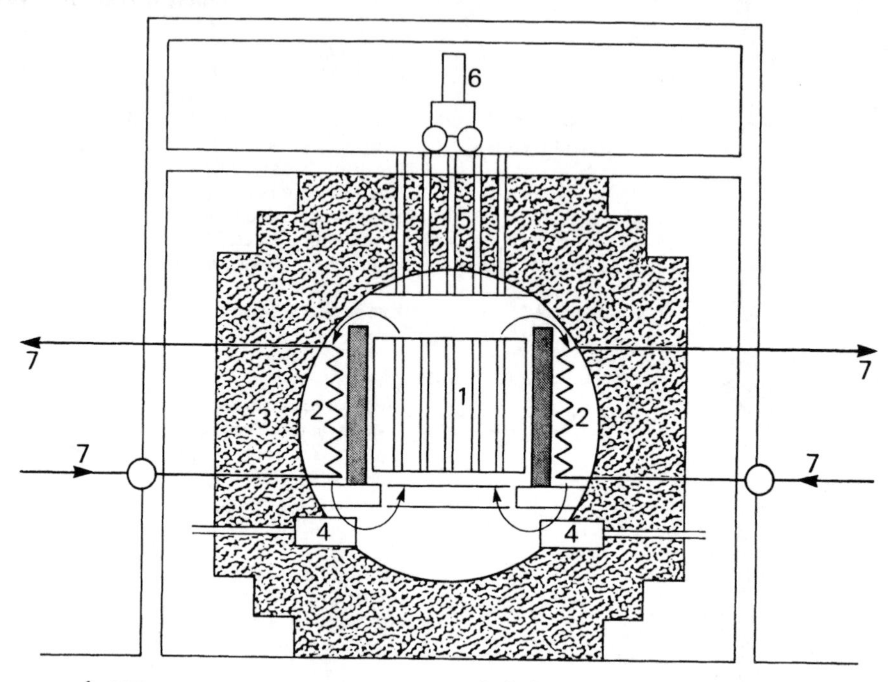

1. core
2. boilers
3. prestressed concrete pressure vessel
4. coolant gas circulators
5. fuel and control rod standpipes
6. charge and discharge machine
7. steam and feedwater pipes

Layout of gas-cooled graphite-moderated reactor (WYLFA) *Figure 2*

nected to it by standpipes (5) passing though the biological shield and pressure vessel. Fuel is loaded and unloaded through these pipes.

The following data refer to the Wylfa power station, which has two reactors:

Core size: 9.15 m high by 17.4 m diameter.
Fuel elements: 28 mm diameter natural uranium rods in finned Magnox cladding; each element 1.07 m long; 8 elements per channel; 6150 channels, each 98 mm diameter, in the core on a 222 mm square pitch; total mass of uranium per reactor: 595 t.
Coolant: CO_2 at 27.6 bar pressure; outlet temperature from reactors: 410 °C.
Pressure vessel: prestressed concrete.
Heat exchangers: once-through boilers producing steam at 395 °C, 48.3 bar pressure.
Thermal output per reactor: 1875 MW.
Electrical output per reactor: 590 MW.
Average rating: 3.2 MW/t of uranium.
Net thermal efficiency: 31%.

The *advanced gas-cooled reactor* (AGR) is the second generation of British gas-cooled reactors, developed directly from the first generation Magnox reactor. Figs. 3 and 4 show the fuel assembly plan and general layout of the AGR. The principal difference between the AGR and Magnox reactors lies in the fuel elements, which in the AGR are assemblies of slightly enriched uranium (2% uranium-235 as uranium dioxide in stainless steel tubes, 14.5 mm diameter (numbered 1 in diagram). Each fuel assembly (2) contains 36 of these fuel rods. The core (3) is graphite, approximately 10 m high by 11 m diameter, and the coolant is carbon dioxide. The core and the heat exchangers (5) are enclosed within a prestressed concrete pressure vessel (4) (which also acts as the biologi-

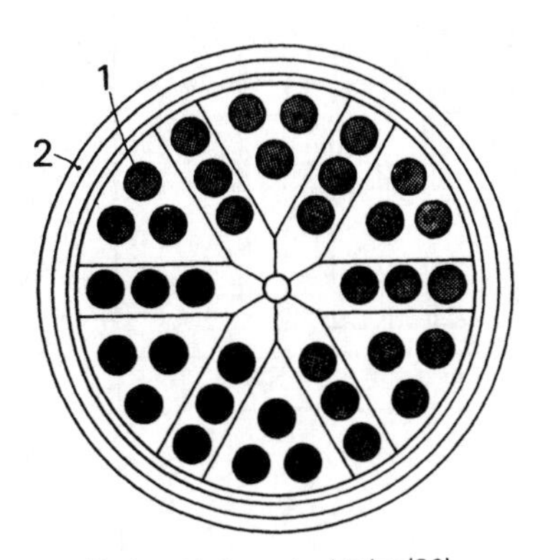

1. UO_2 in stainless steel tube (36)
2. graphite sleeve round fuel assembly

Fuel assembly plan of the AGR *Figure 3*

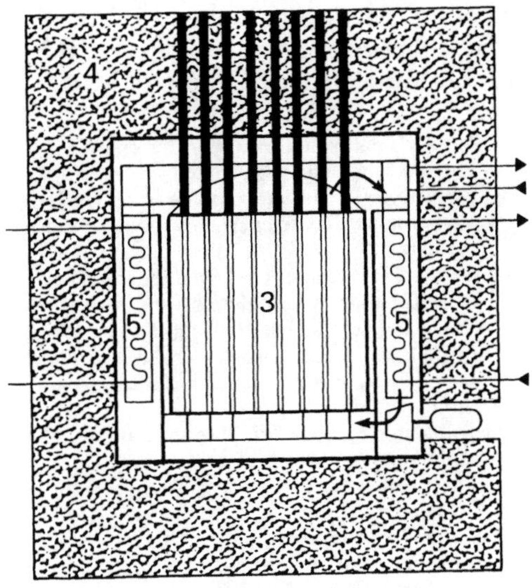

3. core
4. prestressed concrete pressure vessel
5. heat exchangers

Layout of the AGR *Figure 4*

cal shield); the heat exchangers are located in the annular space between the core and pressure vessel or (in the case of one design) in separate cylindrical holes in the pressure vessel. The reactor is designed to be refuelled on load, and the high melting point of the fuel and the stainless-steel cladding permit steam to be generated in the heat exchangers at the same temperature and pressure as in fossil-fuelled power stations.

The design parameters for the latest UK AGR at Torness, E. Lothian are typical of the second generation of this type of reactor (*see* table below).

The *high-temperature gas-cooled reactor* (HTGR) is a high-temperature version of the Magnox (gas-cooled graphite-moderated) reactor with a different type of fuel and with helium as the coolant. Only five have been built in the world. The HTGR is designed at present to operate on the uranium-thorium cycle, producing uranium-233 by *breeding. Initial fuel loadings have been of highly enriched uranium (93% of uranium-235) and thorium, both in the form of the carbide, UC_2 and ThC_2. Future fuel loadings are expected to include uranium-233, currently being

Thermal output (one reactor):	1549 MW
Mass of uranium (one reactor):	113.5 t
Average rating:	13.6 MW/t
Burnup:	18000 MWd/t
CO_2 temperature, inlet to reactor:	288°C
CO_2 temperature, outlet from reactor (mean):	616°C
CO_2 pressure (mean):	41 bar
Steam pressure & temperature at turbine stop value:	160 bar, 538°C
Electrical output gross (two reactors):	1320 MW
Electrical output net (two reactors):	1222 MW
Thermal efficiency:	39.4%

produced. The layout is shown in Fig. 5.

The following description refers to the Fort St. Vrain HTGR, Colorado, USA.

Fuel: very small spherical UC_2/ThC_2 particles, about 0.3 mm diameter, coated with layers of pyrocarbon and silicon carbide to retain fission products; coated particles about 1.0 mm diameter.
Core size: 6 m diameter by 4.8 m high; made up of 1482 hexagonal graphite prisms, each one 36 cm across flats, with holes drilled parallel to their axes; fuel particles are loaded and sealed into some holes, coolant is circulated through others.
Fuel loading: 19.48 t of thorium, 0.88 t of uranium.
Helium temperature at outlet from reactor: 770°C.
Maximum fuel temperature: 2300°C.
Thermal output of reactor: 842 MW.
Net electrical output: 330 MW.
Overall thermal efficiency: 39.2%.

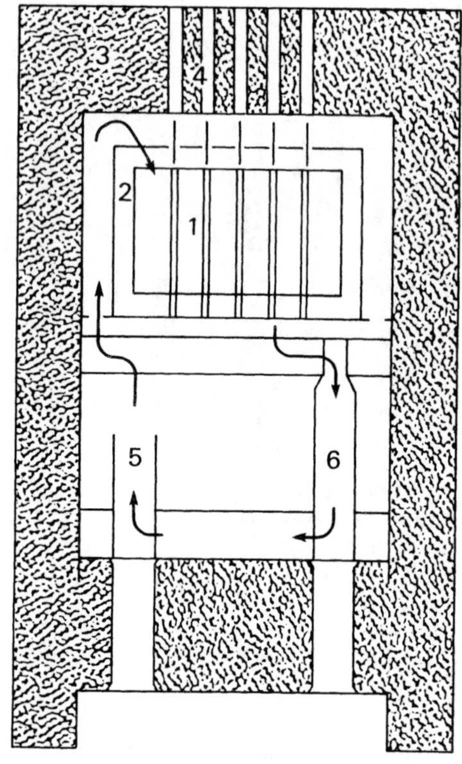

1. core
2. reflector
3. prestressed concrete pressure vessel
4. fuel and control rod standpipes
5. gas circulators
6. heat exchangers

Layout of high-temperature gas-cooled reactor
Figure 5

The very high coolant temperatures, which are made possible by the use of helium as the coolant and graphite as the main structural material, have led to proposals that the HTGR may in future be developed for direct-cycle gas turbines and high-temperature industrial processes such as thermochemical water splitting.

The *pebble-bed reactor* (AVR) is a type of high-temperature gas-cooled reactor developed in West Germany. The fuel is in the form of spherical pebbles approximately 38 mm diameter, each pebble containing the uranium carbide/thorium carbide (UC_2/ThC_2) fuel enclosed in graphite moderator. The reactor is cooled by helium, blown upwards through the reactor vessel, which is filled with the fuel pebbles. The reactor may be refuelled on load by removing irradiated fuel pebbles and replacing them by new ones. Gas-outlet temperatures in excess of 900°C have been achieved in this reactor, which makes it suitable for high-temperature industrial applications.

In the *pressurized-water reactor* (PWR), another type of thermal reactor, water acts as the moderator and coolant. The PWR was first developed in the USA as the power plant for submarines and other naval vessels. It has since been developed for electricity generation to such an extent that it is now the most common type of nuclear power reactor in the world, being used extensively in the USA, and to a lesser extent in Japan, France, West Germany and other European countries.

A principal feature of the PWR is that the water must be maintained at a very high pressure (typically 150 bar) to prevent boiling while at the same time al-

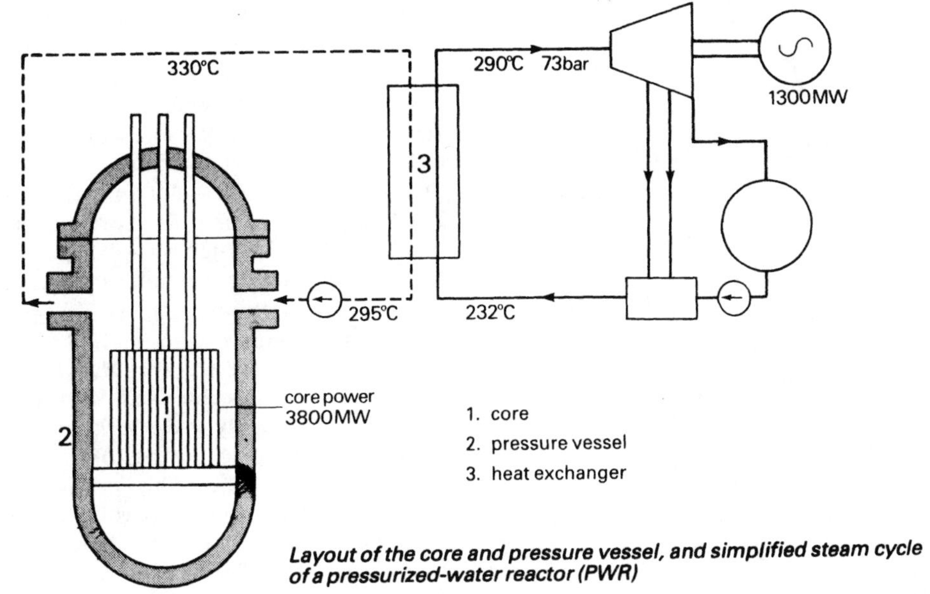

Layout of the core and pressure vessel, and simplified steam cycle of a pressurized-water reactor (PWR)

Figure 6

lowing a fairly high temperature. The coolant temperature (which must be well below the critical temperature of water, 374°C) is rather low for a thermodynamic power cycle, and leads to rather low thermal efficiency. The very high system pressure which is characteristic of the PWR makes it necessary to have a very strong pressure vessel and pipework; the possibility of rupture of these components remains one of the problematical safety features of this type of reactor.

After 30 years of progressive development in the USA, the PWR has reached a stage where most new reactors are very similar (*see* Fig. 6) and the following description and data are typical of present PWRs:

Fuel: uranium dioxide, UO_2, enriched to 2.5–3% of uranium-235 in form of pellets 8 mm diameter by 10 mm long, enclosed in 4 m long Zircaloy tubes of 10 mm outside diameter.
Zircaloy tubes closely packed in square arrays of 20 cm side, 240 fuel tubes per fuel assembly; these square assemblies arranged together to form core (1 in diagram).
Core size: approximately 3.8 m diameter by 3.7 m high, within the pressure vessel (2).
Pressure vessel size: 4.7 m inside diameter by 10 m high.
Mass of fuel in the core: about 100 t.
Water pressure in core and pressure vessel: about 150 bar.
Water temperatures at inlet and outlet: 295°C and 330°C respectively; in heat exchangers (3) steam is generated at 73 bar pressure with a moisture fraction of less than 0.025.
Thermal output of present standardized PWR: 3800 MW.
Electrical output: 1300 MW.
Thermal efficiency: 34%.

The use of water as a moderator and coolant makes it possible to construct a smaller core than is possible using other moderators. This leads to a more compact reactor with higher power density and lower capital costs than other thermal reactors, and offsets the cost of enriched uranium and the low thermal efficiency. Considerable attention must be paid to the integrity of the pressure vessel and pipework, and elaborate *emergency core cooling systems must be installed in the reactor to prevent overheating in the event of a *loss of coolant accident.

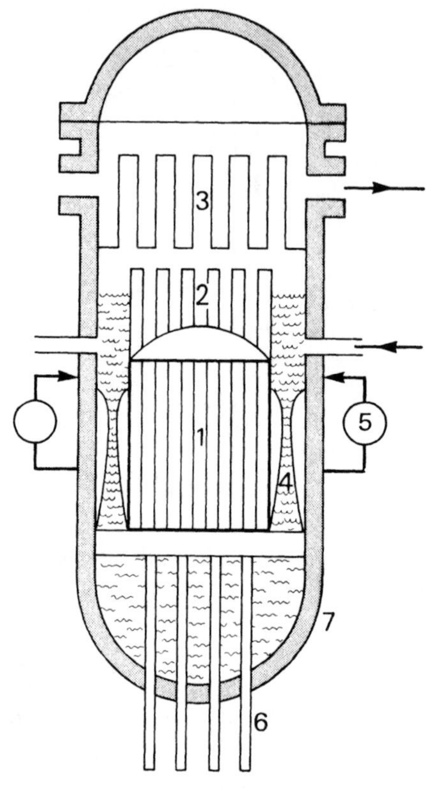

Layout of the core and pressure vessel of a boiling-water reactor *Figure 7*

The *boiling-water reactor* differs from the pressurized-water reactor in that fission energy causes the water coolant to boil in the core. The layout is shown in Fig. 7. The fuel is uranium dioxide, UO_2 (containing approximately 2% of uranium-235). The fuel is in the form of pellets, 1 cm diameter, which are loaded into Zircaloy tubes. A square array of these tubes, typically 8 × 8, makes a fuel bundle which is assembled in a 14 cm square Zircaloy tube. The core (1) is composed of a large number, typically 560, of these Zircaloy clad fuel assemblies; the dimensions of the core are approximately 4.3 m high by 4.4 m diameter. Water, acting as moderator and coolant, circulates through the core at high pressure and in the upper half of the core it boils. Above the core the steam-water mixture is separated (2). Saturated steam passes directly to the turbines and water is returned to the bottom of the core by jet pumps (4). The core is contained in a cylindrical *pressure vessel (7), about 8 m diameter by 22 m high. The space above the core is occupied by the steam separators and driers (3), and the control rods (6) are inserted at the bottom of the core. Provided water purity is maintained, the carry-over of *radioactivity to the turbines is limited to the very short *half-life nitrogen-17 isotope. The parameters shown in the table (*below*) are typical of boiling-water reactors.

CANDU is the acronym given to the Canadian design of natural-uranium fuelled, heavy-water (D_2O) cooled and moderated reactor. This type, developed from the early Chalk River, Ontario, reactors, has become the standard Canadian power reactor for electricity generation. The layout is shown in Fig. 8.

The following description refers to the Gentilly 2 reactor, which is typical of present-day CANDU design. The core vessel is a cylindrical stainless steel *calandria 6 m long and 6.28 m diameter,

Thermal output (1 reactor):	3850 MW
Electrical output:	1200 MW
Mass of uranium:	170 t
Average rating:	21 MW/t
Burnup:	19000 MW d/t
Pressure and temperature at core outlet:	72 bar, 288 °C
Thermal efficiency:	33%

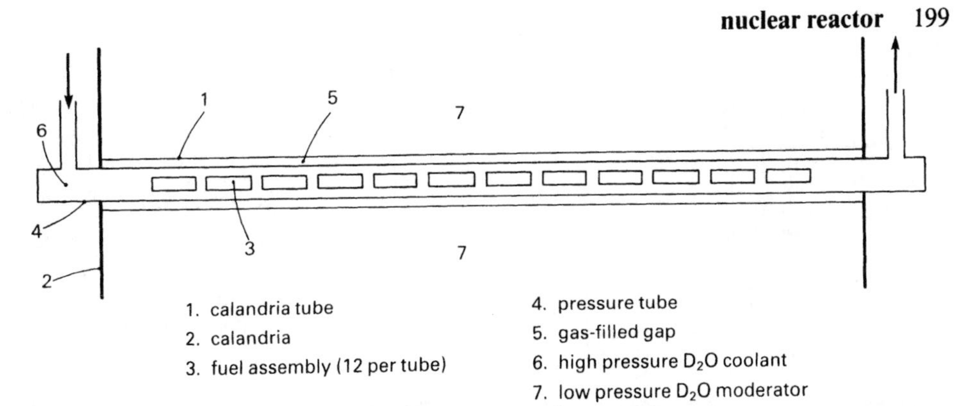

Arrangement of fuel assemblies, fuel tube and calandria tube in CANDU reactor ***Figure 8***

with its axis horizontal and 380 horizontal zirconium alloy calandria tubes (numbered 1 in diagram) passing through it. The heavy-water moderator is contained in the calandria (2) at low temperature and pressure. The fuel is natural uranium dioxide, UO_2, in the form of pellets 12 mm diameter. These are loaded into Zircaloy tubes, and bundles of 37 tubes form one fuel assembly (3). Twelve such assemblies, each about 0.5 m long, are loaded into a Zircaloy pressure tube (4) which is located inside a calandria tube. An annular gas-filled gap (5) between the pressure tube and the calandria tube provides thermal insulation between the high-pressure high-temperature D_2O coolant passing through the pressure tubes (6) and the low-pressure low-temperature D_2O moderator in the calandria (7).

An important characteristic of the CANDU reactor is that due to the very low capture cross-sections of heavy water and zirconium, the *conversion ratio of this reactor is higher than for any other type of thermal reactor. Consequently the fuel can be used to higher burnups, thus giving very good fuel economy.

Performance figures for this reactor are:

Thermal output: 2180 MW.
Net electrical output: 640 MW.
Efficiency: 29.4%.
Fuel burnup: 7500 MW d/t (U).

The *steam-generating heavy-water-moderated reactor* (SGHWR) has been built and operated at the United Kingdom Atomic Energy Authority's research establishment at Winfrith, UK. The layout is shown in Fig. 9. It uses heavy water as the moderator (2) contained in a calandria (1) with vertical pressure tubes (3). Inside these tubes the fuel elements consist of bundles of 36 Zircaloy fuel tubes, 1.5 cm in diameter, containing enriched uranium dioxide pellets (2.3% of uranium-235). The coolant, which passes upwards through the pressure tubes at 67 bar pressure, is ordinary water. It boils in the core and passes to a steam drum (4) where dry saturated steam is separated from the water and passes directly to the turbines (6) at 62 bar pressure. Performance figures are:

Thermal output: 300 MW.
Electrical output: 94 MW.
Thermal efficiency: 31%.

The SGHWR combines the advantages of the heavy-water moderator and pressure-tube design of the CANDU reactor with the direct cycle of the boiling-water reactor. Despite the success of the design and the excellent operating record of the SGHWR, no further reactors of this type have been built.

The *RBMK reactor* is a design which has been developed in the USSR, and reac-

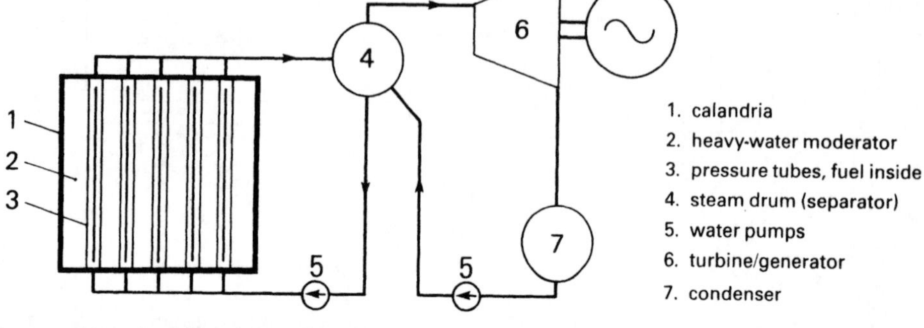

Layout of the calandria and simplified steam cycle of the SGHWR ***Figure 9***

tors of this type operate only in the USSR. The reactor is a graphite-moderated, pressure tube design in which the coolant is water, flowing in 1693 zirconium-niobium pressure tubes, 88 mm diameter, through the core. The graphite core is 7 m high by 11.8 m diameter. The fuel is in the form of assemblies of 18 fuel pins, each containing UO_2 pellets (enriched to 1.8% of uranium-235) in zirconium–niobium cladding 13.6 mm in diameter. The cooling water in it flow upwards through the pressure tubes, boils and passes to separators from which saturated steam at 284°C, 70 bar pressure passes to the turbines. The thermal output of the reactor is 3140 MW, and the electrical output is 1000 MW. (These data refer to the RBMK-1000 reactor).

The *liquid metal cooled fast breeder reactor* (LMFBR) is the most common type of fast reactor in use at present. Liquid sodium is chosen as the coolant because of its non-moderating characteristic and excellent heat-transfer properties, which are necessary for the coolant in a highly rated reactor. One problem associated with the use of liquid sodium as a reactor coolant is that during its passage through the core it captures neutrons to form radioactive sodium-24:

$$^{23}_{11}\text{Na} + ^{1}_{0}\text{n} \rightarrow ^{24}_{11}\text{Na} + \gamma$$

This radioactive sodium must not pass to the heat exchangers in which steam is raised. It is therefore necessary to have an intermediate (secondary) sodium circuit which is shielded from the core by a neutron shield and does not become

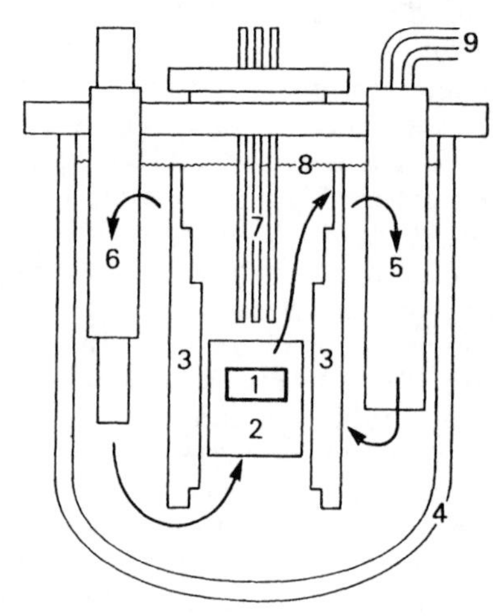

Arrangement of pressure vessel, with core, pumps and heat exchangers

1. core
2. blanket
3. neutron shield
4. stainless steel vessel
5. primary heat exchanger
6. sodium pump
7. control rods
8. sodium level in vessel
9. secondary sodium to steam generators

Prototype fast reactor: Dounreay ***Figure 10***

radioactive. About five large fast breeder power reactors have been built in the world: in the UK, France and

USSR.

The following description applies to the Prototype Fast Reactor (PFR) built at Dounreay, Scotland, the layout of which is shown in Fig. 10:

Fuel: mixture of plutonium oxide and natural uranium oxide (containing 22–30% of plutonium-239); 6 mm diameter fuel pellets are clad in stainless steel.

Core size: 1 m high by 1.8 m diameter; core surrounded by the blanket which contains uranium oxide elements for breeding; neutron shield surrounds core and blanket.

Liquid sodium primary coolant is contained in large stainless-steel vessel, which houses core, primary heat exchangers and sodium pumps; free surface of the sodium is covered by an inert gas atmosphere.

Primary sodium temperatures at inlet and outlet of core: 430°C and 595°C respectively.

Secondary sodium passes through primary heat exchanger, entering at 370°C and leaving at 590°C.

In steam generators, steam is raised at 160 bar pressure and 565°C, with reheat.

Thermal output: 600 MW.

Electrical output: 250 MW. DB

Nuclear Regulatory Commission (NRC). An independent agency of the US Federal Government. Its major role is to ensure that the use of nuclear power for commercial purposes does not adversely affect either the health and safety of the general public or the environment. It is responsible for nuclear reactors and other nuclear facilities, and assures safety in the possession, use, processing, transport, handling and disposal of nuclear materials. It was formed under the provisions of the Energy Reorganization Act of 1974, whereby all licensing and related functions formerly assigned to the Atomic Energy Commission were transferred to the NRC.

Address: 1717 H St. N.W., Washington, D.C. 20555 USA. PH

nuclear steam supply system. A term of US origin used to denote that part of a nuclear power plant, namely the reactor and its heat exchangers, which generates and supplies steam for the conventional turbine-generator part of the power plant. DB

nucleon. A constituent particle of the *nucleus of an atom. Nucleons are thus either protons or neutrons. DB

nucleus. The central part of an *atom, composed of *protons and *neutrons. The nucleus is positively charged due to the positive charges of the protons and constitutes almost all the mass of the atom. The protons and neutrons of stable nuclei are tightly bound together by the short-range nuclear force, known as the strong interaction. DB

nuclide. The term used to identify atomic nuclei with specific numbers of *protons and *neutrons. Nuclides with the same number of protons, *Z*, but different numbers of neutrons, are all *isotopes of the element of *atomic number *Z*. The total number of protons and neutrons in the atom gives the *mass number of the element. A nuclide is identified by the chemical symbol of the element, with mass number as superscript and atomic number as subscript. For example, ${}^{238}_{92}U$ denotes the isotope of uranium which has an atomic number of 92 and a mass number of 238. DB

numeraire. A unit of account for expressing stocks, flows, state variables, rates, etc. In economics the numeraire is normally in money units. Since the value of money is not constant and changes unpredictably, this can pose problems when forecasting. This can be partly overcome by the concept of *real prices. In science and engineering common numeraires are mass, volume, energy or other units. In *energy analysis energy is used as a numeraire and attempts have been made to express *macroeconomic *models in energy terms (e.g., joules). MS

numerator. The multiplier or top line of a division calculation. For example, in the calculation a/b, a is the numerator. MS

nutrients. Chemicals necessary in small quantities for biological growth. For example, potassium is necessary as a trace element nutrient for plant growth. JWT

O

Oak Ridge National Laboratory (ORNL). An independent non-profit-making multidisciplinary research and development organization, owned by the US government and supported by the Department of Energy and other US governmental agencies. It is operated under contract by the Nuclear Division of Union Carbide Corporation for the US Department of Energy. Its principal fields of research are nuclear energy development, biomedical and environmental problems, basic energy sciences and magnetic fusion energy.

Address: Oak Ridge, Tennessee 37830, USA.

PH

ocean dumping. *See* radioactive waste management.

ocean farm project. *See* giant kelp.

ocean thermal energy conversion (OTEC). Energy obtained by virtue of a temperature difference between surface and subsurface layers of seawater. This temperature difference is used to power a volatile-liquid heat engine. Some tropical waters have temperature gradients from the surface of 10°C per 200 m, and may be suitable for generating power. The relatively small temperature difference (about 20°C) available for the generating machines allows only very low thermal *efficiency, as described by the *second law of thermodynamics. Nevertheless the potential for supply is extremely large in tropical waters, especially where hot and cold sea currents meet. Allowing for the thermal efficiency of the machines, a 20°C temperature differential with a pumped water flow of 1 m/s could produce about 2–3 MW per m^3 of water. However this does not allow for the energy requirements of the pumps and the transformation and transport of the energy to land in a convenient form, such as *hydrogen. Research and development programmes, especially in the US, are determining these factors.

JWT

ocean wave power. *See* wave power.

octane. A flammable aliphatic liquid hydrocarbon, C_8H_{18}, obtained from crude petroleum and by chemical reaction in a petroleum refinery. There are in fact 18 *isomers with this molecular formula and they form a major proportion of the mixture of hydrocarbons in gasoline. Iso-octane is used as a reference fuel in the *octane rating test. The gross calorific value of octane is 48.35MJ/kg.

WG

octane number. *See* octane rating.

octane rating. When a mixture of petrol (gasoline) vapours and air is compressed by the piston in the hot cylinder of an *internal combustion engine, the hydrocarbons may be so easily oxidized that the mixture ignites before the spark is passed by the spark plug. This premature ignition is effectively a detonation which produces shock waves, causing the engine to overheat and lose power. The phenomenon is heard as a metallic tapping, known as knock, when the engine is under load.

When the knocking tendency of a petrol is measured on a standard engine, it may be compared with that of standard pure hydrocarbons. Iso-octane has been found to have an excellent antiknock rating (arbitrarily set at 100) and n-heptane a very poor value (set at 0). These so-called reference fuels in mixture are compared with a petrol for an octane number evaluation. The octane number is then the volume percentage of iso-octane in a mixture with n-heptane that has the same knocking tendency as the petrol under test.

Various grades of petrol are sold with octane numbers in the range 92–100. Numbers in excess of 100 may be obtained by the use of the lead alkyl additives, *tetramethyl lead and *tetraethyl lead.

WG

off-peak. Denoting or relating to those periods of time which alternate with periods of high demand for energy in one form or another. In the interests of efficiency and economy, an even distribution of load may be desirable and this has led to financial incentives to energy consumers to depart from the normal diurnal or seasonal pattern of demand in favour of off-peak usage.

TM

off-shore. Not on land, yet not in deep water.

MS

ohm (Symbol: Ω). The SI unit of electrical *resistance. One ohm is the resistance between two points of a conductor when a constant *potential difference of one *volt applied between these points produces in the conductor a current of one *ampere. This relationship between current and potential difference is known as *Ohm's law.

TM

Ohm's law. The relationship established by Georg Ohm, which in simple terms states that the electric *current flowing in a conductor is directly proportional to the *potential difference across it and inversely proportional to its *resistance.

TM

oil. (1) *See* crude oil.

(2) A popular term embracing many forms of *organic liquids which may not necessarily be pure hydrocarbons, for example castor oil. However, this word is most commonly applied to oil from the ground (*crude oil), *fuel oil, and lubricating oil.

MS

oil blowout. A sudden uncontrolled release of oil from a wellhead or from a geological fault close to a drilling operation.

PH

oil burner. *See* burner.

oil equivalent. *See* tonne of oil equivalent.

oil field. A natural underground reservoir of oil, often associated with gas. *See also* caprock; salt dome.

MS

oil seepage. The slow natural release of oil from rocks into the marine or terrestrial environment. The best-known occurrence of natural seepage is along the coast of southern California, especially on the Santa Barbara Channel; there are additional occurrences in the Caribbean Sea and on the coast of Dorset (UK) at Kimmeridge, where there is a well-known exposure of black oil shale. Estimates of the global input of oil to the oceans from natural seepage vary widely but may be of the order of $0.2-6.0 \times 10^6$ tonnes per year.

PH

oil shale. A fine-grained clay containing a high proportion of organic matter which was laid down during the Carboniferous period and has been compressed into a rock. The fossilized organic matter is a waxy substance known as *kerogen. When the rock is broken up and heated in a retort, hydrocarbons with a wide range of molecular weight may be distilled from it in the form of a crude oil. The yield of oil is normally about 10% of the weight of shale. Thus production of the oil on a large scale requires the mining of massive volumes of shale and the disposal of almost the same volume of spent shale.

WG

oil spill. The accidental release of oil into the marine or terrestrial environment. At sea, spills may occur as a result of the collision or grounding of oil tankers (*see Torrey Canyon, Amoco Cadiz*), or may be due to the blowout of underwater oil wells (*see* Ixtoc-1). The majority of spills occur during loading and unloading operations at oil ports.

PH

olefin. An unsaturated hydrocarbon in which one or more pairs of carbon atoms are joined by double valency bonds, as with *ethylene:

$$H_2C:CH_2$$

Owing to this double bond olefins are considerably more reactive than *paraffins and react readily with many chemical elements and other chemical compounds. Under suitable conditions of temperature and pressure they also form *polymers. Olefins with four or fewer carbon atoms in the molecule are gases; those with higher molecular weights are liquids.

WG

oligopoly. An industry in which there are only a few sellers of a homogeneous product and many buyers. Because of the small number of sellers, when firms make output and price decisions they must take account of the response they will induce from their rivals. A variety of oligopoly *models have been established depending on the type of reaction which is assumed, e.g. leader–follower behaviour, *collusion and *cartels. The oil industry is generally taken to be an oligopoly. *See also* market structure.

MC

once-through cooling. The standard method for disposing of waste heat from a power plant. Water is drawn from a lake, river or estuary, passes through the steam condenser where its temperature is raised, and then is returned to the body of water from which it was drawn.

PH

on-load refuelling. Of *nuclear reactors, the removal and replacement of the *fuel in the *core while the reactor is operating. This method of refuelling can be carried out in the following reactors: Magnox, advanced gas-cooled, CANDU and pebble-bed reactors.

DB

open cast mining [US: strip mining]. The winning of coal and lignite from depressions cut in the surface of the earth. Coal seams which are exposed on the surface may be worked by machinery similar to that used in open stone quarries. Much of the coal exposed in this way is of low rank, but has been used in steam raising and in gasification processes for the manufacture of town gas.

Open-cast mining methods usually result in serious disturbances to the surface environment, including erosion, subsidence, scarred landscapes, loss of topsoil, silting of local streams and *acid mine drainage. Extensive reclamation procedures are generally required.

WG, PH

open fuel cycle [throwaway cycle]. The type of nuclear *fuel cycle in which spent fuel is not reprocessed but held instead in prolonged storage. Canada currently bases her nuclear fuel cycle on this concept.

PH

opportunity cost. The cost of doing something, measured in terms of the loss of opportunity of doing the next best alternative activity with the same time or the same resources. Often opportunity cost is measured by the money paid, since such expenditure denies the opportunity to spend the money on something else. The money outlays on the resources input to production measure accurately the opportunity cost of the output only if there is no *market failure. Thus, for example, the opportunity cost of electricity from coal-fired power stations is not measured by the costs incurred in generation unless the electricity utility is either constrained to have sulphur and other emissions at the appropriate level or bears the costs arising from the pollution that excess emissions give rise to, by means of taxes on emissions. *See also* polluter pays principle.

MC

opportunity energy requirement. The energy foregone in order to provide another energy type. For example, a certain amount of fossil fuel energy may be used to construct a device to collect solar energy.

MS

optimal. Denoting a state of affairs which is the best available in the circumstances, according to some particular criterion. For example, for the owner of an oil well the optimal *depletion programme is that which maximizes the *present value of the profit stream arising from operating the well. This depletion programme may not be the one which is optimal according to the criterion of the government of the country in which the oil well is located. Such a divergence would arise, for example, if the oil-well owner and the government used different *discount rates in converting the profit stream over time to a present value.

MC

option. The buyer of an option acquires the right to buy/sell a quantity of a commodity, or of *securities, at a specified price within some specified future period. An option to sell is known as a put option, an option to buy as a call option. With a call option, for example, the buyer will be able to gain if during the period of his option the *spot price rises above the price stated in his option by an amount which would more than cover the price paid for the option. Note that an option is different from a forward contract, which is a firm commitment to buy or sell. Forward contracts are common in the oil market; option trading is most common in securities markets and in markets for *commodities in the special sense of internationally traded raw materials.

MC

organic. Denoting or relating to carbon compounds which form complex structures containing hydrogen and possibly other elements including oxygen, nitrogen, and chlorine. The term was originally confined to compounds produced by living matter, but was extended when it was discovered that some of these could be synthesized in the laboratory. *See also* inorganic.

WG

organic waste. Plant or organic matter for which no use has been discovered and which generally incurs a cost, both energy and economic, before it can be returned safely to the environment. Such wastes may be of agricultural, domestic or industrial origin and, as they have no alternative value, may be considered candidates for *bioenergy production. The term waste is however imprecise since what may be classed as waste in one set of circumstances may be useful in another. Cattle dung, for instance, is a polluting waste in some situations but is a vital energy and fertilizer source in others, especially in developing countries.

CL

Organization for Economic Cooperation and Development (OECD). An organization formed in 1961, replacing the Organization for European Economic Cooperation (OEEC). It originally had 20 members – rising to 23 – with Yugoslavia as an associate member. Its aims are to encourage non-discriminatory trade between its members, to increase members' national income and to coordinate aid to developing countries. In 1974 it set up the *International Energy Agency (IEA) to coordinate policies on energy matters. There is also a *Nuclear Energy Agency dealing specifically with the application and development of nuclear power in member countries.

Address: 2 Rue André Pascal, 75775 Paris.

MC

Organization of Arab Petroleum Exporting Countries (OAPEC). An organization formed in January 1968 by an agreement signed by Saudi Arabia, Kuwait and Libya: by 1972 all the Arab oil-exporting nations were members. The organization enables members to integrate their oil-production policies and to participate in joint ventures in oil-related industrial activity such as oil transport. It was OAPEC's 25% cut in oil production, to be followed by monthly cuts of 5% until the USA adopted (vis-à-vis the Arab–Israeli conflict) a more sympathetic attitude to the Arabs in November 1973, which made effective the oil price increases declared by OPEC. By the end of 1973 the *posted price of OPEC oil had nearly quadrupled compared with its level before the outbreak of the Arab–Israeli war.

Address: PO Box 20501, Kuwait.

MC

Organization of Petroleum Exporting Countries (OPEC). An organization formed in 1960 following initiatives by Venezuela; the other founding members were Iran, Iraq, Kuwait and Saudi Arabia. Subsequently the following nations joined: Qatar (1961), Indonesia (1962), Libya (1962), United Arab Emirates (1967), Algeria (1969), Nigeria (1971), Ecuador (1973), Gabon (1973). The stated aims of the organization were the coordination of the petroleum policies of members, the stabilization of prices for oil, and the safeguarding of the interests of members (in relation to the international oil companies).

Widely regarded as a *cartel, OPEC has been a success from the point of view of most of its members. By 1980 the ownership of oil deposits located in OPEC member countries had effectively been shifted from the oil companies to the member countries, though explicit nationalization has been relatively rare. OPEC provides loans to non-OPEC developing countries for the finance of *balance of payments deficits and development projects. Such developing countries have been especially badly affected by the rising price of oil.

Address: Obere Danaustrasse 93, A-1020 Vienna, Austria.

MC

orientation. The direction, relative to the compass points, faced by any particular part of a building envelope. Orientation in relation to the glazing in the building envelope is a major determinant in the *passive solar heat which can be collected by the building.

TM

Orsat apparatus. A portable apparatus for the analysis of gas mixtures. The volume of gas to be analysed is measured over water before contacting it in turn with a series of absorbant solutions specific for the various components of the gas mixture. After each absorbtion is complete, the volume of gas remaining is measured to enable the composition to be calculated. The unit is particularly convenient for the analysis of flue gas.

WG

***Oscillatoria*.** A *blue–green alga capable of *nitrogen fixation and of acting as a *biofertilizer in argriculture.

CL

Oslo Dumping Convention. Short for the Convention for the Prevention of Marine Pollution by Dumping from Ships and Aircraft, which was concluded in Oslo in 1972 and came into force in 1974. It applies to the high seas and territorial waters off the nations in the north-east Atlantic and North Sea regions. It divides dumping at sea into three categories: that subject to approval, that subject to permit and that completely prohibited. *See also* London Dumping Convention.

PH

osmotic energy. *See* saline energy.

Otto cycle. *See* heat engine.

output. The end products of the processes by which *factors of production, *natural resources and *intermediate products are combined to produce goods and services. Not all outputs are positively valued: the wastes arising in production are an output of the production process. *See also* external cost; joint production.

MC

oven. A heated enclosure designed to provide even heating of its contents so that a process may take place. This may range from a bakery oven in which bread is baked to a *coke oven where coal is carbonized. Heating may be achieved by oil or gas flames or by electric elements.

WG

oxidation. The combination or reaction of an element or compound with oxygen. Thus carbon burning in oxygen is oxidized to carbon dioxide. Similarly, as the converse of *reduction, the removal of hydrogen from a compound is considered to be an oxidation. More generally, oxidation is considered to take place in a reaction when electrons are removed from an element or ion.

WG

ozone. An allotrope of oxygen, O_3, which may be prepared by passing an electric discharge through gaseous oxygen. It is a poisonous gas and a powerful oxidizing agent. It finds use as a sterilizing agent and in the purification of water. Ozone is present on the atmosphere in small concentration and is formed at high altitude by the action of short wave length ultraviolet light on oxygen.

WG

P

Paasche index. An *index number which uses the current year variables as weights. For example, let p^0 and p^t represent initial-period and current-period prices respectively, and similarly for quantities q^0 and q^t. Then a Paasche *price index is given by:

$$P_p = \sum_{i=1}^{n} w_i' p_i^t = \frac{\Sigma p_i^t q_i^t}{\Sigma p_i^0 q_i^t}$$

where $w_i' = q_i^t / \Sigma p_i^0 q_i^t$

A Paasche quantity index is given by:

$$Q_p = \sum_{i=1}^{n} w_i'' q_i^t = \frac{\Sigma p_i^t q_i^t}{\Sigma p_i^t q_i^0}$$

where $w_i'' = p_i^t / \Sigma p_i^t q_i^0$

The Paasche index is less widely used than *Laspeyres index due to the need for the continual updating of the weighting system.

MC

pair production. The process whereby *gamma radiation of energy greater than 1.02 MeV interacts with an atom: an *electron and a *positron are formed and the gamma-ray photon disappears. This is an example of a process in which energy (of the gamma photon) is converted to mass (of the electron and positron). The positron is short-lived and combines with an electron; both particles are annihilated and two gamma photons, each of 0.51 MeV, are formed. In this annihilation process mass is converted to energy.

DB

panemone. A wind turbine which is able to turn with the wind from any direction. In practice all such machines have a vertical axis. It is usually only used for *drag-type machines.

JWT

parabolic reflector. A mirror which can be used in solar *focusing collectors to concentrate radiation onto an absorbing surface. The mirror has the cross-section of a parabola, which enables it to focus the parallel light to one focal position.

JWT

paraffin. (1) [alkane] Any saturated hydrocarbon with a straight or branched-chain molecule, e.g. butane (C_4H_{10}) or cetane ($C_{16}H_{34}$). The name was given to such hydrocarbons in recognition of their small affinity or reactivity when contacted with powerful oxidizing agents. Paraffins with four or less carbon atoms in the molecule are gases, those with 5 to 16 carbons in the molecule are liquids and those with 17 or more carbons are solids, at normal room temperature.

(2) A term used in the UK for domestic *kerosene.

WG

paraffin base crude. A crude petroleum from which paraffin wax separates when cooled. The hydrocarbons present in the crude are in general straight-chain, i.e. *paraffins, with the minimum of cyclic hydrocarbons. The crude when distilled thus provides poor gasoline; good kerosene, good diesel oil and a range of good lubricants.

WG

paraffin wax. A mixture of high molecular weight *paraffins which is solid at ambient conditions and is used in the manufacture of candles. It can also be cracked to provide fractions suitable for use in the formulation of fuels for internal combustion engines.

WG

parameter. An arbitrary but constant quantity or multiple in a mathematical expression. Often the term is more loosely used to imply a factor influencing a situation.

MS

Pareto optimality. *See* efficiency.

partial equilibrium analysis. An approach used in economics to consider the conditions for the existence and nature of an *equilibrium in a single market considered in isolation, ignoring *feedback linkages with other markets in the economy. Thus, for example, it studies the effects on the price and quantity of oil of the imposition of a tax on oil, without considering how the lower oil consumption will affect output elsewhere in the economy and the implications of such output effects for the oil market itself. Obviously, the usefulness of the method depends on the validity, in a particular case, of the assumption that feedback effects are small. The method is popular because, compared to the alternative *general equilibrium analysis, it is relatively easy to handle and has manageable information requirements.

MC

partial oxidation. A process operated in a petroleum refinery in which hydrocarbon gases are burned in a restricted amount of air so that carbon monoxide and hydrogen are formed instead of carbon dioxide and water vapour. The gases produced are then used to synthesize other chemicals such as alcohols.

WG

particulate material. In the context of air pollution, any material, either solid or liquid, of which the individual particles are larger than about 0.002 micrometres. Particulate material is released into the atmosphere by many natural processes, such as volcanic eruptions and wind erosion of surface soils, and also by man-made processes, such as the incomplete combustion of fossil fuels. More than a third of the total particulate material attributed to human activities comes from the combustion of fossil fuels (mainly coal) in power stations and from the exhaust of motor vehicles. *See also* particulate pollution.

PH

particulate pollution. *Pollution caused by the release of *particulate material into the atmosphere by man-made processes. Airborne particulates can have a wide range of deleterious effects on human health, plants, materials and the climate.

As far as human health is concerned, particulates enter the respiratory tract during inhalation. The size of the particles inhaled determines the extent of penetration into the respiratory system; for example, particles over 5 micrometres in diameter are stopped by the hairs lining the nasal passages and are deposited mainly in the nose and throat. Particles ranging in size from 0.5–5.00

micrometres may penetrate as far as the narrowest passages (bronchioles) in the lungs where most are deposited on the walls in mucus, which is in turn removed by tiny hairs (cilia) within a few hours. Particles less than 0.5 micrometres in diameter reach the air sacs deep in the lung tissue and may settle there, some being absorbed into the blood.

The effect of particulates on human health depends on their nature, and on *synergistic actions with other air pollutants present. The particles may be intrinsically toxic because of their chemical or physical characteristics, for example asbestos and beryllium, or the particles may carry adsorbed molecules which are irritating or toxic to lung tissue, for example *aromatic hydrocarbons such as *benzopyrene, or *tetraethyl lead from motor vehicle exhaust.

In plants the deposition of particulates on leaf surfaces interferes with *photosynthesis by reducing incident solar radiation, resulting in reduced growth rate.

In materials particulates may accelerate corrosion of metals, especially in the presence of sulphur-containing compounds. They may damage and soil buildings, sculpture, and other structures, and react with painted surfaces. Deposition of particulates on textiles may require more frequent cleaning which weakens materials.

Climate is also affected by particulates in the atmosphere which influence the amount and type of solar radiation reaching the earth's surface as a result of light scattering and absorption by the particulates. Particulate pollution may reduce visibility and decrease the incident solar radiation, creating a greater need for artificial illumination. The formation of clouds may be influenced by particulates when they act as nuclei upon which water condensation can take place. Patterns of precipitation over cities and adjacent countryside show a correlation with atmospheric particulate levels. Worldwide climatic change may also be related to particulate pollution; the reduction in the solar radiation reaching the earth's surface may be disturbing the heat balance of the atmosphere by causing an atmospheric temperature decrease which may more than compensate for the *greenhouse effect of increased atmospheric carbon dioxide levels.

PH

partitioning. When a production process yields two or more outputs, in order to arrive at an *energy requirement for each output the energy inputs must be partitioned between the outputs. There is at present no universally agreed convention. Possible bases are value, mass, potential enthalpy of combustion (useful only for fuels) and stoichiometric balance (for chemicals).

MS

pascal (Symbol: Pa). The recommended *SI unit of pressure, equal to a force of one newton acting over an area of one square metre. Standard atmosphere pressure is 101,325 pascals. One pascal equals 0.145×10^{-3} lb per square inch.

JWT

passive solar energy. Heat trapped within the structure of a building, usually by the absorption of radiation in rooms through windows. A conservatory or greenhouse attached to the sun-facing side of a house is a passive device. *See also* Trombe wall.

JWT

pay-back time. (1) The period, usually in years, required for the capital cost of an energy-generating or energy-saving system to be recovered. The prediction of pay-back time is liable to considerable error due to unknown factors such as actual capital costs, interest rates, future price of energy, government or other incentives or subsidies and secondary benefits such as employment, depreciation and maintenance.

(2) The period during which a device for capturing *renewable energy must operate before it has captured and delivered as much energy as was used to construct the device. The concept has been extended to non-renewable energy devices, such as power stations, oil rigs or nuclear reactors. The numbers so derived for the renewable and non-renewable cases are not comparable. In the former, an energy investment is resulting in a supply of energy with no further consumption of non-renewable energy. In the latter case the system continues to consume non-renewable energy; in this sense then, it can never 'pay back' the energy being extracted from the ground.

JWT, MS

peak [peak load]. The maximum occurrence, over some significant time interval, of demand or load on an energy-supply system. The magnitude of the peak will determine the sizing of the supply plant and distribution network and may also determine the supply tariff. Typically maximum demand may occur on winter evenings in cold countries and during exceptionally warm weather in hot countries. Some electricity supply systems have capacity far in excess of maximum demand (over-capacity), while some have under-capacity and have to disconnect consumers or lower the voltage of supply. *See also* blackout; brown-out.

TM, MS

peak-load pricing. A pricing policy which recognizes that demands for a non-storable (or largely non-storable) product, e.g. electricity, is not constant throughout the day or throughout the seasons. Different prices are then charged during different time periods to reflect demand and costs associated with such periods, particularly where, as in the case of electricity, the meeting of the peak load requires that the supplier maintains a large stock of *capital equipment.

MC

peat. A deposit of partially decomposed vegetable matter, generally associated with a bog, in which the surface layer continues to grow and the lower layers are buried ever deeper and become subject to bacterial decomposition. It is thus considered to be the first stage in the process of forming a *coal. Peat may be cut from the bog as a semi-solid brown mass, which has largely lost its vegetable character, and air-dried in heaps for use as a domestic fuel. Modern mechanical methods have also been used to win peat from such deposits. The material may be dried and ground to provide a pulverized fuel for steam generation in power-station boilers.

WG

pebble-bed reactor (AVR). *See* nuclear reactor.

Peltier effect. The thermal response at the junctions of an electric circuit made up of dissimilar metals when a direct current flows in it. One of the bi-metal junctions is raised in temperature, the other cooled. If the direction of the current is reversed, the junction which was formerly heated will be cooled and the formerly cooled junction will be heated.

TM

Pelton turbine. A rotating machine driven by jets of high-pressure water impinging on cups at the outer circumference. The cups are shaped to deflect the jet backwards and sideways. Pelton turbines are efficient for relatively high pressure *heads of water and are used in *hydroelectricity generation.

JWT

penstock. A pipe or channel supplying water, e.g. to a water *turbine from a dam.

JWT

pentane. A flammable highly volatile liquid hydrocarbon, C_5H_{12}, which is present in gasoline and other light petroleum spirits. It has a gross calorific value of 50.51 MJ/kg.

WG

per capita. Measured per head. Thus a nation's per capita energy use is total energy used in the nation divided by population size.

MC

perfect competition. A *market structure in which sellers can sell as much as they wish at the ruling price, which they are unable to affect by any action of their own. The conditions necessary for sellers to be in this position are that there are many sellers and many buyers of a homogeneous product, and there are no barriers to entry to or exit from the industry by firms. Perfect competition is usually also characterized by the condition of perfect information, regarding all relevant circumstances, on the part of all buyers and sellers. It is a conceptual and analytical device rather than an attempt to describe any actual market situation. Economists see its usefulness as in providing a point of comparison for other market structures; perfectly competitive firms operate where *marginal cost equals price and hence behave as is required for *efficiency in the allocation of resources. By comparison, a *monopoly (as an example of *imperfect competition) would operate where marginal cost was less than price, thus failing to satisfy the efficiency requirement.

MC

perimeter induction system. A system for *air conditioning a building in which dehumidified air and either hot or cold water are circulated throughout the building from a centrally located plant room. In each space which is to be air conditioned, a perimeter induction unit is sited. The dehumidified air is delivered from the unit at high pressure thus entraining air into the unit from the room and causing it to pass over a *heat exchanger through which the hot or cold water flows. In summer, cold water will be supplied with a changeover to hot water in winter.

TM

peroxyacetyl nitrate (PAN). *See* photochemical smog.

petrocurrency. The surpluses of foreign currency, especially US dollars, that are accumulated by oil-exporting countries – especially the members of OPEC. These are largely deposited in Europe and become part of the *eurocurrency market.

MC

petrol. The UK name for *motor spirit. *See* gasoline.

petroleum. A complex mixture of *hydrocarbons which is found underground in the pores of sedimentary rocks, the mixture varying considerably with locality. The *crude oil separated from the rock by the creation of an oil well may vary from a pale yellow volatile liquid to a black viscous liquid or semi-solid. It is considered that the organic matter from which petroleum was derived was laid down under marine conditions and is derived from marine organisms such as shellfish, coral, algae and plankton. Settlement of such material on the sea beds led to its decay and embedment in fine-grained muds. Further deposition caused compression of the lower layers to form petroleum compounds by processes not yet fully understood, and at temperatures of less than about 200°C.

It is likely that the liquids so formed have migrated through the rocks due to hydrostatic pressure until further progress is barred by some impermeable layer of rock. A reservoir of petroleum is thus formed. During the long periods of migration and of rest in reservoir rock, further chemical changes are likely to take place in the liquid. These changes determine its final physical and chemical characteristics as a crude oil when it reaches the surface in an oil well. On average, it has a *calorific value of 45 MJ/kg, a carbon content of 83–87%, a hydrogen content of 11–14% and a sulphur content of 1–6%. There is also a negligible amount of ash.

By distillation and chemical conversion processes operated in the petroleum refinery, the petroleum may be separated into a number of fractions of different boiling range. From these may be derived *liquified petroleum gas, *gasoline, *kerosene, *diesel fuel, *fuel oil, lubricants, *paraffin wax and *bitumen.
WG

petroleum coke. A coke produced by the carbonization of high molecular weight hydrocarbons; a solid porous mass of carbon left in a petroleum refinery plant after the *thermal cracking of petroleum residues. It may be removed and cut up by high-pressure jets of water. The coke may be utilized as a solid smokeless fuel, but owing to its low ash content it is used largely in the preparation of graphite electrodes for the aluminium industry.
WG

petroleum fraction. An arbitrary division of the mass of a *crude oil by a physical separation process. In distillation processes the fractions are divided by boiling point with an eye to the end use of each fraction, e.g. as gasoline, kerosene or gas oil. In solvent extraction, solubility in the solvent determines the separation. Fractions are also obtained by adsorption processes and by selective crystallization.
WG

petroleum refining. A combination of industrial processes which separate crude *petroleum into a number of fractions with properties suitable for their use as fuels and lubricants. The processes available include:

(a) physical separation, such as *fractional distillation or solvent extraction.

(b) chemical conversion, such as *cracking and *reforming.

(c) treatment processes in which the petroleum fractions are contacted with materials which will remove undesirable substances present in low concentrations, e.g. colouring matters or corrosive components.
WG

petroleum spirit. *See* benzine.

petroleum tar. Dark-coloured and viscous liquid residues from refinery processes such as *distillation, *acid treatment, etc.

pH. A scale of acidity and alkalinity. In pure water there exists a concentration of 10^{-7} g/litre of hydrogen*ions. This concentration increases with acidity and declines with alkalinity. The pH scale utilizes the negative exponent of hydrogen ion concentration to express this. Thus pH 7 is neutral, whereas lower values are acid, higher values alkaline.
MS

phenol [carbolic acid]. The compound C_6H_5OH, which is the simplest of a wide range of more complex hydroxybenzenes. It is a colourless crystalline solid with a characteristic sweet odour. It is both corrosive and highly toxic. Obtainable from the tar acid fraction in coal tar distillation, it is now manufactured by petrochemical processes involving the oxidation of aromatic hydrocarbons. It is used in the manufacture of resins and other polymers.
WG

photoautotrophic organisms. *See* autotrophic organisms.

photobiological. Denoting or relating to reactions in living organisms brought about by light. Examples are *biophotolysis, the biological splitting of water molecules into hydrogen and oxygen in the presence of light, and *photosynthesis, in which light energy is converted to chemical energy, ultimately in the form of sugar. Water-splitting is carried out in plant photosynthesis, but bacterial photosynthesis requires other compounds such as organic acids or sulphides.

CL

photocell. *See* photovoltaic conversion.

photochemical oxidant. *See* photochemical smog.

photochemical smog. An atmospheric haze which can occur over heavily industrialized or urbanized areas when *hydrocarbons and *nitrogen oxides occur together in the presence of sunlight. Photochemical smog is characterized by high temperatures, bright sunlight, low humidity and an eye-burning haze. Secondary pollutants are formed from the hydrocarbons and oxides of nitrogen, including ozone, oxidized hydrocarbons and organic nitrates such as PAN (peroxyacetyl nitrates); these are referred to as photochemical oxidants, and are responsible for health effects, such as irritation to the eyes and respiratory tract, and damage to vegetation.

The air pollution of Los Angeles is the classic example of photochemical smog, but many metropolitan areas in which there is heavy use of motor vehicles, such as Tokyo, Rome and Denver, are similarly afflicted.

PH

photoelectric effect. The process whereby gamma radiation of low energy interacts with the electrons in atoms. All the energy of the gamma photon is transferred to an electron which is ejected from its orbit. The gamma photon disappears and may be considered to have been absorbed.

photolysis. Chemical reactions induced by light, e.g. direct production of hydrogen gas.

JWT

photon. The basic energy packet of *electromagnetic radiation. The photon energy, E, is related to the frequency of radiation, ν, by

$$E = h\nu$$

where h is the Planck constant, equal to 6.63×10^{-34} J s. Thus a photon of green light has energy 36×10^{-20} joule (2 electronvolts).

JWT

photorespiration. The light-stimulated oxidation of carbohydrates to carbon dioxide and water by plants. Along with *dark respiration it lowers net *photosynthesis and hence reduces *biomass yields. Whereas dark respiration provides energy to the plant cell, photorespiration apparently serves no function. Photorespiration occurs at much higher rates in *3-carbon than in *4-carbon plants. The rate may be approaching 50% of net photosynthesis in C_3 species, resulting in a great loss of biomass potential. Therefore the introduction of a blocking mechanism into the photorespiratory reaction sequence, either through genetic mutation or chemical inhibition, would greatly increase plant productivity, particularly in temperate C_3 species selected as energy crops.

CL

photosynthesis. The conversion of light energy into chemical energy in a plant. Most photosynthesis occurs within the *chloroplasts of green plant cells and the overall process may be divided into two sets of reactions, the first requiring light and the second not requiring light. Photosynthesis may be simplified to the equation:

$$CO_2 + 2H_2O + \text{light} \rightarrow (CH_2O) + H_2O + O_2$$

where (CH_2O) represents one-sixth of a glucose molecule. Light energy is initially absorbed by the *chlorophyll and other chloroplast pigments, removing electrons from water molecules and eventually liberating molecular oxygen in a water-splitting reaction. The displaced electrons are conveyed through an elaborate electron transport system to ultimately reduce pyridine nucleotide chemical compounds to $NADPH_2$ (hydrogenated nicotinamide adenine dinucleotide phosphate). In order for each electron to be displaced from water, two *photons of light energy are required. The $NADPH_2$ molecule provides the reducing power for the reduction of CO_2 to sugar. 'High-energy' adenosine triphosphate molecules are also formed during the transfer of electrons in two *photosystems so that stored energy is available for the subsequent dark phase CO_2 reduction reactions. The overall reaction can be portrayed as:

$$NADP + H_2O + 2ADP + 2Pi + light\ (2e^-) \rightarrow NADPH_2 + 2ATP + \tfrac{1}{2}O_2.$$

Thus two electrons are required to form one molecule of $NADPH_2$ plus $\frac{1}{2}O_2$. Therefore four electrons, and hence two $NADPH_2$ molecules, are needed to complete the overall reduction of CO_2 to (CH_2O) and O_2. Since the displacement of four electrons from H_2O necessitates the absorption of eight photons, then at least eight photons of light energy need to be sequestered for the photosynthetic reduction of one CO_2 molecule and the liberation of one O_2 molecule.

The two main mechanisms of plant photosynthesis are described under *3-carbon plants and *4-carbon plants; bacterial photosynthesis is mentioned under *bacteriorhodopsin; the synthetic route is covered under *photosynthetic artificial membrane.

CL

photosynthetic artificial membrane. A synthetic structure which mimics plant *photosynthesis and seeks to overcome the inefficiencies and instability factors that are inherent in the natural process. The concept entails the use of a photoelectric membrane modelled on the natural photosynthetic membrane of the plant *chloroplast, and utilizing *chlorophyll as the primary element to absorb *photons of light (*see* photoelectric effect). Electron transfer generates a potential difference and develops a current within the external circuit, resulting in the production of electricity. This coupling of the primary photochemical reactants to electrodes should constitute an appreciable gain in efficiency because it bypasses the need to remove and store hydrogen. Nevertheless, hydrogen gas could be liberated for storage if desired. Calculations predict a practical limit of 13% maximum efficiency of solar energy conversion to chemical energy. This compares favourably with the 5–6% maximum theoretical efficiency of natural photosynthesis, which does not include the further losses of subsequent biological energy conversion of the biomass to a higher-grade fuel.

CL

photosynthetic efficiency The proportion of incident solar energy captured by organisms during *photosynthesis. The efficiency of solar-energy capturing systems is an important parameter since the lower the efficiency the greater the land area required to collect a given quantity of energy. From the simplified photosynthetic equation (*see* photosynthesis), where (CH_2O) represents one-sixth of a glucose molecule, then the Gibbs *free energy (ΔG) stored per gram mole of CO_2 reduced to glucose is 477 kJ. It has been shown that at least eight *photons of light are required for this overall reaction to occur, with the usable energy input equivalent to that of monochromatic light of wavelength about 575 nanometres (nm). Eight photons of 575 nm light have an energy content of 1665 kJ, thus giving a theoretical maximum photosynthetic efficiency of

$$477/1665 = 28.6\%.$$

The actual efficiency may well be less

than this because laboratory experiments have demonstrated that up to ten photons are needed in practice. However, since only light of wavelengths from 400 to 700 nm can be utilized in plant photosynthesis, and this photosynthetically active radiation constitutes only about 43% of the total incident solar radiation, then the photosynthetic efficiency falls to

$$28.6\% \times 0.43 = 12.3\%.$$

In the case of land plants, the optimally arranged leaf canopy can at most absorb 80% of active radiation while respiration, needed for translocation and biosynthesis, accounts typically for one-third of the energy stored by photosynthesis, leaving just 66.7%. Thus, combining the photosynthetic efficiency with absorption and respiration factors gives an overall possible efficiency for the conversion of solar energy into stored chemical energy of

$$12.3\% \times 0.8 = 0.667 \times 6.6\%.$$

This is an approximate figure and many scientists give a lower value of about 5.6%. Under optimum field conditions 3–5% efficiencies are possible for limited periods, but at present typically annual conversion efficiencies are 0.5–1.3% for temperate crops and 0.5–2.3% for subtropical and tropical plants. There is therefore definite room for improvement here.

CL

photosystem. A biochemical system involved in *photosynthesis. The natural photosynthetic process involves two main photosystems, called photosystems I and II, that act sequentially in the production of each oxygen molecule.

JWT

photovoltaic conversion. Production of electricity directly from light or *solar radiation in a solar cell or photocell. Ultraviolet, visible and near-infrared *photons separate electrical charges in certain semiconductor materials, such as silicon, cadmium sulphide and gallium arsenide, thereby producing an electric potential or voltage. Each cell when illuminated is equivalent to a low-voltage *battery producing *direct current; for example, a silicon cell operates at about 0.5 volt producing about 0.01 watt per cm^2. Higher voltages are produced with cells connected in series, and higher current with cells in parallel.

The complete photocell will consist of the photovoltaic surface as a flat layer encapsulated in a sealed container with a transparent cover. The surface may be a slice cut from a large single crystal, a thin layer perhaps grown as a ribbon, a layer deposited by condensation from a vapour, or an amorphous substrate. The encapsulation has to be of a high standard to benefit from the (potentially) extremely long lifetime of the photovoltaic surface; the cost of the final device is therefore limited by the cost of the encapsulation, however cheap the surface may become. At present, photovoltaic surfaces are not cheap: an array of cells that produces 1 watt peak power under optimum conditions will cost about $25. The cost of the active surfaces is steadily decreasing as the technology improves and commercial sales increase.

The theoretical maximum efficiency of photovoltaic conversion surfaces is about 25%, depending on the particular material. This efficiency in solar radiation decreases with increase in temperature, so cooling may be important. In practice the efficiency of commercial cells is considerably less than the theoretical maximum. For terrestrial use, silicon cells reach about 12% efficiency, cadmium sulphide 7%, gallium arsenide about 4% and amorphous silicon about 5%. Since the solar intensity in bright sunshine is about 1 kW/m^2, the maximum electrical output in practical devices is about 80 W/m^2 or less. The maximum electrical output per day may be about 0.3 kWh/m^2 on a fixed orientation array of cells, but can increase to about 0.5 kWh/m^2 if optimum orientation is arranged by tracking the sun's path. Reflecting concentrators or *Fresnel lenses may be used to increase output per unit cell, but overheating and damaging electric-current variations may make this unadvisable.

Arrays of photovoltaic cells produce electricity in both *direct and *diffuse solar radiation, are lightweight, have very long lifetimes and are almost maintenance-free. These excellent properties are marred solely by price. The present price is about $20 per peak watt for a commercial array, many times greater than hydro- and nuclear-generated electricity. Present use is limited to low-power electricity use in remote locations, e.g. space, telecommunication repeater stations, areas remote from grid electricity with oil-supply difficulties. The steadily reducing costs however are producing more varied uses. Some experimental houses now use photovoltaic converters for all their electricity supplies. The pumping of water for domestic use in tropical Third World countries may be the first extensive use of photovoltaics, especially if acid programmes finance purchases.

JWT

phurnacite. A manufactured solid smokeless fuel. Coal with low volatile matter is mixed with about 6% by weight of pitch and the mixture formed into small egg-shaped briquettes. These are carbonized in special ovens at 1000 °C for four hours. The pitch used is a by-product of the carbonization process.

WG

Pigovian taxes. Taxes intended to make the generator of a detrimental *externality have a perception of the cost of his activities which coincides with their social cost. As *emissions taxes, Pigovian taxes are distinguished by being set at rates designed to secure *efficiency in allocation. A typical situation in which a Pigovian tax would be appropriate is the emissions into the atmosphere associated with the burning of coal to produce electricity. The tax rate would not, other than in exceptional cases, be set at such a level as to reduce emissions to zero, since the objective is to reduce pollution to the extent that people are willing to pay for such reduction. Pigovian taxes are very rarely applied in practice due to the difficulty of measuring such *willingness to pay.

MC

pinch effect. *See* fusion reactor.

pinking. *See* knock.

pit. An underground working of a deposit of coal which the workers (miners) reach by travelling down a vertical shaft by an elevator or lift from the surface. At the pit bottom horizontal or slightly sloping roadways or tunnels enable the miners to reach the seams being worked.

WG

pitch. The black residue, a solid at normal temperatures, from the distillation of *coal tar.

WG

Planck constant. *See* photon; Planck's law.

Planck's law. The law stating that the energy of *electromagnetic radiation can be regarded as being confined in indivisible packets, known as *photons, the energy of a photon being directly proportional to the frequency of the radiation. If the frequency is equal to ν, then the energy, E, of each photon is given by

$$E = h\nu$$

where h is the Planck constant. The absorption or emission of radiant energy involves whole numbers of photons, i.e. a discrete amount of energy, being absorbed or emitted by an atom.

VI

Planck's radiation distribution. The radiant energy emitted from a heated material has a spectrum of wavelengths with maximum *intensity at an intermediate value. The shape of this intensity distribution with wavelength and the position of the maximum is explained by Planck's equation of the distribution. The equation also predicts how the wavelength of the maximum intensity varies inversely with the temperature of the material, originally put forward as Wien's law, and how the total emitted intensity varies as the 4th power of the absolute temperature, the *Stefan–Boltzmann law.

JWT

plasma. A very high temperature gas, the temperature being greater than one million kelvin. At this temperature the individual atoms of the gas have such a high energy that they are ionized.

DB

plastics. *See* polymer.

plenum ventilation. A system of effecting *airchange and environmental control by putting the ceiling void under positive pressure and introducing conditioned air to the occupied space through a ceiling diffuser.

TM

plume dispersion. *See* atmospheric dispersion.

plutonium. A *transuranic element (atomic number: 94; symbol: Pu), which does not occur naturally but which can be produced by nuclear reactions. Some of its *isotopes can be used as nuclear fuel. The isotope plutonium-238, whose half-life is 86.4 years, emits 5.5 MeV alpha particles. It forms a light, compact and reliable power source which has been used for applications in space and for heart pacemakers. The isotope plutonium-239 is produced by neutron capture in uranium-238 as a result of the following processes:

$$^{238}_{92}U + ^{1}_{0}n \rightarrow ^{239}_{92}U$$

$$^{239}_{92}U \rightarrow ^{239}_{93}Np + ^{0}_{-1}e$$

$$^{239}_{93}Np \rightarrow ^{239}_{94}Pu + ^{0}_{-1}e$$

Plutonium-239 is radioactive, with a very long *half-life, 24,400 years, and it is *fissile, hence its importance as a source of energy and for nuclear weapons. At present it is being produced in quite large quantities in all operating reactors. Other isotopes of plutonium, ^{240}Pu, ^{241}Pu and ^{242}Pu are also produced in operating reactors, but in smaller quantities; of them, ^{241}Pu is fissile but it is not as important a source of energy as ^{239}Pu. Plutonium is highly toxic and this together with its radioactivity makes it an extremely hazardous substance. *See also* breeder reactor; breeding.

DB

plutonium oxide. A compound, PuO_2, with a very high melting point which renders it more suitable than pure metallic plutonium for reactor *fuel elements. Present-day fast reactors have fuel elements of mixed plutonium oxide and uranium oxide, UO_2. It is a powder and is fabricated into fuel pellets by powder metallurgy.

MS

pneumoconiosis. A general term describing occupational diseases of the lungs found in miners and others who work with dusty materials. It is characterized by a chronic inflammation of the lungs, which may develop into tuberculosis, *emphysema or bronchitis.

PH

pollutant. An agent or effect which causes *pollution.

PH

polluter pays principle. The idea that the generator of pollution should bear the cost to which it gives rise. The principle would exclude, for example, government grants to polluting firms to pay for the installation of pollution-reducing equipment. The principle would admit *emissions taxes generally, and *Pigovian taxes in particular. Since the *incidence of taxation is not necessarily the same as the point at which tax is levied, care needs to be taken in assessing the implications of the principle. If the polluting firm is not a *price taker, it will pass some of the tax on to its customers in the form of a higher price. For example, the imposition of a tax on the emissions into the atmosphere from coal-burning electricity generating stations would lead to a higher price charged for electricity.

MC

pollution. The accidental or deliberate contamination of the environment as a result of human activities. It may be caused by a variety of agents or effects, known as pollutants, which include sewage, waste chemicals, waste gases, waste heat and excess noise. Pollutants may adversely alter the environment if they interfere with human health, comfort, amenities or property values. They may also interfere with *food chains, change the growth rate of species or be toxic to particular species.

Wastes arising in the activities of production and consumption are eventually returned to the natural environment. If this produces an environmental effect which is perceived as adverse by some people, then, and only then, does pollution exist in an economic sense. It is not the case in economic terms that the *optimal level of pollution, so defined, is zero, since typically the total elimination of some particular kind of pollution would cost more than the damage avoided would be valued at. *See also* materials balance principle; polluter pays principle.

PH, MC

polymer. A chemical compound composed of a number of identical repeated molecular groups. It may be formed by a chemical reaction which links the individual molecular units (monomers) into chains and other networks. A structure consisting of two different sets of units is known as a copolymer. Some polymers are semiplastic. These are organic substances which soften when heated and may be moulded into a desired shape and then cooled to harden again. They constitute the so-called plastics used for a wide variety of purposes formerly served by wood, metal, etc. The basic building blocks of polymers are today provided by derivatives of crude *petroleum, but they can also be produced, though more expensively, from *biomass sources.

WG

positron. The antiparticle of the *electron, i.e. the particle whose mass is the same as that of the electron and whose charge is positive. Positrons are produced in certain types of radioactive-decay processes (*see* radioactivity) and in the *pair production process. The positron is very short-lived and reacts with an electron; the two particles are annihilated and two gamma photons, each of 0.51 MeV energy, are formed. This is an example of the complete conversion of mass to energy.

DB

posted price. The tax reference price for oil, i.e. the price which serves as the basis for calculating the tax to be paid by an oil company to the nation in which it was operating an oil concession, per barrel of oil extracted. The posted price is not generally the same as the market price, with the latter sometimes above and sometimes below the posted price. The posted price is now of very little significance since in most OPEC countries the oil extraction industry is effectively nationalized; this means that the *rent element in the oil price accrues directly to the government without the need for complex taxation arrangements.

MC

potential difference (Symbol: V). The scalar interval of a magnetic or electric field intensity between two points, measured relative to some accepted zero potential, e.g. earth. The unit of potential difference is the *volt (V), defined as the difference of electric potential between two points of a conductor carrying a constant current of one *ampere when the power dissipation between the points is one *watt. The term can also be used where any potential energy state differs from another.

TM

potential energy. Energy that may be produced by a change in state of an object or system. It is particularly applied to energy obtained by virtue of a mass raised to a height, e.g. by water able to flow down a hill. A mass m raised to a height h when the acceleration due to gravity is g (9.8 m/s^2) has potential energy equal to mgh. Chemicals that may react to produce heat have a chemical potential energy. *See also* kinetic energy.

JWT

potentiometer. An instrument for the measurement of *potential difference, *current and *resistance in an electric circuit.

TM

pour point test. A standard test which determines the lowest temperature at which an oil will flow when cooled in a test tube periodically tilted through 90 degrees.

WG

power (Symbol: P). The energy transferred per unit time or alternatively the rate at which work is done. The universal measure of power is the *watt, an SI unit. *See also* conversion tables; energy flux.

MS

power alcohol. *See* gasohol.

power coefficient (wind). The ratio of the *power obtained from a wind turbine, to the power (energy passing per unit time) in the airstream impinging on the swept area of the turbine blades. In the figure the ratio is P/P_t. The power coefficient, C_p, is therefore the efficiency of wind-power extraction. Because air must be able to leave from behind a wind turbine and so take away energy, the maximum value of C_p is much less than 100%. By the *Betz criterion the maximum is 59%. An efficient wind turbine might have C_p less than or equal to 40%.

JWT

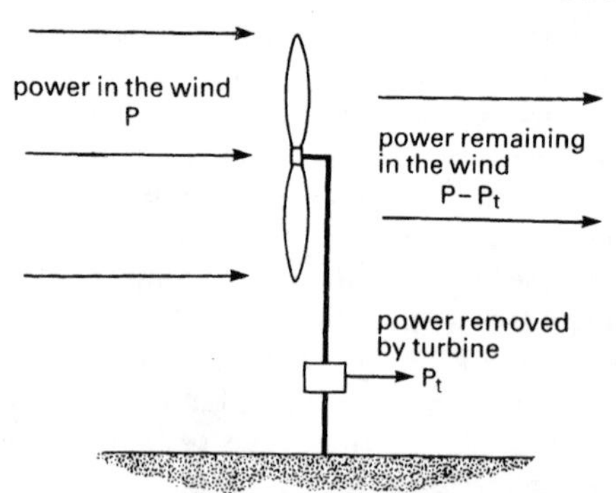

power coefficient of reactivity. A characteristic of a *nuclear reactor which expresses the change in *reactivity of the reactor caused by a change in its power. If an increase in power leads to an increase in reactivity which in turn causes the power to increase further, the power coefficient of reactivity is positive and an unstable operating condition exists. If an increase in power leads to a decrease in reactivity which in turn causes the power to drop, the power coefficient of reactivity is negative and a stable operating condition exists. The latter condition is desirable from the point of view of safe reactor operation. DB

power factor. The proportion of current in an alternating current supply which can be used as energy. The current supplied can be divided into two components: the active component which is in phase with the voltage and can do useful work, and the reactive component which, to some degree, is out of phase with the voltage and can do no work. Power factor correction can be achieved by the introduction of electrical *capacitance into the circuit. TM

predatory pricing. *See* dumping.

present value. The current value of a monetary value arising at some future date, found by using the *discount rate. Thus the stream of money sums V_1, V_2, ..., V_T arising at times 1, 2, ..., T becomes a stream of present values at time 0 which are

$$\frac{V_1}{1+\rho}, \frac{V_2}{(1+\rho)^2}, \frac{V_3}{(1+\rho)^3}, \ldots, \frac{V_T}{(1+\rho)^T}$$

where ρ is the discount rate. *See also* cost–benefit analysis; project appraisal. MC

pressure vessel. A steel or prestressed concrete vessel which withstands high internal pressure. In a *nuclear reactor it encloses the core, its primary function being to allow the coolant to be circulated through the reactor at a pressure greater than atmospheric. This is particularly important in the case of pressurized-water reactors and boiling-water reactors, since to achieve optimum thermal efficiency the water pressure and temperature in the reactor should be as high as possible. In gas-cooled reactors the circulation of the coolant at high pressure increases efficiency by reducing the pumping power required to circulate the coolant.

pressurized-water reactor (PWR). *See* nuclear reactor.

prestressed concrete. A structural system used, for example, for the *pressure vessels of certain types of nuclear reactor, notably *advanced gas-cooled reactors. Its use instead of steel enables higher pressures to be achieved with greater integrity. The concrete pressure vessel also provides shielding for the reactor. DB

Prevention of Oil Pollution Act 1971 (UK). An Act which makes it an offence for British ships to discharge oil in any part of the sea in any part of the world. It also prohibits British and foreign ships from discharging oil specifically in British national waters. PH

price. What has to be given up, usually expressed in terms of money as *numeraire, in exchange for one unit of a good or service. Where a market exists, the price is determined by the interaction of demand and supply. *See also* absolute price; relative price. MC

Price–Anderson Act (US). An Act passed in 1957 to provide for damages arising from a nuclear accident in the USA. It requires a utility which operates a nuclear power plant to purchase the maximum amount of liability insurance it can obtain ($60 million in 1957) while the Federal Government provides additional insurance to a maximum of $560 million for public liability claims. The Act was extended in 1965 and again in 1975 after congressional hearings which investigated the need for such protection. The 1975 extension provided for phasing out government indemnification, permits the $560 million limit to float upwards, and extends coverage to certain nuclear incidents that may occur outside the territorial limits of the USA. PH

price index. An index designed to show how the average price of some group of commodities, such as fuels, has moved over time. *See also* index numbers; Laspeyres index; Paasche index.

MC

price mechanism. The means by which competitive markets allocate resources, as between the production of different commodities, in response to the signals arising from movements in *relative prices; this coordinates the activities of many individual decision-making units. Thus, insofar as rising energy prices lead to energy-saving responses by energy users and energy-producing responses by actual and potential energy suppliers without governmental direction, the price mechanism is coordinating the appropriate responses to increased energy scarcity.

MC

price taker. A producer who has no control over the price of his output and must accept it as given. He can sell at or below this price but sells nothing if he tries to sell at a higher price. Firms operating in an environment of *perfect competition would be price takers. Oil-producing nations not members of *OPEC and having an output small in relation to total world output, approximate to being in the position of price takers. The UK is a case in point, not due to the fact that different types of crude oil are not perfect substitutes one for another, transport costs, etc.

MC

primary air. Air mixed with a combustible gas as it enters a burner or introduced under the grate when coal is burned as a fixed or fluidized bed. It is also the air used to atomize liquid fuels or to inject pulverized fuel.

WG

primary energy. A *non-renewable energy resource in the ground which, by extraction, refining and delivery, can become an economically useful energy or fuel. It is rare that a primary energy source can be used without some refining before use. Natural gas and coal can, on occasion, be used without treatment.

MS

primary productivity. The net yield of dry plant matter produced by *photosynthesis within a given area and period of time. *Photosynthetic efficiency is the main determining factor in primary productivity, with tropical *4-carbon plants generally more productive than temperate *3-carbon plants. The annual global primary productivity is from $100-125 \times 10^9$ tonnes on land plus $44-55 \times 10^9$ tonnes in the world's oceans. Most of this biomass (44.3%) is formed in forests and woodlands, with 35.4% in oceans, 9.7% on grassland, 5.9% on cultivated land, 3.2% in freshwater and 1.5% in desert and semi-arid regions. Tropical rain forests alone account for $22-27 \times 10^9$ tonnes each year, thereby supplying the main energy supply to rural areas of the Third World.

CL

process analysis. An analysis of mass and energy balance made on a process or a part thereof. Such analysis is often used to establish the *energy requirement of making a product, and is subject to a number of rules and conventions. *See also* energy analysis.

MS

process energy requirement (PER). The energy required to drive one part of a process. For example, having prepared the gas, ethene, one might compute the PER to turn it into polythene. If the PER is computed for every step of the process from energy source in the ground to final product, the sum of all the PERs is the *gross energy requirement (GER) of the product.

MS

producer gas. A relatively low-grade fuel gas generated by blowing air through red-hot coke. It is essentially a mixture of carbon monoxide and nitrogen. However, since reaction is not complete and since some water vapour is injected to control the temperature of the generator, the final gas also contains some carbon dioxide and hydrogen. Non-caking coals may be used in place of coke. Typical *calorific values are as follows:

producer gas from coke: GCV $=4.8\,MJ/m^3$
producer gas from coal: GCV $=6.0\,MJ/m^3$

WG

product differentiation. A situation which exists when each of the firms comprising an industry sells a product with characteristics that make it different from, though generically the same as, the products of its competitors. It is an essential feature of *monopolistic competition. Where it exists, firms are not *price takers. The retailing of petroleum products is marked by product differentiation; the retailing of electricity, being run by a monopoly, is not.

MC

production function. The relationship between output and input levels for a production activity, given that there is technical efficiency in the operation of the process. Where just two inputs are distinguished, the properties of a production function can be represented graphically by means of the *isoquants. The properties of principal interest are the *returns to scale exhibited and the *elasticity of input *substitution. It is the properties of the production function relationship which determine, for example, how the operation of the process for producing some commodity will respond to higher energy prices.

MC

production possibility frontier. A conceptual device designed to illustrate the nature of the fundamental problem of economics: choice in conditions of *scarcity. In the diagram, OB represents the amount of agricultural output which

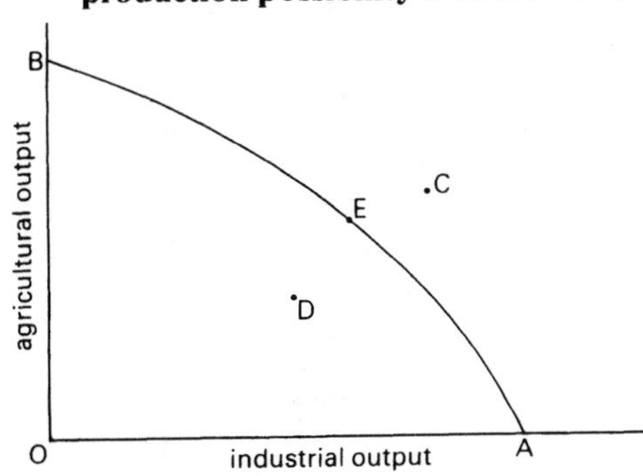

could be produced if all available energy, say, were used in agricultural production, while OA is the amount of industrial output which could be produced if all energy were used there. The production possibility frontier is the curve joining A and B; it represents the locus of possible combinations of agricultural and industrial output available, with all the available energy used in one line of production or the other.

The particular shape taken by the curve will depend on the *production functions for agriculture and industry: the shape shown in the diagram reflects the assumption that in both lines of production the energy input is subject to *diminishing marginal returns. Output combinations lying outside the production possibility frontier, C for example, cannot be produced given the total amount of energy and other resources available. Output combinations lying inside AB can be produced, but to do so would mean leaving some available energy and other resources unused. A move from D to E, for example, would represent a move to an efficient use of resources, since it is only along AB that it is impossible to increase one output without decreasing the other output; this situation characterizes *efficiency in allocation. Where there is no *market failure, a system of competitive markets will ensure that the economy ends up at a point on the production possibility frontier.

MC

productivity. Output per unit of factor input. For example, labour productivity is the output per unit of labour (i.e. the average product of labour). Generally it is difficult to separate changes in labour productivity from changes in *capital productivity in practice.

MC

profit. In an accounting sense, the difference between the revenue from sales by a firm and the outlays on the costs of running the firm, including the purchases of inputs to production, in the relevant period. In economics the definition of cost is wider and includes the full *opportunity costs of all factors used in production; costs therefore include the return to risk taking and the cost of using capital equipment, known as *user cost. Profit may be measured *gross or *net of *depreciation.

MC

profit maximization. The assumed objective of private sector firms in conventional economic analysis. The *supply function for some commodity, for example, is derived by assuming that the firms in the industry concerned respond to market changes affecting their output and their inputs in a way conditioned by their continual striving after the highest level of profit available in given circumstances. Increasingly the empirical validity of this assumption has been questioned by some economists, especially as it relates to large firms where the owners are a distinct group from the managers of the firm.

MC

progressivity. A property of a system of taxation such that the proportion of income taken in tax rises with the level of income. The property is considered desirable by those who favour economic policies which promote an egalitarian outcome. Progressivity is usually discussed in the context of direct taxation of income, where it is achieved by having levels below which no tax is paid and/or marginal tax rates which increase with income. It can also be a feature of *indirect taxation if, for example, such tax applies only or predominantly to commodities purchased by people with higher incomes. The effect of the indirect taxation of energy commodities is generally taken to involve *regressivity rather than progressivity.

MC

project appraisal. A special case of *cost–benefit analysis where the project lies in the private sector of the economy, where *cost is measured solely in terms of actual money outlays arising from the project, where *benefit is measured solely in terms of money receipts arising from the project, where the *discount rate used is the appropriate interest rate, and where the decision on whether or not the project should go ahead is to be taken according to the criterion appropriate in the private sector of the economy, i.e. long-term *profit maximization. With respect to the opening up of a new coalfield, for example, a project appraisal by the coal mining firm would not necessarily lead to the same decision as a cost benefit analysis.

MC

Project Salt Vault. An experimental programme conducted by *Oak Ridge National Laboratory in the USA between 1963 and 1968 to study the feasibility and safety of the storage of solid radioactive waste in a bedded salt deposit. A disused salt mine near Lyons, Kansas, was chosen for the project. A pilot demonstration plant at this site was abandoned in 1972 largely because of evidence that water could in the future enter the repository via abandoned oil or gas wells in the area. *See also* radioactive waste management.

PH

proliferation. An increase, as in the number of countries producing nuclear weapons, always associated with programmes to develop nuclear energy. JWT

prompt criticality. The condition of a *nuclear reactor which is in a state of *criticality by virtue of the *prompt neutrons only. Without the effect of *delayed neutrons to retard changes in the power of a nuclear reactor, such changes can take place very rapidly, making control difficult if not impossible, and putting the reactor into a very unsafe condition. It is a cardinal rule of reactor operation that prompt criticality is avoided. *See also* delayed neutrons.
DB

prompt neutrons. Neutrons produced at the instant of *fission, as distinct from *delayed neutrons. DB

propane. The systematic name for the *alkane with three carbon atoms. Thus it is a saturated hydrocarbon with the formula C_3H_8. It is found naturally in small proportions in *natural gas and gas associated with *crude oil. Separated and compressed to a liquid, it is sold as a *bottled gas, particularly suitable for low temperature environments as it boils at 231.09 K (−42.07 °C). *GCV 50.25 MJ/kg.

propene. The systematic name for the *olefin propylene, C_3H_6, a flammable gaseous hydrocarbon with one double bond in the molecule. It is frequently a constituent of refinery gases. It has a gross calorific value of 49.00 MJ/kg. WG

proportional counter. A *gas-filled radiation detector whose output pulses are proportional to the energy of the radiation being detected, enabling radiations of different energies from two or more sources to be identified and counted separately. DB

propylene. *See* propene.

protection. The act of imposing a *tariff, *quota or non-tariff barrier on imported goods as a means of securing the domestic market. It may be for economic reasons, e.g. the protection of new industries or an anti-dumping measure, or it may be non-economic, e.g. to secure self-sufficiency for strategic reasons. Thus in 1932 the USA introduced a tariff on imports of crude and heavy oil principally to serve the economic interests of domestic producers. The oil-import restrictions introduced in the USA in 1959 had the effect of serving the same interests, but were justified in terms of the strategic need for self-sufficiency.
MC

proton (Symbol: ^1_1p, p). A particle of matter having a mass of approximately one *atomic mass unit and a charge equal in magnitude to that of the *electron but opposite in sign, i.e. positive in value. The proton is one of the constituent particles of the atom: together with *neutrons, protons form the atomic *nucleus, their combined number in a nucleus being equal to the *mass number, *A*, of the element; the number of protons in a nucleus gives the *atomic number, *Z*, of the element. The proton is thus equivalent to the nucleus of a hydrogen atom. DB

proton recoil detector. An *ionization chamber for detecting *fast neutrons. The detector usually has a lining of a hydrogenous material, such as polythene or paraffin wax. Fast neutrons react with hydrogen atoms in the lining by a *scattering process, and hydrogen nuclei are ejected as protons, which in turn cause ionization in the detector.
DB

Prototype Fast Reactor (PFR). The liquid-metal-cooled fast breeder reactor built at Dounreay, Scotland. *See* nuclear reactor.
DB

proximate analysis. The analysis of a finely divided representative sample of a coal in terms of the percentages of moisture, ash, volatile matter and fixed carbon, all determined by standard procedures specified by the *British Standards Institution. It is effectively a coal assay and is employed, together with gross *calorific value, to determine the most suitable application for the coal.

WG

PSALI. The acronym for *p*ermanent *s*upplementary *a*rtificial *l*ighting of *i*nteriors, i.e. the magnitude of the shortfall between the necessary illumination required within a building and that available as daylight. TM

psychrometry. The study of how the properties of moist air change as a result of *air conditioning processes. Tables and charts of psychrometric data are available: in a typical psychrometric chart, most of which are to an atmospheric pressure base of 1013.25 millibars, the main axes would be moisture content and dry bulb temperature, with other parameters – wet bulb temperature, enthalpy, etc. – superimposed. TM

public good. A *good or service for which one individual's consumption is not at the expense of consumption by others. It follows that markets cannot regulate the consumption of public goods since units of such are not ownable in the sense of purchase of a unit securing exclusive rights to use. The services of a police or defence force are examples of public goods. Air pollution, arising for example from the generation of electricity by burning coal or the emissions from motor cars, is a public 'bad': increased inhalation of polluted air by one individual would not reduce the pollution suffered by others.

public sector. All departments and agencies of central and local government plus publicly owned *firms. The size of the public sector varies between nations. In the UK as of 1988 the electricity supply industry and the coal industry are in the public sector, whereas in the USA most firms in these industries are privately owned with the electricity utilities subject to regulation by public authorities. In the 1980s there has in many economies been a move to reduce the size of the public sector. In the UK energy sector, for example, the gas industry has already moved from the public to the private sector, and the electriciity supply industry is expected to do so.

MC

pulverized fuel. Coal or other solid fuel which has been ground to a small particle size so that it may be fed with a stream of air to a burner and consumed in a flame similar to that produced by a liquid or gaseous fuel. Pulverized fuel is used to generate steam in the large combustion chambers of the boilers of coal-fired power stations.

WG

pumped storage. Water pumped up to a reservoir at times of excess electricity production for subsequent *hydroelectricity generation (*see overleaf*). Systems are usually able to accumulate water slowly from excess *base load generation for rapid discharge at peak supply periods. Overall efficiency is 65–80%.

JWT

purchasing power parity. An economic theory which states that the *exchange rate between two countries is determined by their relative domestic purchasing power, i.e. the exchange rate equals the ratio of the two countries' price levels. Although short-run exchange rates are determined by many factors, it is thought that purchasing power parity has a strong influence on their long-run determination.

MC

Purex process. A liquid–liquid solvent extraction method used in the nuclear industry for the separation of *irradiated fuel into its constituents: uranium, plutonium and *fission products. The organic solvent tributyl phosphate (TBP) is used to separate plutonium from uranium in aqueous nitric acid solution.

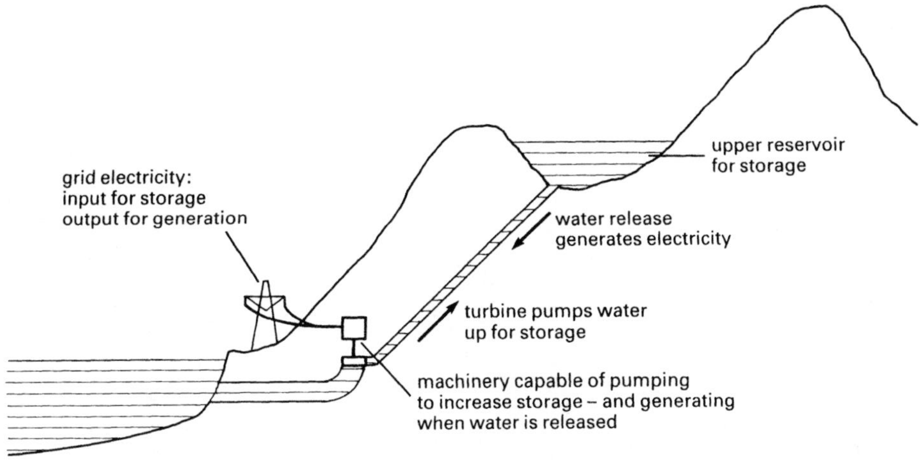

DB

pyranograph. An instrument with a pen and chart for measuring and recording total radiation, including solar radiation and far-infrared radiation.

JWT

pyranometer. An instrument for measuring total or *global radiation from all sources (including solar radiation and far-infrared radiation) at all directions from the hemisphere above. The instrument measures the radiation flux falling onto a *flat plate absorbing surface without a cover.

JWT

pyrheliometer. An instrument for measuring radiation from a limited angular cone, including radiation directly from the disc of the sun. The instrument therefore can measure the radiation flux incident on a *focusing collector.

JWT

pyrites. A mineral which is found in coal and is composed of a mixture of metallic sulphides, largely iron pyrites (FeS_2). It appears in the form of cubic brassy crystals which range from hand-size masses (known as coal brasses or fool's gold) to tiny individual crystals visible only under the microscope. The presence of pyrites in coal leads to the formation of gaseous *sulphur oxides in the products of combustion when the coal is burned. Thus the prior removal of the pyrites reduces atmospheric pollution.

WG

pyrolysis. The thermal degradation of organic substances. Processes involving pyrolysis include the *carbonization of coal and the *thermal cracking of petroleum hydrocarbons.

WG

Q

quad. A unit equal to one thousand million million (10^{15}) *British thermal units, often used for national power supplies. US total energy consumption is about 80 quad per year.

JWT

quality factor. *See* radiation dose.

quantity theory. Any of several theories concerning the demand for money, forming the essential theoretical underpinning of *monetarism. The basic idea is that a change in the *money supply will lead to *disequilibrium in the money market, with demand not equal to supply, the effects of which will work through to other markets in the economy. If, for example, the money supply is increased when the economy is operating at full capacity, the attempt by people to spend the additional money on goods and services will not result in extra output, but only in higher prices. The *inflation will continue until *equilibrium is restored in the relationship between the demand for and supply of money.

MC

quantum. The minimum amount by which the energy of a system can change. The quantum of electromagnetic radiation is the *photon.

JWT

quantum efficiency. The fraction of light quanta or *photons utilized in a photon-sensitive device or process, such as *photovoltaic conversion or *photosynthesis. In photosynthesis this fraction is also known as the *photosynthetic efficiency.

JWT

quota. A limit imposed on the quantity of goods purchased, as in the case of imports, or produced, as in planned economies and some *cartels. An import quota may be imposed as a means of *protection or as a means of reducing a *balance of payments deficit.

MC

Q-value. Of a nuclear reaction, the energy released by that reaction. It is positive for *exothermic reactions and negative for *endothermic reactions. It should be noted that in chemical reactions the sign convention is the opposite. Q is represented by ΔH, *enthalpy change, which for exothermic chemical reactions has a negative sign.

DB

R

rad. The unit of absorbed *radiation dose, recently superseded by the SI unit, the *gray (Gy). 1 rad is equal to 0.01 gray. The use of the rad is likely to continue for some time.

DB

radiation. Energy passing through space in the form of *electromagnetic radiation, such as light or ultraviolet radiation, or as a stream of particles, e.g. electrons or protons. Thermal radiation is energy transmitted as electromagnetic waves from hot surfaces. The wavelength distribution of the radiation follows *Planck's radiation distribution law, from which the peak intensity can be derived, and the total emitted energy is given by the *Stefan-Boltzmann law. Surfaces with temperatures less than red hot will emit in the infrared region of the *electromagnetic spectrum. Radiation arriving at surfaces may be absorbed, reflected or transmitted. Absorbed radiation causes the absorbing material to increase in temperature and hence increase its own emission of radiation. The energy exchange between two surfaces thus depends on the *net radiation transmitted between them. *See also* ionizing radiation.

JWT

radiation damage. Changes produced in the molecular or crystalline structure of a substance as a result of prolonged exposure to *ionizing radiation.

DB

radiation detector. An instrument used for the detection and measurement of nuclear radiation. Most operate on the basis of one of two processes: the *ionization of a gas by charged particles or the production of scintillations (tiny flashes of light) in a phosphor by high-energy electromagnetic radiation. *See also* gas-filled radiation detector; Geiger–Muller tube; ionization chamber; scintillation detector.

DB

radiation dose. A general term for quantity of *ionizing radiation. Four more specific terms are in current use to describe the dose of radiation absorbed by an organism or part of an organism, or by a population.

(1) Absorbed dose is a measure of the energy imparted to a unit mass of matter such as tissue. It is expressed in a unit called the *gray (Gy):

$$1\,\mathrm{Gy} = 1 \text{ joule per kilogram (J/kg)}$$

Absorbed dose was formerly expressed in *rads.

(2) Dose equivalent is the absorbed dose of ionizing radiation multiplied by a numerical factor, the quality factor,

which allows for the different effectiveness of the various types of radiation in causing biological damage. It is expressed in a unit called the *sievert (Sv). The factor for gamma rays, X-rays and beta particles is 1, for neutrons is 10, and for alpha particles is 20. This unit was previously expressed in *rems.

(3) Effective dose equivalent (often shortened to dose) is the quantity of radiation dose obtained by multiplying the dose equivalent to various tissues and organs by the risk weighting factor appropriate to each, and summing the products. It is expressed in sieverts.

(4) Collective effective dose equivalent (or collective dose) is the quantity of radiation dose obtained by multiplying the average effective dose equivalent by the number of persons exposed to a given source of radiation. It is expressed in man-sieverts or person-sieverts.

See also radiological protection standards.

PH

radiation dose limits. *See* radiological protection standards.

radiation exposure pathway. The route taken by a particular *radionuclide from its point of discharge into the environment to the point of exposure to the human population. For example, the radionuclide iodine-131 released in the gaseous discharge from a nuclear fuel reprocessing plant may be deposited on grass which may in turn be ingested by cows. The radionuclide is then transferred to milk which may be consumed by members of the human population.

PH

radiation sickness. Sickness and ill-health induced in humans as a result of overexposure to *ionizing radiation.

DB

radiative capture. *See* absorption (neutron).

radiator. Any body which emits radiant heat energy. The term is commonly, but inappropriately, applied to those parts of a *central heating system in which the circulating hot water or steam gives up its heat to the rooms of a building; in such a case, the main exchange of heat is in fact convective, not radiative. The term is more appropriately applied to an electrical room heater in which the heating element is incandescent.

TM

radioactive decay. *See* radioactivity.

radioactive decay chain. If the *daughter product of a radioactive decay process is itself radioactive, it in turn will decay, possibly to yet another unstable daughter product. The sequence of decays is known as a radioactive decay chain, and in some such chains, e.g. the decay of naturally occurring uranium-238, several radioactive daughter products are formed before eventually a stable daughter product ends the chain. Uranium-238 eventually decays to lead. *See also* radioactivity.

DB

radioactive waste. The byproduct of reprocessing *irradiated fuel from nuclear reactors, and to a much lesser extent other discarded radioactive materials. It is separated into different categories according to its activity and *half-life, and disposed of or stored accordingly. *See* radioactive waste management; *also* high-level waste; low-level waste.

DB

radioactive waste management. Liquid, solid and gaseous radioactive wastes are produced at all stages of the nuclear *fuel cycle. Three basic philosophies are applied to the management of these wastes.

(a) Dilute and disperse: some of the low-level liquid and gaseous wastes produced at nuclear facilities are released directly into the environment where they are diluted into bodies of water or air; an example is the low-level liquid effluent discharged by Windscale into the Irish Sea.

(b) Delay and decay: some radioactive wastes containing radionuclides with short *half-lives are stored until their *radioactivity has decayed to relatively innocuous levels before further treatment or handling; for example, *irradiated fuel assemblies are stored under water in cooling ponds at the reactor or reprocessing plant prior to reprocessing.

(c) Concentrate and contain: highly radioactive wastes are treated by various chemical and mechanical means to produce a high-level liquid waste fraction, which then requires special treatment for long-term storage and disposal.

Several methods of disposal are in use or have been proposed. Ocean dumping of intermediate and low-level radioactive wastes has been practised by many countries since the 1940s. Between 1946 and 1970 the US Atomic Energy Commission licensed the dumping of about 95,000 curies into the Atlantic and Pacific Oceans. This practice was phased out in 1971 because of the prospect of much less expensive land-based disposal alternatives. From 1951 to 1966 Britain dumped about 45,000 curies into the Atlantic Ocean. Since 1967 Britain and seven other European countries have continued to dump under the auspices of the Nuclear Energy Agency. The current dumping site is about 900 km SSW of Lands End, UK, at an average depth of 4.5 km. The dumping procedures followed are based on recommendations of the International Atomic Energy Authority within the terms of the *London Dumping Convention (1972).

In land geologic disposal, sites must be completely isolated from circulating ground water. The rock must have very low penetrability for water, and geological faults and fractures should be absent. The area should not have a history of significant seismic activity, and should be far removed from major drainage basins and from bodies of surface water to avoid the possibility of flooding. The most suitable geological media are considered to be: rock salt, in thick beds or stable domes; hard crystalline rocks such as granite and basalt; limestone, a sedimentary rock formed by compression of shells and dead aquatic organisms; shale, a sedimentary rock produced by consolidation of clay and mud.

Seabed disposal is possible in areas of the deep seabed which are geologically and seismically stable, as well as being remote and biologically relatively unproductive. Solidified packaged high-level waste (HLW) could be placed in seabed sediments or in the underlying basaltic bedrock. The major oceanic regions considered as possible sites for disposal are continental margins, mid-oceanic ridges and the ocean floor. The latter at depths of 5–11 km contains some of the most promising areas for disposal.

Another proposed method is polar ice sheet disposal. Ice sheets are thick permanent layers of ice overlying polar land masses, especially in Greenland and Antarctica. Due to their low temperature they have a natural capability to dissipate the heat of radioactive decay. Several concepts for HLW disposal have been considered. For example, in the meltdown concept a container of waste would be inserted in a hole 50–100 m deep drilled in the ice. The decay heat would cause the ice to melt and the container would sink slowly under gravity until it reached bedrock at a depth of 3–4 km within 5–10 years.

For extra terrestrial disposal of HLW some of the space trajectories that have been considered are very high earth orbit, solar orbit other than that of the earth and planets, solar impact and escape from the solar system.

PH

radioactivity. A phenomenon exhibited by one or more *isotopes of most elements. These radioactive *nuclides, or radionuclides, are unstable even in their ground state of energy and can spontaneously change to form another nuclide. This transformation is known as radioactive decay. With very few exceptions, radionuclides do not exist naturally but are produced by nuclear reactions, e.g. by neutron capture. The rate at which a radionuclide decays is characterized by its *half-life. Several different decay processes are possible, the following types being the most important:

*alpha particle emission, e.g.

$$^{238}_{92}U \rightarrow ^{234}_{90}Th + ^{4}_{2}He$$

*beta particle emission, e.g.

$$^{60}_{27}Co \rightarrow ^{60}_{28}Ni + ^{0}_{-1}e$$

*positron emission, e.g.

$$^{22}_{11}Na \rightarrow ^{22}_{10}Ne + ^{0}_{1}e$$

*K-capture, e.g.

$$^{7}_{4}Be + ^{0}_{-1}e \rightarrow ^{7}_{3}Li$$

Radioactive decay processes are *exothermic, most of the energy released being carried by the emitted particle. In the case of beta particle and positron emission, some of the energy is carried by the *neutrinos which are also emitted. In most decay processes the product or daughter nuclide is not formed at its lowest or ground state of energy but at an excited state, from which it decays immediately by the emission of *gamma radiation. This gamma radiation emitted by the daughter product often characterizes and identifies the original radioactive nuclide. *See also* radioactive decay chain.

DB

radiological protection. The science and practice of limiting the harm to the human population from radiation.

PH

radiological protection standards. Standards designed to protect the general public and radiation workers from the harmful effects of radiation. The central principles of radiological protection as expressed by the *International Committee on Radiological Protection (ICRP) are:

1. no practice shall be adopted unless its introduction produces a positive net benefit;
2. all exposures shall be kept 'as low as reasonably achievable, economic and social factors being taken into account';
3. the *radiation dose equivalent to individuals shall not exceed the dose limits recommended for the appropriate circumstances by the ICRP.

The dose limits are designed to limit the incidence of radiation-induced effects, such as cancers and hereditary damage. For a radiation worker the dose limit is 50 millisieverts per year (i.e. 5 rem per year), and for a member of the public 5 millisieverts per year (500 millirem per year). These limits exclude medical exposure.

Further secondary standards are recommended by the ICRP to control the maximum amount of *radioactivity allowed in the environment or in the bodies of individuals. These include the following:

Maximum permissible concentrations (MPCs) for various radionuclides in air or water.

The maximum permissible body burden (MPBB), which defines the amount of a radionuclide in the body of an individual which would cause him or her to be irradiated just to the level of the recommended dose limit. MPBBs are recommended for 240 different *radionuclides.

Annual limits of intake (ALI), which are standards which have been introduced more recently to define the maximum yearly intake for individual radionuclides. They correspond to a radiation dose equivalent that is equal to the appropriate dose limit set for workers or for members of the public.

Derived working limit (DWL), which is the maximum permissible concentration of a specific radionuclide in a certain critical material in a *radiation exposure pathway that can be permitted without exceeding ICRP dose limits. For example, a DWL is set for the concentration of caesium-137 in trout flesh from Lake Trawsfynydd, Wales, where a nuclear reactor is located.

PH

radionuclide. An unstable or radioactive *nuclide.

DB

radon. A radioactive element belonging to the inert gas group, the most stable radioactive isotope of which is radon-222. Radon-222 is formed by the radioactive decay of radium-226 and itself undergoes decay to form further radioactive products called radon daughters. It emits *alpha particles with a *half-life of 3.8 days.

Since radium is present in uranium ores, the atmosphere within uranium mines contains radon-222. The presence of this gas has been responsible for a high incidence of lung cancer among miners working in inadequately ventilated and poorly managed mines. Radon-222 is also released from discarded uranium tailings; in parts of the southwestern USA, where huge piles of tailings have accumulated, radon-222 constitutes a significant public health hazard, especially in situations where tailings have been used as infill for houses and schools.

Symbol: Rn; atomic number: 86.

PH

radwaste. Short for radioactive waste.

rank. Of a coal, the extent of *coalification of the coal in the transition from peat to anthracite. It is a measure of the geological maturity rather than the geological age, and is indicated by the results of *proximate and *ultimate analysis.

WG

Rankine cycle. *See* heat engine.

Rankine scale. *See* absolute temperature scale.

Rasmussen Report (Reactor Safety Study, Wash-1400). A detailed quantitative assessment of accident risks and consequences in US commercial nuclear power plants, published in October 1975. The final draft contained comments and criticisms from government agencies, environmental groups, industry, professional societies, etc. A group directed by Professor N. C. Rasmussen (MIT) prepared the study, which compares estimated risks from accidents at nuclear reactors with risks that society faces from both natural events such as earthquakes and hurricanes and from non-nuclear man-made accidents such as dam failures and air crashes. The probability of an accident leading to the melting of the fuel core of a reactor was estimated to be 1 in 20,000 reactor-years of operation, or, for 100 operating reactors, one chance in 200 per year.

In July 1977 the *Nuclear Regulatory Commission organized a Risk Assessment Review Group to review the Rasmussen Report, and its report was published in 1978.

PH

rated output capacity. The predicted power from a generator under specified conditions. It is often used for wind generators under conditions of strong wind force.

JWT

rate of return. A proportional measure of the earnings arising from the outlay on some project, most usually applied in the context of *investment expenditure by a company. Thus, for example, in considering expenditure on measures to reduce energy use, such as better instrumentation and control equipment, a company will have to decide whether such investment will earn an acceptable rate of return. A wide variety of measures of rate of return are used, but most suffer from problems of definition regarding outlay and earnings and ignore the time pattern of such. The measure preferred in economics is the *internal rate of return, which does take account of the timing of outlay and earnings. *See also* project appraisal.

MC

rating. (1) In general, the accepted safe load of a device or part thereof under prescribed operating conditions.

(2) Of a nuclear reactor, a measure of its thermal power output per unit mass of fuel. Typical values for current reactors are:

Magnox reactors	3–5 MW/t
advanced gas-cooled reactors	10–15 MW/t
pressurized-water reactors	30–40 MW/t

TM, DB

reactivity. Of a nuclear reactor, a measure of its ability to sustain a *chain reaction, i.e. its *criticality. It is defined as:

$$(k-1)/k$$

where k is the *multiplication factor. The reactivity of a critical reactor is zero. Positive reactivity implies a supercritical reactor, and negative reactivity implies a subcritical reactor. Changes of reactivity occur spontaneously in an operating reactor as a result of burnup of fuel and the production of *fission product poisons. In a reactor operating at constant power for long periods, these changes of reactivity must be compensated by control rod movement in order to maintain criticality. *See also* temperature coefficient of reactivity.

DB

reactor. *See* chemical reactor; nuclear reactor.

reactor poisons. *Fission products or their *daughter products that have very high neutron capture *cross-sections. The two most important reactor poisons are xenon-135 and samarium-149. The build-up of these nuclides in an operating nuclear reactor can have a significant effect on the *reactivity and operating characteristics of the reactor.

DB

reactor rating. *See* rating.

real income. Income measured after making any necessary corrections for change in the prices of goods and services during the relevant period. If between two years, for example, income measured in *nominal terms increases by 10% while the general level of prices of goods and services increases by 5%, then the increase in real income is just 5%. The conversion from nominal to real measurement is achieved by using a *price index for the relevant goods and services. *See also* income.

MC

real price series. *See* constant price series.

real wages. Wage earnings measured in terms of the goods and services they can buy, rather than in *nominal terms.

Changes in real wages are calculated by dividing changes in nominal wages by the appropriate *price index.

MC

reciprocating engine. A device in which the reciprocating motion of a system of pistons within cylinders is converted into rotational motion. The reciprocating motion, as in the case of an *internal combustion engine, may be provided by the expansion of the hot gases of *fossil-fuel combustion; in the case of a *steam engine it is produced by a drop in pressure of the steam. Reciprocating engines are commonly used in the small-scale generation of electricity or to provide locomotion.

TM

reclamation. The restoration of land, e.g. in the vicinity of surface or underground mines, to its original form, or the conversion of marshland or desert to a state suitable for agriculture. Such procedures include grading of the land, drainage, fertilizing and revegetation.

PH

rectifier. A device, without moving parts, used for converting *alternating current into *direct current.

TM

recuperator. A *heat exchanger used to recover heat from hot waste gases and to preheat air for a combustion process. The heating and cooling gases pass at a steady rate on either side of a metal or refractory wall through which the heat is transferred.

WG

recurring costs [running costs]. Expenditure incurred over the lifetime of equipment, plant, buildings, etc., associated with maintenance and energy consumption, as distinct from the *capital cost of acquisition. Accountancy techniques allow capital and recurring costs to be combined into *costs-in-use, thus facilitating *cost benefit analyses.

TM

recycling. The practice of taking wastes arising in economic activity and feeding them as inputs to some production process rather than discharging them to the natural environment. Recycling is worthwhile commercially only where the 'virgin' product's price is higher than the recycled product's price. Even where this is not the case, recycling may be socially desirable insofar as it reduces the *external costs associated with discharge of residuals into the environment. Thus the imposition of *emissions taxes could increase the amount of recycling. Energy cannot be recycled, but the extent to which wastes are recycled affects the use of energy in an economy: the use of a recycled rather than a virgin product in a process frequently reduces the energy inputs required. For example, the direct energy input to the manufacture of glass containers is reduced if waste glass is used instead of virgin raw materials.

MC

reduction. In chemistry, the addition of an electron to an atom or an ion. Since hydrogen is a convenient source of electrons the term has come to mean reaction with hydrogen in either of two ways:

(a) hydrogen is added to the molecule, as in the hydrogenation or reduction of an alkene to an alkane, e.g.

$$\underset{\text{ethene}}{C_2H_4} + \underset{\text{hydrogen}}{H_2} \rightarrow \underset{\text{ethane}}{C_2H_6}$$

(b) hydrogen removes oxygen from the molecule and forms water, as in the reduction of a metallic oxide to the metal, e.g.

$$\underset{\text{iron oxide}}{Fe_2O_3} + 3H_2 \rightarrow \underset{\text{iron}}{2Fe} + 3H_2O$$

Other substances able to remove oxygen

are thus also termed reducing agents, e.g. carbon monoxide in the following reduction of iron oxide:

$$Fe_2O_3 + 3CO \rightarrow 2Fe + 3CO_2$$

WG

redundancy. Capacity within a system to deliver an output greater than normally required. Redundancy may be provided to facilitate maintenance or ameliorate the consequences of breakdown of part of the system, or to meet unpredictably high demand for output from the system. TM

Redwood viscometer. An instrument employed to measure the *kinematic viscosity of liquid fuels and lubricants. It consists of a temperature-controlled cup with a small standard orifice in the base. The time for a standard volume of the liquid to flow through the orifice is noted and recorded in Redwood seconds at that particular temperature. WG

reference profile. A time-based record of the output of a system under a certain set of conditions referred to as the base or reference condition. This may then be used to compare the performance of other operating conditions.

MS

refinery gases. Mixtures of low molecular weight gaseous hydrocarbons which arise in an oil refinery as a byproduct of the various operations carried out there. If they cannot be sold as fuel gas they are burned within the refinery to heat furnaces and raise steam. Gas which is excess to requirement may be disposed of by burning in a *flare. WG

reflectance. The ability of a surface to reflect light. Reflectance is measured as that proportion of the light falling on a surface which is reflected. The reflectance of a surface, which is high for light-coloured surfaces and low for those of a dark colour, affects the observer's impression of brightness. *See also* absorptance; transmittance. TM

reflectivity. The ability of a surface to reflect radiant heat energy. Reflectivity is measured as that proportion of incident radiation which is not absorbed or transmitted but is reflected. TM

reflector. That which reflects incident *radiation in a thermal *nuclear reactor, material surrounding the *core of the reactor in order to scatter neutrons escaping from the core back into it. The desirable properties of the reflector material are low *mass number, high scattering *cross-section and low capture cross-section. Materials used as *moderators for thermal reactors are also suitable as reflectors, namely water, heavy water and graphite. The reflector is usually an outward extension of the core which does not contain any fuel. DB

reforming. A series of chemical reactions which cause the rearrangement of the molecular structure of hydrocarbons. The reactions are utilized in the petroleum refinery to produce aromatics and to increase the *octane rating of naphtha fractions and thus make them suitable for blending into gasoline.

The chemical reactions involved include the *hydrocracking of high molecular weight paraffins, the *isomerization of paraffins to branched-chain isomers, the *dehydrogenation and *cyclization of paraffins to aromatics, and the dehydrogenation of naphthenes to aromatics. The reactions are carried out over a platinum catalyst in the presence of hydrogen. The process temperature, pressure and hydrogen concentration must be carefully selected to achieve the correct balance of hydrogenation/dehydrogenation reactions. *See also* catalytic reforming; thermal reforming.

WG

refrigerant. The fluid used in *absorption chillers and *vapour compression chillers which, in the course of evaporation, provides the cooling or refrigerating effect. The refrigerating effect of the variety of available refrigerants, each of which is identified by a refrigerant number (water is 718, ammonia is 717), is measured in kJ/kg.

TM

refrigeration. The production of a cold(er) environment. Since heat must be taken from a system at low temperature and rejected to one at a higher temperature (the environment), work must be expended. There are several systems for creating refrigeration. Some require a heat source but are very inefficient in *second law efficiency terms; others use motive power and are more efficient.

MS

regenerator. A *heat exchanger which consists of a large refractory brick chamber through which hot waste gas from a furnace is passed. When the chamber reaches a suitable temperature, the gas flow is switched to heat a second such chamber while cold combustion air for the furnace is preheated by being passed through the first regenerator. When the chamber has cooled, the hot waste gas is again passed through. This cyclical system of heat recovery has been used with *blast furnaces and *coke ovens.

WG

regression analysis. A technique widely used in science and *econometrics, by means of which the numerical values for the parameters of an equation can be determined from available data on the variables of the equation, typically according to the least squares criterion for goodness of fit. Suppose it is desired to find numerical values for a and b in the equation $y = a + bx$, where, say, y and x are respectively the quantity of oil sold and the price of oil so that the equation is a simple *demand function for oil. In the diagram crosses represent data points for y and x, and a possible regression line is shown as $\hat{y}_i = \hat{a} + \hat{b}x_i$. The least squares criterion involves taking as the estimates for $\hat{a}$ and $\hat{b}$ those values for a and b which are such that the sum of the squared residuals about the regression line is minimized, i.e. $\hat{a}$ and $\hat{b}$ such that

$$\sum_{i=1}^{T} e_i^2 = \sum_{i=1}^{T} (y_i - \hat{a} - \hat{b}x_i)^2$$

is minimized, where i indexes data points of which there are T in number. Conceptually, by varying the values of $\hat{a}$ and $\hat{b}$ the position and slope of the regression line are varied until the smallest possible value for Σe_i^2 is found. In practice there exist general rules for calculating the required values of $\hat{a}$ and $\hat{b}$ directly from the data on y and x, which rules can be used to write an algorithm. Most computer installations now have standard packages mounted for regression analysis.

MC

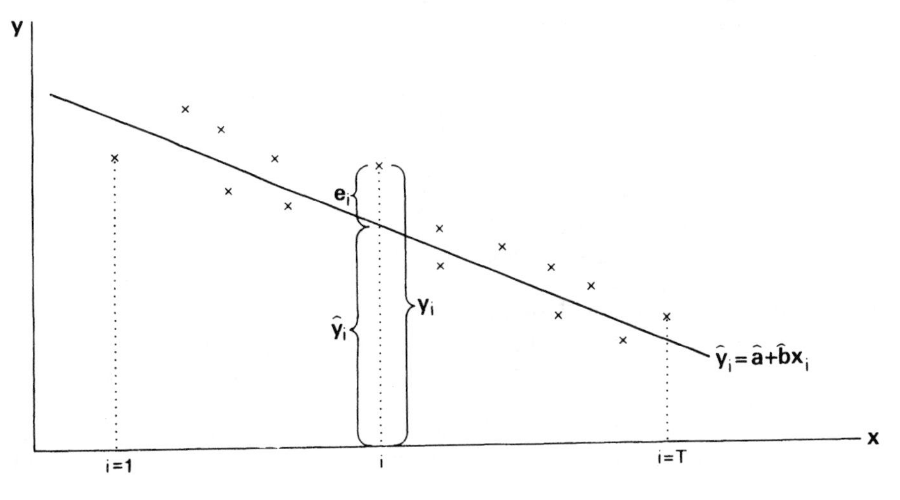

regressivity. A property of a system of taxation such that the proportion of income taken in tax falls with rises in the levels of income. It is the opposite of *progressivity. A system of *indirect taxation may be progressive or regressive in its overall impact, depending on the commodities taxed and the relationship between income levels and household-spending patterns, and on the rates of tax. The effect of indirect taxes on energy commodities, or of higher energy prices for whatever reason, is generally held to be regressive since low-income households typically spend a larger proportion of their income on energy commodities than do high-income households.

MC

relative density [specific gravity]. For a solid or liquid, the ratio of its density to that of water at specified temperatures. The temperature selected may be the same throughout, e.g. the relative density could be the ratio of the density of the substance at 20°C to that of water at 20°C; alternatively the ratio may be stated with reference to water at 4°C. Since at that temperature the density of water is 1.0 gram/cm^3, the value of relative density then becomes numerically equal to density in grams/cm^3.

For a gas, relative density is the ratio of its density to that of air at the same conditions of temperature and pressure, e.g. 0°C and 1 atmosphere. The usual abbreviation is 'rel d wrt air = 1'.

In technological calculations involving fuels, the average molecular weight of air may be taken as 29 grams and the relative density of a gas obtained by calculating the ratio of its average molecular weight to that of air. For example, the molecular weight of methane is 16 grams thus the relative density of methane at 0°C and 1 atmosphere is 16/29 = 0.55.

WG

relative humidity. The ratio, expressed as a percentage, of the amount of water vapour in the air to the amount that would saturate the air at the same temperature.

TM

relative price. The relative price of energy, say, is the quantity of some other good which has to be given up to obtain a unit of energy. There are thus many relative prices for energy according to which other good is used for comparison. If the *absolute price of a unit of energy is $2 and the absolute price of beer is $1 per unit, then the relative price of energy in terms of beer is 2. It is changes in relative prices which are relevant for decisions about resource allocation and for the way the *price mechanism works. Thus if the price of energy increases at the same rate as all the other prices, its relative price is not changing and no *substitution of other inputs for energy in production is called for. It is only when the absolute price of energy is rising faster than all other absolute prices that the relative price of energy is rising. In doing *cost–benefit analysis, for example, it is important to correct for *inflation and consider only changes in relative prices.

MC

rem. The unit of dose-equivalent, recently superseded by the SI unit, the sievert (Sv). 1 rem is equal to 0.01 Sv. The use of the rem is likely to continue for some time. *See* radiation dose.

DB

renewable energy. Energy sources that do not depend on the extraction of *fossil or *fissile resources, but on naturally occuring energies flows. Although these may also be declining in

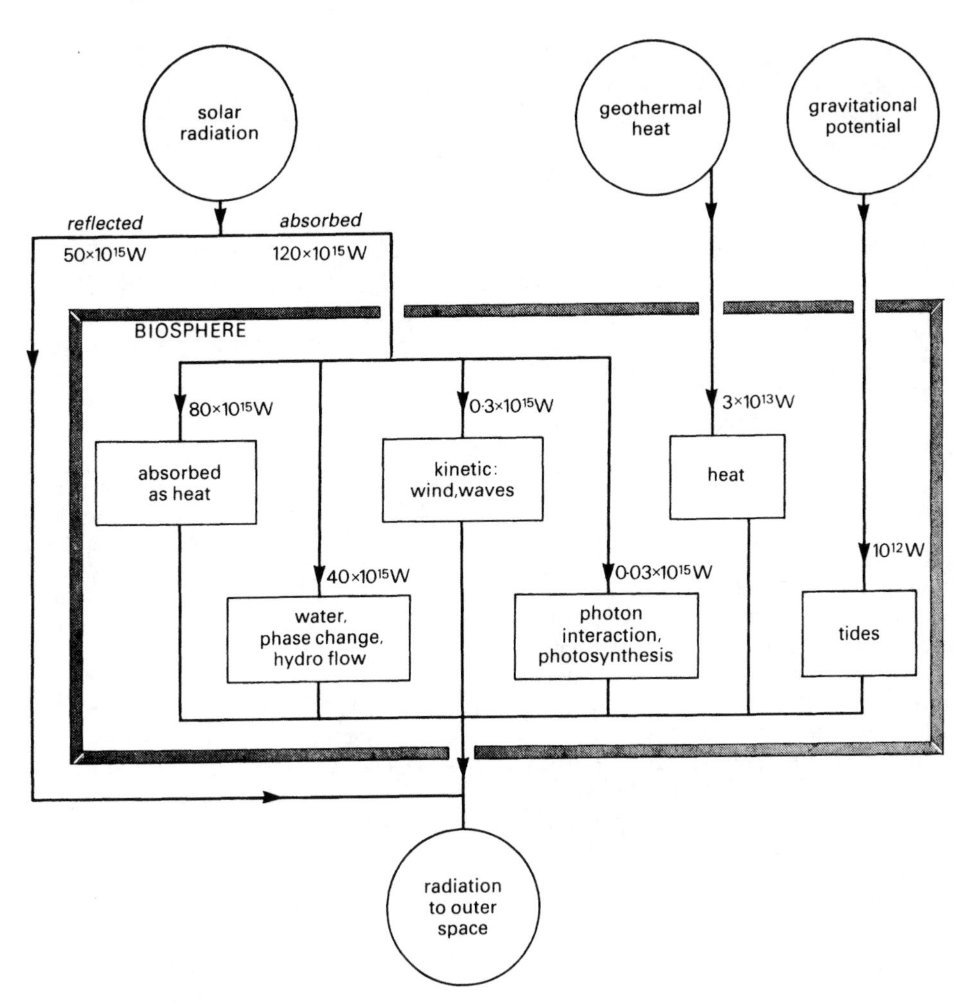

Figure 1

vigour as time passes, the time scale is so great that it may effectively be ignored. Examples are incident radiation from the sun, *geothermal energy, and the rise and fall of the tides. Although these energy flows are some 10,000 times current (1987) world energy use (*see* fig. 1), harnessing them and converting them into economically useful energy sources can require considerably more capital per unit of output than conventional fossil fuel sources. An exception is *biomass (wood, crops, plants, etc.), where the *solar energy is captured through *photosynthesis. In the past biomass provided the single major source of energy for all civilizations, but today is restricted to the less-developed countries, many of whom are consuming forests at a higher rate than they are being replaced, creating serious problems of soil erosion and climate change in addition to the impending fuelwood crisis facing them.

There are two essential problems in harnessing renewable energies:

(a) They occur at low *energy flux, typically at not more than 0.1 kw/m^2.
(b) Some, such as solar radiation, are not continuous, so that they must be integrated with means of *energy storage.

Considerable efforts have been made to develop technologies for the conversion of renewable energy sources into storable high-quality energy sources. *Hydroelectric systems are one example from rain water, while conversion of high-yielding *C_4 plants to *ethanol is another.

MS

rent. In economic analysis generally, a payment to a *factor of production in excess of what is needed to keep it in its present use. In the context of *natural resource economics, the excess of the per unit price of the extracted resource over the *marginal cost of extraction. For example, if Saudi Arabian crude oil has a world price of (say) $25 and the marginal cost of extraction is (say) $1 per barrel, the rent element in the price is $24. This differential arises from the absolute *scarcity of Saudi crude. The history of relations between the international oil companies and the countries where oil deposits are found can be interpreted as a struggle over who benefits from the rent element in the price of oil. For example, OPEC members have generally moved from attempting to capture some of the rent by means of taxation to a situation where the oil extraction operations are effectively nationalized so that the rent accrues directly to the government. *See also* depletion programme; natural resource taxation; severance taxes.

MC

replacement energy requirement. The energy needed to replace an energy-transforming device or system or (as in renewable sources) an energy-producing device or system, taking into account differences of energy quality. *See* gross energy requirement; pay-back time.

MS

reprocessing (nuclear). The chemical or metallurgical treatment of *irradiated fuel designed to separate the *uranium, *plutonium and *fission products from each other.

DB

reserve currency. A foreign currency which a country is prepared to hold as part of its international *reserves. This has usually been sterling and the US dollar.

MC

reserves. (1) With respect to natural resources, there are several classifications of reserves. The most important distinction is between proven reserves and ultimately recoverable reserves. Proven reserves of oil, coal, etc., are usually defined as the amount of the resource which is commercially worth extracting from known deposits at today's prices, or which will be commercially worth extracting at today's expected future prices, with today's technology. Ultimately recoverable reserve estimates refer to amounts which it is calculated will be eventually extracted, allowing for discoveries of new deposits and changes in technology. Reserves figures for any non-renewable natural resource are always in the nature of estimates and should be treated with caution.

(2) In a financial context, receipts not spent but put aside by a company or country for some specified use. An example is a country's foreign exchange reserves, which usually comprise gold,

*special drawing rights and US dollars.

MC

reservoir rock. A sandstone or limestone rock underground, the pores of which are filled with petroleum and above which there is a layer of rock impermeable to the upward movement of hydrocarbons. *See also* anticline; cap rock.

WG

residence time. *See* retention time.

residue. The high-boiling material remaining in a distillation unit after more volatile components have been distilled off. In a petroleum refinery the distillation residues are normally dark-coloured and viscous. If not suitable for further treatment, they may be sold as *heavy fuel oil or blended with a *heavy distillate to provide a fuel oil of lower *viscosity.

WG

resinite. The fossilized remains of plant resins in the shape of small, round inclusions in *vitrain. They show a yellow–orange colour under transmitted light and are mainly found in *bright coal.

WG

resistance (1) (electrical; symbol: R). The property of a material which resists the flow of electric current when an *electromotive force is applied to it. Resistances may be connected in series with each other, in which case the total resistance is equal to the sum of the individual resistances; when connected in parallel the total resistance is equal to the inverse of the sum of the inverses of the individual resistances. The unit of electrical resistance is the *ohm (Ω); the inverse of resistance, *conductance, is measured in *siemens.

(2) (thermal; symbol: R). The property of a material which resists the conductive flow of heat when a temperature difference exists across it. Resistances may be connected in series with each other, in which case the total resistance is equal to the sum of the individual resistances; when connected in parallel the total resistance is equal to the inverse of the sum of the inverses of the individual resistances. Thermal resistance is measured in square metres kelvin per watt (m^2 K/W).

TM

resistivity (1) (electrical; symbol: ρ). The electrical *resistance of a material per unit length. The unit of electrical resistivity is the ohm metre (Ω m). The reciprocal of resistivity is *conductivity, measured in siemens per metre.

(2) (thermal; symbol: k). The thermal *resistance of a material per unit length. Thermal resistivity is measured in m K/W. The reciprocal of resistivity is *conductivity, measured in W/m K.

TM

resonances (neutron). High values of neutron *cross-sections which occur at certain well-defined neutron energies. Resonances are most important in heavy elements such as uranium and for neutrons of energy in the range 1–1000 eV, as shown in the diagram. *See also* Doppler broadening.

DB

resources. In economics, a general term used to refer to the totality of the *factors of production available for use in an economy. If some of the economy's *capital stock is not being used, for example, available resources are not being fully utilized. A situation of Pareto optimality is one in which all available resources are being utilized in such a way that it is not possible to rearrange the pattern of use so as to make somebody better off, except at the cost of making some other individual(s) worse off. *See also* natural resources.

MC

Resources for the Future (RfF). A US think tank which has published many reports on energy questions. Its purpose is to advance research and education in development, conservation and use of *natural resources, and in the improvement in the quality of the environment.

Address: 1755 Massachusetts Avenue N.W., Washington, D.C. 20036, USA.

MC

respiration. Consumption of oxygen to provide energy for metabolic processes in living organisms. Carbon dioxide is given out as a product of respiration.

JWT

response factors. Computed values which characterize the thermal behaviour of, for instance, the walls and roof of a building, under the influence of a time-series of temperature change.

TM

retention time [detention time; residence time]. The length of time necessary for the *fermentation of a given quantity of substrate to an end-product. In the case of anaerobic digestion to *biogas, the total time period for virtually complete digestion may be up to 90 days when operated as a batch fermentation. In continuous or semicontinuous operation, a given volume of input feed is introduced into the vessel during the course of each day and an equal volume of digester contents withdrawn so that once the fermentation is established the rate of gas production is steady. The retention time then equals the volume of the digester divided by the feed rate (volume/time). The loading rate (con-

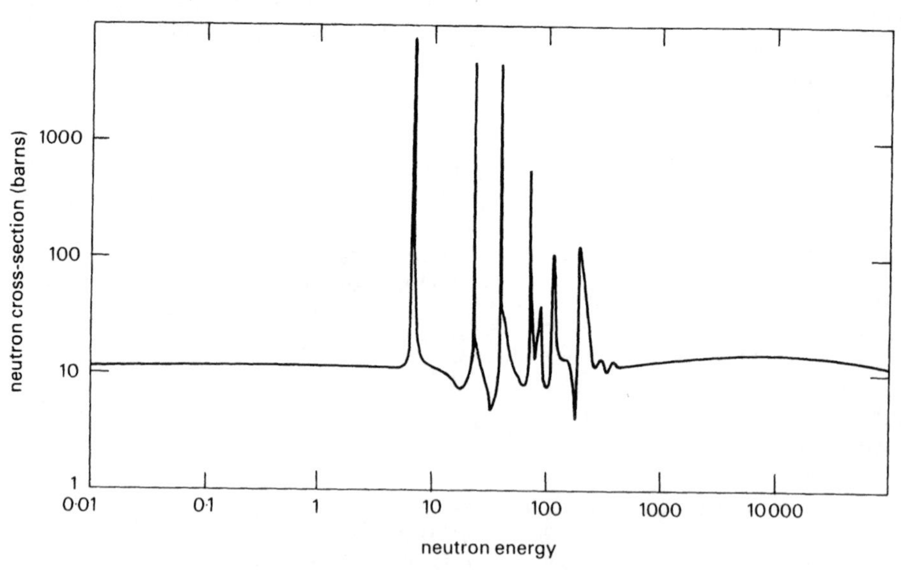

The total cross-section of uranium 238

The total cross-section of uranium-238

centration) is particularly important since at high loading rates the retention time is reduced.

In a truly continuous culture the loading rate (and hence the retention time) is adjusted to precisely match the growth rate of the slowest growing microbes essential to the process – the *methanogenic bacteria. However, in the case of *Gobar gas plants operating under varying conditions of temperature, pH, agitation, etc., in the Third World, such fine control is impracticable. Retention times of 30–50 days are common under tropical conditions where cow dung is the main feed to the digester and where the digestion is *mesophilic. *Thermophilic digestion at temperatures around 55°C can reduce the retention time of a mesophilic digestion by 50% or more.

CL

retort. A unit in which solid fuel is carbonized on a commercial scale. It is designed for the production of coal gas or for the production of gas and a smokeless fuel such as a semicoke. The carbonizing units are tall and have a number of narrow vertical chambers of circular section. Retorts have been used for the carbonization of coal, lignite and oil shale.

WG

returns to scale. The increase in output that results from increasing all inputs in the same proportion. If output increases by more than the proportion by which the factor inputs were increased there is increasing returns to scale; there can also be constant returns to scale; and decreasing returns to scale. The situation with respect to returns to scale is an important property of the *production function. The car manufacturing industry is a standard example of a situation where increasing returns to scale have been operative.

MC

revaluation. The opposite of *devaluation. *See also* appreciation; depreciation.

MC

Rexco. A proprietary solid smokeless fuel prepared by carbonizing coal in a vertical cylindrical chamber about 7.5 metres high. Carbonization is achieved by burning fuel gas within the top of the chamber and passing the hot products of combustion downwards through the charge of coal.

WG

Reynolds number (Symbol: *Re*). A dimensionless number whose value can distinguish between streamline and turbulent flow around objects in a fluid stream, or of fluids in conduits and pipes. It is the ratio of the inertial to the viscous forces acting on the flowing fluid. *Re* is calculated from

$$Re = \frac{vd\rho}{\mu}$$

Where d is a characteristic dimension of the object, v is velocity; ρ is density of the fluid and μ is dynamic viscosity of fluid.

MS

***Rhizobium*.** A bacterium existing in *symbiosis with leguminous plants such as lucerne and the soyabean. It can fix atmospheric nigrogen in the root nodules of the associated plants, the nigrogen compounds formed being utilized by the legume, which in turn supplies energy-rich sugars to the bacteria via photosynthesis. The efficiency of *nitrogen fixation can often be improved by the inoculation of more effective strains than those naturally colonizing the plant.

CL

rich mixture. A mixture of fuel gas or vapour with less air than that required for complete combustion.

ring (chemical). A closed formation of atoms which in a homocyclic ring are all of one kind, and, in a heterocyclic ring, are of different kinds. In organic chemistry the most stable homocyclic ring appears to be that composed of six carbon atoms, as found in benzene (C_6H_6) and cyclohexane (C_6H_{12}). Common heterocyclic rings are those in which one or more of the carbon atoms have been replaced by a nitrogen, sulphur or oxygen atom.

WG

Ringelman chart. A simple device used for classifying the apparent density of smoke by reference to four different shades of grey printed on a piece of transparent plastic.

PH

risk. A phenomenon arising in a situation where the actual outcome of a decision cannot be known when the decision has to be taken, but where objective probabilities can be assigned to the possible outcomes on the basis of past experience. For example, an individual who has to decide whether to enter a particular occupation supposed to be dangerous cannot know when he will die if he does take up the occupation. But mortality statistics mean that the probabilities that he will die at particular ages can be objectively calculated. An alternative example is the decision on whether or not to drill exploratory oil wells in a particular area: probabilities for finding deposits of particular sizes can be calculated from past experience in similar areas. Decision makers will usually require compensation for taking decisions to go ahead with projects involving risk: their expected *rate of return will have to be higher than the rate of return that would have given rise to a decision to go ahead in the situation where the outcome was known with certainty. *See also* uncertainty.

MC

risk assessment. A technique for estimating the risk of death or injury to the human population from exposure to a particular environmental pollutant, technology, occupation, etc. For example, it has been estimated that the average annual risk of death in the UK from smoking 20 cigarettes a day is 1 in 200, while that from being exposed to 1 millisievert of radiation is 1 in 80,000.

PH

risk factor. An estimate of the risk of fatal cancer or serious hereditary defect as a result of exposure to radiation. It is usually expressed as risk per unit dose equivalent, i.e. risk per *sievert or per *rad. For example, the risk factor for leukaemia is estimated to be about 1 in 500 per sievert. This means that if a person receives a dose equivalent of 1 sievert to the bone marrow there is a 1 in 500 chance that he will eventually die of leukaemia. *See also* radiation dose; radiological protection standards.

PH

Risley. An establishment of the United Kingdom Atomic Energy Authority which provides engineering-design services and technical and economic assessments and has engineering laboratories. Risley is also the headquarters of British Nuclear Fuels Limited.

Address: Risley, Warrington, Cheshire, England.

DB

rock bed. A mass of rocks loosely piled with air spaces. Hot air passed through the bed gives up heat for storage. Later cold air may be passed through and so itself became heated. In a *solar energy heating system, rock beds can be used to store heat for days, weeks and occasionally months.

JWT

rose. In wind power technology, (1) a diagram of lines pointing from a central point, with each line representing the magnitude of the average wind speed in the direction of the line;

(2) a turbine with many blades positioned on a circular framework.

JWT

rotor. The rotating part of a machine. In wind turbines for example it is the rotating blades and hub.

JWT

roughness length. The height within vegetation, trees, buildings, etc., below which the wind has unpredictable behaviour. It is technically defined by mathematical equations describing the increase in wind speed with height.

JWT

ROV. Remote-operated vehicle, as used, for example, in undersea operations of the oil industry.

MS

Royal Commission on Environmental Pollution (UK). A standing Royal Commission appointed on 20 February 1970 to 'advise on matters, both national and international, concerning the pollution of the environment, on the adequacy of research in this field, and the future possibilities of danger to the environment. The Commission has published six reports: the sixth report entitled 'Nuclear Power and the Environment' was published in 1976 under the chairmanship of Sir Brian Flowers, and is often referred to as the 'Flowers Report'.

PH

royalty. In the context of *natural resource extraction, a payment, in the form of a levy on the value of output, to the owner of the site where extraction occurs by the extractor of the resource under the terms of some license or concession agreement between the site owner and the extractor. The landlord or site owner may be either a government or a private individual or company. In the UK all mineral rights are vested in the crown so that all royalties are effectively taxes on mineral extraction paid to the government. In the USA, mineral rights are often vested in private ownership so that royalties are often paid to private landlords.

MC

running costs. *See* recurring costs.

run-of-mine coal. Coal as mined, varying in size from large lumps to fine dust and mixed with mineral matter from rocks adjacent to the coal seam.

WG

run of the river. A descriptive term used for *hydropower generation from the immediate flow of a water supply and not from a stored volume of water.

JWT

run of the wind. The distance travelled by wind in a specified time, equal to the average wind velocity multiplied by the time specified. Measurement is given by certain types of *anemometer that continually add to a dial reading.

JWT

ruthenium-106. A radioactive isotope of the metallic element ruthenium with a *half-life of 369 days. It is a *fission product present in the low-level liquid radioactive effluent from a nuclear-fuel reprocessing plant such as Windscale. Prior to 1972 it determined a *critical pathway for radiation exposure of the human population resulting from the operation of this plant. The edible seaweed *Porphyra umbilicalis* accumulates ruthenium-106 from sea water. This seaweed used to be collected from the Cumbrian shore and processed into a foodstuff called laverbread, mainly consumed in South Wales. PH

S

safety rods. In a *nuclear reactor, rods containing materials such as boron, cadmium or hafnium, which have very high neutron capture *cross-sections. During the normal operation of a reactor they are held out of the *core, but in the event of any emergency or malfunction of the reactor they are dropped into the core. The resulting increased neutron absorption in the rods makes the reactor subcritical (*see* criticality) and the fission *chain reaction is rapidly shut down.

DB

saline energy [osmotic energy]. Energy obtained by means of the difference in amount of dissolved matter between sea water and fresh water. The mechanisms suggested for tapping this energy source include using the saline liquid in a *battery, perhaps with a membrane structure, or in a mechanical system where the larger density of saline water would produce hydrostatic pressure.

JWT

salt dome. Rock salt tends to flow plastically when subjected to high pressures underground and may be forced up through overlying strata to form a plug or dome. Since salt is impervious to oil, porous sediments which are bent up by the intrusion of the dome create a trap or reservoir for crude *petroleum.

WG

samarium. A metallic element whose stable isotope, samarium-149, is produced in nuclear reactors by the decay of the fission product promethium-149. Samarium-149 has a very high neutron capture *cross-section and acts as a *reactor poison.

DB

Sankey diagram [Sankey chart; spaghetti diagram]. A diagram which depicts the sources and uses of energy within a given system, whether it be a factory or an entire economy. The width of the supply lines are an indication of the proportion used of each type of energy source, while the delivery of energy shows lines proportional to each principal type of end-use.

MS

saturated hydrocarbon. A hydrocarbon in which each carbon atom is linked to the next one in the molecular chain or ring by only one valency bond. Examples include the *naphthenes and *alkanes.

WG

saturation. The state of thermal equilibrium at the interface between a gas and its parent liquid.

TM

saving. That part of *income not consumed and hence an addition to the stock of *wealth.

MC

Savonius rotor. A vertical-axis wind machine with specially shaped panels which cause rotation by *drag forces. The Savonius rotor starts in moderate winds but does not develop high rotational frequencies. JWT

scarcity. A situation in which the resources available for producing *output are insufficient to satisfy wants. Economics is the study of choice in conditions of scarcity, without the existence of which there would be no need to think about the best allocation of available resources. With human wants virtually unlimited, scarcity is pervasive. Absolute scarcity refers to the situation which arises with non-renewable *natural resources such as fossil fuels, where use must imply eventual exhaustion. MC

scattered radiation. Light or other *electromagnetic radiation bent away from its initial direction by small particles or molecules in a gas or liquid. Various processes cause this scattering. For example the blue of the sky is caused by blue light in solar radiation being scattered more than red light. Scattered light forms part of the diffuse solar radiation. JWT

scattering (neutron). An interaction between a *neutron and an atomic *nucleus in which the neutron can be considered to bounce off the nucleus, an interaction which is analogous to the collision of two hard spherical balls. If the neutron initially has a high energy (greater than the kinetic energy of the nucleus), the effect of the scattering collision is to reduce the neutron's energy. The nucleus acquires some energy but is otherwise unchanged, i.e. no nuclear transmutation takes place as would be the case in a neutron capture reaction (*see* absorption). DB

scenario. An expression of the future based on explicit or implicit assumptions. Although scenarios may be evaluated through the medium of a *model, they are only as good as the quality of the inbuilt assumptions, and hence cannot be used for rigorous *forecasting. MS

Scholler process. A process developed in Germany during the First World War to hydrolyse woody cellulosic materials using dilute sulphuric acid to form fermentable sugars for subsequent fermentation and distillation to *ethanol. The acid was present at a concentration of 0.5–1% and the hydrolysis took place at 180°C for 12 hours, when 352 kg of fermentable sugar was the average yield from 1 tonne of wood. Several plants using the Scholler hydrolytic process were operating in Europe up to and during the Second World War, with the end-product used mainly as a transportation fuel. *See also* Madison process. CL

scintillation detector. A *radiation detector in which the radiation is absorbed in a phosphor, e.g. a crystal of sodium iodide, and produces tiny flashes of light. These flashes are converted at the photo-cathode to electron pulses which are amplified in a photomultiplier tube. The output pulses at the anode are proportional to the energy of the radiation absorbed in the phosphor. DB

scram. The rapid shut-down of a nuclear reactor in the event of a malfunction or an emergency. DB

seabed disposal. *See* radioactive waste management.

seasonal adjustment. The removal from time series data of repetitive seasonal variations, so that in the seasonally adjusted or de-seasonalized data any underlying trends or changes are more clearly revealed. For example, if monthly data on domestic coal or electricity purchases is to reveal anything about the effects of fuel prices on domestic fuel consumption, it must first be seasonally adjusted. Many officially published data series, e.g. energy consumption, are provided with and without seasonal adjustment.

MC

secondary air. Air added over the top of a bed of burning solid fuel or around the flame of liquid or gaseous fuels to complete the combustion process and to control the size and shape of the flame. WG

secondary recovery. The process of recovery of oil from an oil well after it ceases to flow under its own internal pressure.

MS

second best problem. If any one of the conditions for *efficiency in allocation cannot be fulfilled, the second best theorem states that the best attainable situation can then only be realized if all the other conditions for efficiency in allocation are violated. The problem which arises is that it follows that intervention according to standard prescriptions to meet one source of *market failure is not justified if all other sources of market failure are not either absent or corrected for. Consider the case of a *monopoly supplying electric power where the burning of coal gives rise to *external cost and *pollution. Correcting the pollution problem by, for example, the imposition of an *emissions tax will have the effect of reducing the output of electricity. But this output is already lower than is required for efficiency in allocation by virtue of the fact that the electricity is produced by a monopoly. Imposing the emissions tax alone is as likely to reduce *welfare as it is to improve it. What is required is an emissions tax together with a measure which would on its own induce an expansion of electricity output. The inappropriateness of a piecemeal approach to problems of market failure makes it difficult to design appropriate market intervention strategies for the achievement of efficiency in allocation.

MC

second law efficiency. (1) (work) The ratio of the work effected by a unit of energy at a given temperature compared to the work that could be theoretically obtained from an ideal heat engine at the same conditions (*see* Carnot cycle). Practical heat engines may have second law efficiencies between 50% and 80%.

(2) (heat) When a high-temperature energy source is used to produce heat for a low-temperature duty (e.g. an oil-fired furnace used to produce space heating at 20°C) the work potential of the degraded heat is much less than that of the original heat in the furnace. Thus the second law efficiency of a domestic oil-fired heating system may be as low as 8%, even though the *first law efficiency may be 70% or better.

MS

second law of thermodynamics. The physical law of energy processes explaining the maximum amount of heat energy that can be transformed into work energy by a *heat engine. Since energy cannot be destroyed (*see* first law of thermodynamics), the second law also explains the minimum amount of heat that is rejected by a heat engine, e.g. in exhaust gases. The law can be quantified by reference to the *Carnot cycle for a heat engine, since no heat engine can produce more than a Carnot cycle engine working at the same temperatures. If a heat engine takes in energy Q_1 at temperature T_1 kelvin and ejects waste heat at T_2 kelvin, then the maximum amount of work produced is

$$W = Q(T_1 - T_2)/T_1.$$

In practice, real engines are far less efficient than a Carnot engine and also the so-called waste heat can be usefully used (*see* combined heat and power; negentropy).

JWT

securities. Financial *assets that give entitlement to income, such as *stocks, *shares, *bills and *bonds. It is a broad term covering many financial assets of short-term, medium, and long-term duration.

MC

Seebeck effect. The phenomenon in which an *electromotive force (emf) is produced when two wires of different composition are joined at their ends to form a circuit, as long as the two junctions are at different temperatures. The emf drives a current around the circuit. The magnitude of the emf increases with increasing temperature difference. *See also* Peltier effect, thermocouple. JWT

selective surface. A surface that absorbs radiation well at solar frequencies but emits poorly in the infrared at the temperature of the surface. Thus the selective surface will be at a higher temperature than a *black body surface when absorbing solar radiation. JWT

Sellafield (originally Windscale). An establishment of the United Kingdom Atomic Energy Authority and British Nuclear Fuels plc in Cumbria, England. It was the site of the first British plutonium-producing *nuclear reactors. commissioned in 1952 and closed down in 1957 after a serious fire in one of them. More recently the experimental advanced gas-cooled reactor (AGR) was built and operated at Sellafield. This reactor, the Windscale AGR, is now closed down and is the subject of trials to develop decommissioning techniques for nuclear reactors. Sellafield is also the site of the *irradiated fuel reprocessing plant and waste storage facilities of British Nuclear Fuels plc. The *Calder Hall nuclear power station is adjacent to Sellafield.
Address: Sellafield, Cumbria, England.

DB

semicoke. A coke formed by a low-temperature carbonization process. It contains about 10% volatile matter and may be burned smokelessly on an open grate or in a closed stove. Owing, however, to its relatively low density, the stove requires to be refuelled more frequently than with *anthracite.

WG

semiconductor. A material having an electrical *resistivity approximately midway between that of a good conductor and that of a good insulator. TM

sensible heat. The heat energy stored in a substance as a result of an increase in its *temperature. It can be contrasted with *latent heat, which is an *isothermal property associated with phase change.

TM

separative work. The energy required to enrich a dilute component in a mixture. The term is usually applied to uranium *enrichment. The separative work requirement depends on the percentage of uranium-235 in the feed, the product and the tailings. Different enrichment processes have different energy requirements (or costs) per unit of separative work, the *gaseous diffusion process requiring approximately ten times as much as the centrifuge process. DB

services. In economics, *commodities other than *goods, i.e. outputs from productive activity which are not material substances, such as education, banking, insurance, policing, etc. The relative size of the services sector is increasing in advanced economies. *See* engineering services.

Sesbania. A genus of shrubs and trees which are *legumes, i.e. able to fix atmospheric nigrogen. *Sesbania grandiflora* is a multipurpose tropical tree which holds much promise for reforesting eroded and grassy wastelands. It also combines well with agriculture in areas where trees are not usually grown, thus becoming an important source of *firewood. This tree has long been used for firewood in SE Asia, though the wood is not very dense, having a relative density of 0.42. Up to 3000 stems can be planted per hectare, and wood yields of 20–25 m^3/ha.y are commonly achieved in Indonesian plantations – the equivalent of around 200 GJ/ha.y. It is believed that this figure can be improved quite substantially.

CL

severance taxes. Taxes on the output of firms extracting *natural resources, such as oil. They may be *ad valorem taxes or specific taxes with the rate set per unit of physical output, as with a 'barrelage tax' on crude oil production. Typically such taxes are intended, on equity grounds, to enable the government to capture some of the *rent arising in the extraction of the natural resource. Except in ideal circumstances they have the effect of changing the *depletion programme in the direction of less rapid depletion, i.e. a given deposit will be more slowly exhausted with severance taxes in operation than without.

MC

severity index. A measure of climatic stress experienced by a building. The number of *degree days is a simple severity index which fails, however, to embody the full range of relevant factors.

TM

shade. An area of partial darkness caused by an object that obstructs rays of light from the sun or other source of illumination. The degree to which the *glazing on the envelope of a building is shaded – by the building itself, by other buildings, by trees, etc. – greatly affects its thermal response and hence its energy consumption. In locations where *insolation is great, shading devices may be fitted above the windows of a building to reduce solar heat gain.

TM

shade ring. A metal ring of width sufficient to shade a solar radiation *pyranometer from the direct rays of the sun. The instrument thereby records only the diffuse radiation from the sky, clouds and reflecting objects not in the path of the sun.

JWT

shadow price. An imputed *price for a good or service, used, for example, in *cost–benefit analysis. In some cases shadow prices are used because no market price exists, as with public health services or environmental quality. In other cases shadow prices are used because market prices do not properly reflect *opportunity costs. For example, in a region with a high rate of unemployment the going wage rate is unlikely to reflect the opportunity costs of employing labour there. Again, if a country has a *tariff on or a *quota for oil imports, the domestic price for oil will be above the world price, and in evaluating oil conservation measures oil saved should be valued at a shadow price which is the world price.

MC

shale oil. An oil produced from oil shales by heating the shale to about 600 °C in a specially designed retort. It is a dark-coloured liquid similar to crude petroleum and may be distilled to create a number of fractions, including a spirit, a wax-free oil and a wax-bearing oil. The first two are refined to produce gasoline and diesel fuel, and paraffin wax may be separated from the third. The diesel fuel is of high quality but the octane number of the motor spirit can be as low as 58.

WG

shape factor. A measure of the geometry of an electrical or heat conductor. The variety of geometric shapes which pass the same electric current in response to a given voltage drop, will have the same shape factor; analogously, the variety of geometric shapes which have the same steady-state conductive heat flow in response to a given temperature difference, will have the same shape factor.

TM

shares [equities]. A particular type of financial *security issued by companies. The holder of a share is a member and part owner of the company to the extent of the nominal value of his share certificate, and participates in profit through dividend payments. Shareholders are risk bearers in that, in contrast to the holders of *bonds, they do not receive fixed or guaranteed dividend payments.

MC

shelter. Protection from an inclement environment. Building design in different parts of the world reflect the type of shelter from climate influences required by the occupants.

TM

shield heating. Heating of the concrete *biological shield of a *nuclear reactor as a result of the absorption of *gamma radiation. If the temperature rise is high enough it may lead to thermal stresses and cracking of the concrete. *See also* thermal shield.

DB

shielding. The material surrounding a radioactive source to protect personnel in its neighbourhood from any radiation emitted. Shielding materials are normally chosen for their high density, these being the most effective for absorbing highly penetrating radiation such as *gamma radiation. Lead is used as the shielding material for small sources of radiation in hospitals, laboratories, etc. Concrete and steel are used for the shielding of nuclear reactors, which are intense sources of radiation. Boron is also used as a neutron shield in reactors since it has a high capture *cross-section for neutrons.

DB

shift reaction. The reaction of *carbon monoxide with water vapour to produce carbon dioxide and hydrogen. It is used to adjust the ratio of carbon monoxide to hydrogen in *synthesis gas so that it is suitable for a particular reaction, such as the production of *methane.

shift reaction:

$$CO + H_2O \rightarrow CO_2 + H_2$$

methanation:

$$CO + 3H_2 \rightarrow CH_4 + H_2O$$

The carbon dioxide may then be removed from the gas mixture by absorption in an alkaline solution.

WG

shipping cask. A container used for the transportation of irradiated fuel assemblies from a nuclear reactor to a reprocessing plant or storage facility. *See* transportation of nuclear materials.

PH

short-flame coal. A semi-bituminious coal which contains 10–15% volatile matter and burns with a short flame. It was thus preferred for use in steam boilers with combustion chambers of limited size, as found in ships and railway locomotives.

WG

short-rotation forestry. The cultivation and harvesting of forest trees on a time scale significantly shorter than normal for a given species. This has several advantages for *biomass production over the traditional tree-growing operation of planting widely spaced trees, using species with long growing periods of 30 to 80 years or more prior to harvesting. These advantages include higher yields per unit land area, a shorter time span from initial investment in stand establishment to positive cash flow from the harvestable crop, the capacity of many species to regenerate by *coppicing, so reducing regeneration costs, and the ability to take quick advantage of cultural and genetic advances. The actual short-rotation period may be 20 years or less, with 3–5 years a common target for very fast-growing species. Biomass, and hence wood energy yields, can be up to five times greater using short-rotation crops over conventional forest species, though establishment and management costs will also be higher.

CL

short run. A period of time long enough to alter output by increasing the use of the variable *factors of production but not long enough to change the fixed factors. Usually *labour is treated as the variable factor and *capital and/or *land as the fixed factor. *See also* long run.

MC

short-term paper. *See* bill.

short-wave radiation. Visible light and *infrared radiation present in *solar radiation. It forms a region of the *electromagnetic spectrum comprising radiation of wavelengths emitted by surfaces at the sun's temperature (approximately 6000°C), i.e. of wavelength range 0.7 to about 2 micrometres. It contrasts with *long-wave radiation in the far infrared region. Short-wave radiation can be transmitted through glass.

JWT

shroud. A structure to deflect and concentrate wind into a *wind turbine.

JWT

shut-down. The reduction in the power of an engine or other heat or power plant to zero. *See also* scram.

DB

siemens (Symbol: S). The SI unit of electrical *conductance (G). One siemens is the conductance between two points in a conductor when a constant *potential difference of one volt applied between these points produces in the conductor a *current of one ampere.

TM

sievert (Symbol: Sv). The SI unit of dose equivalent. *See* radiation dose.

PH

silica. The dioxide of silicon, SiO_2, a hard colourless crystalline substance which is found in the earth in several different forms. These include quartz, flint and silica sands. It may be fused at a high temperature and formed into bricks and other shapes which are used as a heat-resistant (refractory) material in the construction of furnaces.

WG

silicate. A salt of one of a number of silicic acids, which are best described as a mixture of hydrated forms of silica,

$$SiO_2 \cdot nH_2O$$

Thus orthosilicic acid, H_4SiO_4, may be written as

$$SiO_2 \cdot 2H_2O$$

and the salt is M_4SiO_4, where M is a monovalent metal. A wide variety of silicates occur in minerals and rocks.

WG

silicic acids. *See* silicate.

silt. A fine earthy sediment, especially a deposit of mud or fine soil from running water. An area can become filled or choked up with silt, particularly behind a dam or other place of retarded water flow. This process is known as siltation or silting.

PH

silviculture. The cultivation of trees. The term is synonymous with forestry, but in the sense of energy tree plantations it is more closely associated with *coppicing and *short-rotation forestry.

CL

single duct system. A system of *air conditioning a building in which dehumidified air at an appropriate temperature is circulated throughout the building in a single branching duct from a centrally located plant room. The volume of air delivered to any space within the building may be controlled by thermostatically operated dampers on the duct outlet.

TM

singles. Coal with an approximately uniform lump size, the pieces of which are about 25 mm in diameter.

WG

sink. A device, volume, etc., with a capacity to absorb, virtually without change, a rejected product of some process. For example, the earth's atmosphere and oceans are a sink for waste heat and discarded chemicals. A sink can be over-loaded, and is then regarded as polluted.

MS

sink temperature. The temperature of the environment surrounding a *heat engine, to which the engine can reject low-grade heat.

MS

SI units [from Système Internationale d'Unités]. The international set of units adopted by the General Conference of Weights and Measures. The base units are the metre (m) for length; the kilogram (kg) for mass; the second (s) for time; the ampere (A) for electric current; the kelvin (K) for absolute temperature; the mole (mol) for amount of substance; the candela (cd) for luminous intensity.

All other units can be expressed in these base units. For example, the unit of energy, the joule (J), is equivalent to 1 kg m^2/s^2. *See also* conversion tables.

JWT

Sizewell. The site of a *Magnox reactor (Sizewell A) at Suffolk, England. However, the name has become associated with one of the longest and most detailed public enquiries ever held in the United Kingdom as to the merits of building PWR reactors. In 1987 the government gave the go-ahead in the face of significant public protest.

MS

slag. A mixture of inorganic oxides, silicates, phosphates, etc., formed when the non-metallic components of a metallic ore are separated and float on top of the molten metal in a furnace. They form a liquid which prevents access of oxygen to the metal. On cooling, slag solidifies to a stone-like material.

WG

slagging gasifier. A unit for gasifying coal in which the ash constituents are heated to such a high temperature that they melt to form a liquid slag. The slag collects in the bottom of the reaction chamber, from which it can be conveniently run off.

WG

slag-top furnace. A pulverized-fuel combustion chamber in which the flames are directed to concentrate the heat near the base of the chamber and raise the temperature above the fusion point of the ash. The latter collects as a molten slag in the base of the chamber and can be run to waste from time to time. This technique reduces the dust content of the waste gases and is particularly suitable where only low-ash fusion point coals are available.

WG

slurry. A concentrated mixture of low-grade coal particles and water of a paste-like consistency. It is the refuse from a coal-washing plant. With specially designed handling plant and combustion units it may be burned to make steam for power generation.

WG

'Small is Beautiful'. *See* intermediate technology.

Smithsonian Agreement. A multilateral agreement made at the Smithsonian Institute in Washington in December 1971 which realigned the major world currencies with respect to the dollar and increased the margin of fluctuation around these central rates to $\pm 2.5\%$. It was an attempt to salvage the *Bretton Woods Agreement, which came under pressure in 1971, but could not deal with the international turmoil of 1973 associated with the oil price increases.

MC

smog. *See* photochemical smog.

smoke. A suspension of finely divided

solid particles in gaseous products of combustion. It appears as a grey–black plume rising from a chimney when combustion of the fuel has not been completed in the furnace. The particles are largely amorphous carbon but inorganic dust from fuel-ash components may also be present. In an attempt to promote pollution control, smokes may be classsified according to the density of their colour when compared with standard charts.

WG

smokeless fuel. A solid fuel which may be burned with the formation of little or no smoke. As the smoke-forming propensity is proportional to the volatile-matter content of the fuel, smokeless fuels are largely forms of *coke prepared by carbonizing coal to suitable temperatures. If carbonized to about 600°C a *semicoke with about 12% volatile matter is formed. This is sufficiently reactive to burn in an open fire and should produce no smoke.

WG

smoke point test. A standard test to determine the quality of a *kerosene [UK: paraffin]. The smoke point is the height to which the flame on a standard lamp may be raised before smoke is produced. Premium grade kerosene has a somewhat higher smoke point than regular grade.

(SNG). *See* synthetic natural gas.

WG

social cost. The *opportunity cost properly considered from the point of view of all the individuals in society, rather than the private cost as perceived by a particular firm or individual. If there is no *market failure, then private and social cost coincide, otherwise in general they do not. The social cost of coal-fired electricity, for example, exceeds the (private) cost perceived by the electricity-supply industry since it does not bear the health costs arising from the atmospheric pollution to which its activities contribute. *See also* external cost.

MC

social welfare function. A statement of a society's economic objectives, a performance criterion according to which alternative allocations of resources and outputs can be ranked. Some economists take the view that it is not a useful concept, it being impossible to formulate a social welfare function which is derived from the preferences of the individuals who constitute the society and which satisfies certain 'reasonable' conditions. Others take the view that it is a useful concept insofar as it is interesting to examine the implications of maximizing a given social welfare function for patterns of resource allocation. For example, a concern for the welfare of future generations leads to a pattern of use for *natural resources, such as fish stocks or oil deposits, which is quite different from that which would arise if only the welfare of people alive now were to be of social concern.

MC

sodium. A very reactive soft metallic element. Its melting point is 97.8°C, and as a liquid it has excellent heat-transfer properties. It is therefore used as a *coolant for fast breeder *nuclear reactors. It is also non-moderating, an important point in a fast reactor. Two problems are associated with the use of sodium as a reactor coolant. Firstly, its chemical reactivity with air and water requires very high engineering standards in the integrity of pipes and heat exchangers. Secondly, in a nuclear reactor sodium captures neutrons to form the radioactive isotope sodium-24:

$$^{23}_{11}\mathrm{NA} + ^{1}_{0}\mathrm{n} \rightarrow ^{24}_{11}\mathrm{Na}$$

It is therefore necessary to circulate the liquid sodium through two coolant circuits with an intermediate heat exchanger. The primary circuit passing through the reactor contains radioactive sodium, and the secondary circuit which goes to the steam-raising units contains non-active sodium.

Symbol: Na; atomic number: 11; atomic weight: 23.0; density: 970 kg/m^3.

DB

sodium–potassium alloy (Na–K alloy). An alloy of sodium and potassium with a lower melting point than pure sodium. The eutectic (i.e. lowest melting point) Na–K alloy has a melting point of −11°C, and can be used as a *coolant for fast breeder *nuclear reactors; it is not at present used as much as sodium.

DB

soft coal. A bright glossy coal with an obvious banded structure. It tends to be friable and liable to shatter in coal-handling plant. It is composed largely of the coal constituents *clarain and *fusain.

WG

soft energy. Energy use and technology drawing from *renewable energy sources that does not harmfully disrupt the natural environment or the social structures of the people involved. *See also* appropriate technology.

JWT

softwoods. Coniferous trees, generally slow-growing initially and not suitable for *short-rotation forestry. They do not include *coppicing species. However, many softwoods such as pine and cedar are useful *firewood sources in various parts of the world. In *ethanol manufacture, softwoods are less easily hydrolysed to fermentable sugars than are *hardwoods, owing to their higher lignin content. They do have the advantage, nonetheless, of a lower proportion of pentose sugars in their molecular make-up. Pentose sugars cannot be fermented by the *Saccharomyces cerevisiae* yeasts used in ethanol *fermentations and must therefore be removed prior to fermentation.

CL

sol-air temperature. *See* temperature.

solar cells. Solar-energy devices producing electricity directly from sunlight. *See* photovoltaic conversion.

JWT

solar constant. The average solar radiation flux (*insolation) in space at the average distance of the earth from the sun. The solar constant has been accurately measured in satellites as 1373 watts per square metre.

JWT

solar energy. Energy incident in the form of radiation from the sun. This restricted use of the phrase is common in Europe, in contrast with the USA where derived forms of the sun's energy, e.g. wind and *biofuels, are also called solar energy. Radiant solar energy (sunshine) is incident in the direct solar beam and also in *diffuse radiation scattered from clouds, the atmosphere and the neighbouring environment. *See also* photovoltaic conversion; renewable energy.

JWT

Solar Energy (Information) Unit. An information unit to encourage the use of solar energy in the UK. Funded by the UK Department of Industry, it publishes a regular magazine, *Helios*.

Address: Dept. of Mechanical Engineering and Energy Studies, University College, Newport Road, Cardiff, Wales.

JWT

solar input. The energy entering a device or system in the form of *solar radiation. The maximum flux of solar radiation on earth in direct sunshine is about 1 kW/m^2.

JWT

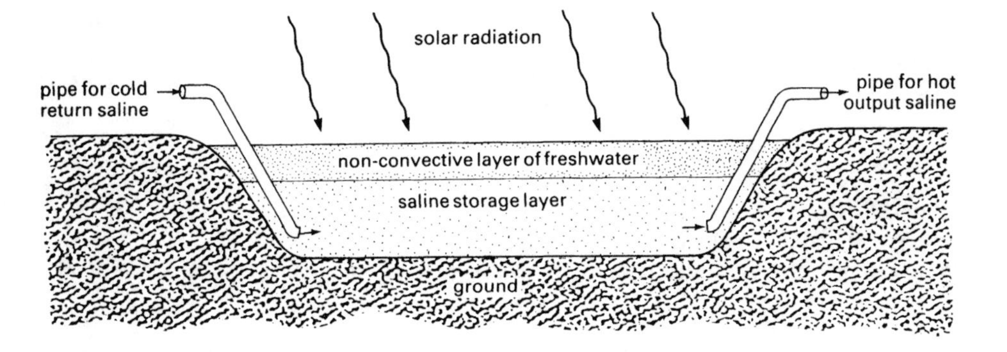

solar pond. A volume of saline water covered by a less dense non-convective layer of fresh water. Solar radiation is absorbed in the low layers, which increase in temperature. The heat is trapped by lack of circulation and may be removed for space heating or to power a generator (*see* figure). With careful control, solar ponds reach an efficiency of 15% trapped *solar radiation at temperatures above ambient of 30°C or more. Some experimental solar ponds have reached boiling point in the saline layer.

JWT

solar radiation. The *electromagnetic radiation, primarily ultraviolet, visible and infrared radiation, emitted from the outer layers of the sun. The intensity of the radiation in space is given by the *solar constant. The earth's atmosphere absorbs certain wavelengths, especially high-energy ultraviolet, so the intensity at the earth's surface is reduced. *See also* short-wave radiation.

JWT

solar spectrum. The distribution of *solar radiation with wavelength. The sun emits electromagnetic radiation ranging in energy from gamma rays to radio waves. The major emission occurs over a continuous range of wavelengths from 0.3 to 2.5×10^{-6} m, covering the ultraviolet, visible and infrared regions. Peak intensity occurs at visible wavelengths but about 50% of the total intensity is distributed in the infrared region.

JWT

solar still. A distillation device energized by solar energy. It is used particularly to obtain fresh water from brackish or saline water. Simple solar stills are enclosed tanks with a sloping glass cover facing the sun. The brackish or saline water is thus heated and evaporated, i.e. fresh water condenses on the underside of the glass to trickle into a gutter leading to the fresh-water container.

JWT

solar system. The group of celestial bodies comprising the sun and the bodies orbiting the sun, i.e. the planets, their satellites, the asteroids, etc. The term does not imply an energy system powered by solar energy.

JWT

solar tower. A complex device in which a *heat exchanger is placed at the top of a tall tower, around which are placed numerous mirrors. The mirrors focus and concentrate the sun's rays upon the heat exchanger. The array of mirrors may cover an area of more than one kilometre square, thus beaming onto the exchanger a large amount of energy. Since solar energy is of high quality, it is possible to achieve high temperatures and thus obtain steam or other vapour from the heat exchanger. This can drive a turbine to produce electricity. Solar towers are being tested in the USA, Sicily and Spain.

MS

solenoid. A coil, usually of helical form, through which an electric current is passed to produce a magnetic field.

TM

solidity. In *wind-power machines, the ratio of the total blade area (perpendicular to the wind direction) to the swept area. Thin bladed high *tip speed devices have low solidity, e.g. 0.05. Wide bladed high *torque devices have high solidity, e.g. 0.6.

JWT

somatic effects. *See* biological effects of radiation.

soot. The black solid product of the incomplete combustion of a fuel, found adhering to the inner surfaces of furnace flues and chimneys. It consists largely of tiny particles of amorphous carbon bound together with tarry matter.

WG

sour gas. Petroleum gas containing hydrogen sulphide and low molecular weight organic sulphur compounds, and thus having an offensive smell. The gas may be used as a source of sulphur for sulphuric acid manufacture. *See also* sweetening.

WG

South of Scotland Electricity Board (SSEB). A UK government-appointed board responsible for electricity generation and transmission, and the construction of power stations of all types in the southern part of Scotland. With a total capacity of approximately 7800 MW, it operates one of the largest proportions of nuclear power in the world: 17%.

Address: Cathcart House, Glasgow G 44, Scotland.

MS

space heating. *See* central heating; domestic heating.

spaghetti diagram. An informal term for *Sankey diagram.

spark ignition engine. An *internal combustion engine in which the mixture of gasoline vapour and air is ignited by a high-tension spark produced between the electrodes of a spark plug set into the end of the combustion chamber. The timing mechanism allows the spark to occur close to the point at which the flammable mixture has been compressed to the greatest extent by the piston.

WG

special drawing rights (SDR). An international *reserve *asset issued by the *International Monetary Fund (IMF) and first introduced in 1969. Its purpose is to boost the supply of international reserves and to replace gold. It is the basic unit of account of the IMF and is now a reserve asset for many countries and international institutions.

MC

specific gravity. An earlier name for *relative density.

specific heat capacity. *See* heat capacity.

speculation. The purchase (or sale) of a *commodity, *security or *currency in the expectation of making a profit from its resale (purchase) when its price rises (falls). It is distinct from *arbitrage and *hedging in that it does not take advantage of different rates nor is it a means of protection. It is based on the accurate forecasting of future prices.

MC

spent fuel. *See* irradiated fuel.

spent shale. The solid inorganic residue left after the destructive distillation of oil shales in a retort for the recovery of shale oil. WG

spot price. A price specified in a contract for immediate delivery, as opposed to a *forward price. In the context of oil prices, use of the term also generally indicates that reference is being made to a price which is the outcome of competitive bidding as opposed to a price set by an OPEC member or an oil company, i.e. an 'official price'. MC

stack effect. The movement of warm air upwards within a chimney, cooling tower, building, or the like, and its replacement at low level by cool air. TM

stainless steel. An alloy of iron, chromium and nickel. It is widely used as the *cladding and structural material in nuclear reactors because of its high strength and corrosion resistance, reasonable cost and fairly low neutron capture *cross-section. DB

standard atmosphere. A unit of pressure equivalent to 1.01325 *bar.

standard cubic foot (SCF). A once-common volume measure for gas in the gas industry. It is the amount of gas in a cubic foot when the gas is under one atmosphere pressure and at 20°C. MC

stand-by system. The additional plant or system capacity held in reserve against the possibility of failure of the regular plant or system to meet the imposed load demand, due either to breakdown or inadequate capacity.

TM

starter. A variable *resistance device in circuit with the *armature of an *electric motor to reduce the instantaneous electrical and mechanical load on the motor at start-up.

TM

stationary state. An economy in a stationary state would experience no economic growth: the level of *national income would remain constant over time and there would be no *net investment, so that the size of the *capital stock would remain constant. It is advocated as the proper objective of *macroeconomic policy by some economists who take the view that economic growth is no longer an appropriate objective for the advanced nations of the world. This advocacy reflects a view that in such countries further economic growth will lead to the excessive depletion of *natural resources (from the viewpoint of later generations) and/or that economic growth is no longer synonymous with increased *welfare.

MC

steady state. A state within any defined system where the net flux of mass or energy is constant over time. This simplifying assumption is commonly used in the computation of the heat flow through a membrane, where it is assumed that the temperature conditions on either side of the membrane remain constant. The assumption that steady-state conditions prevail inside and outside a building gives rise to significant errors in the estimation of energy consumption and comfort, hence the introduction of more sophisticated approaches – the *admittance method, *response factor methods and *finite difference methods – which attempt to model variation in the influences of climate and weather.

TM

steam. The vapour into which water turns when its *temperature is raised to boiling point. Steam raised in a *boiler by the combustion of *fossil fuel has been widely used in *reciprocating engines or in *turbines to provide mechanical power for the generation of electricity or the movement of people and goods. It is also used for the transmission of heat.

TM

steam coals. Coals suitable for steam raising for boiler installations on ships and land. They are *free-burning and produce little smoke. The most valuable for this purpose are coals whose analyses lie between *anthracite and strongly caking *bituminous coals, known in the UK as semi-bituminous or carbonaceous coals. They contain 10–18% *volatile matter and have a low ash content and a high calorific value. A typical gross calorific value on a dry ash-free basis is 36.8 MJ/kg.

WG

steam distillation. A distillation process in which steam is injected into the liquid being vaporized. The presence of the water vapour helps the material to vaporize at a lower temperature and thus reduces the possibility of its chemical breakdown due to elevated temperatures.

WG

steam-generating heavy-water-moderated reactor (SGHWR). *See* nuclear reactor.

steam reforming. The reaction of steam with hydrocarbon vapours in the presence of a catalyst to form carbon monoxide, hydrogen and hydrocarbons of small molecular weight. The process may be employed to turn naphtha into town gas.

WG

steam turbine. *See* turbine.

Stefan–Boltzmann constant [Stefan's constant]. *See* Stefan–Boltzmann Law.

Stefan–Boltzmann law. A law enabling the radiant energy emitted from a surface to be calculated. It states that the energy E emitted per unit area of a surface at *absolute temperature T into the hemisphere above the surface is given by

$$E = \varepsilon\sigma T^4$$

where σ is the Stefan–Boltzmann constant, equal to 5.663×10^{-8} J/m^2.K^4, and ε is the *emissivity of the surface. For an ideal black surface or body ε is equal to one by definition, but in general it is less than one.

JWT

stellarator. A magnetic containment system developed by the Princeton Plasma Physics Laboratory in the USA for plasma containment and *fusion studies. The stellarator is of toroidal shape with external helical multipole windings. *See also* fusion reactor.

DB

sterilization. The act of governments nullifying the effects on the domestic *money supply arising from a deficit or surplus on the *balance of payments. A deficit (surplus) on the balance of payments will generally lead to a fall (rise) in the money supply which may be deliberately offset, in part or in whole, by government action.

MC

Stirling cycle. The theoretical analysis of the thermodynamic operation of a Stirling engine. This engine operates with a closed volume of air working in a twin piston arrangement and powered from a hot but constant-temperature source of energy, such as a flame or solar *absorber. The engine is suitable for generating mechanical work from low-quality sources of heat or controlled regular combustion. *See also* heat engine.

JWT

Stirling engine. *See* Stirling cycle.

stock. (1) An alternative term for *inventory.

(2) A particular type of financial *security, as in stocks and shares, with stocks being distinguished from shares in that whereas the latter are issued by companies as fixed nominal amounts, the

former are issued by governments in a consolidated form and can be held in any amount.

(3) Generally, a variable without a time dimension, as in the stock of *capital, as opposed to the flow of *income.
MC

stock exchange [stock market]. A market in which *securities are bought and sold. These securities can come from central and local government and from companies. The market deals with the buying and selling of existing securities as well as new issues. It does not, however, deal in short-dated *bills (such as Treasury bills or commercial bills). A company must satisfy a set of conditions in order to have its securities quoted on the stock exchange. The stock exchange is an important institution of the *capital market.
MC

stockpile. To form a store of ready-to-use material, or the accumulated store itself.
MS

stoichiometric. A chemical term used to describe the proportions of each chemical species in a chemical reaction. Thus in the combustion of methane (CH_4) with oxygen the stoichiometric ratio of methane to air for complete combustion is given by the equation $CH_4 + 2O_2 \rightarrow 2H_2O + CO_2$, i.e. it requires 2 volumes (*moles) of oxygen per mole of methane. If air is used in place of oxygen, since air contains 21.8% oxygen the ration would be 2/ 0·218 = 9.17 volumes of air.
MS

stoker. A mechanical device used to feed coal at a controlled steady rate into the combustion chamber of a steam boiler furnace. With an overfeed stoker, coal is thrown onto the top of the fire; an underfeed stoker brings fresh coal up through the grate from below. A chaingrate stoker introduces coal on the fire on a continuous belt of steel links. A ramfeed stoker uses a ram to push coal at intervals into the front of the fire bed.
WG

storage (of energy). *See* energy storage.

straight-chain hydrocarbons. *Hydrocarbons composed of molecules in which the carbon atoms are linked in a single chain.
WG

straight-run. A *petroleum fraction produced only by a *fractional distillation process.
WG

straw. The dried stems of grain crops remaining after harvest. Straw is commonly regarded as waste material and large quantities are burnt in fields after harvest. However it is a form of *biomass and may provide a useful energy source by *combustion and certain other methods after processing.
JWT

strip mining. US term for *open-cast mining.

strontium. An element whose radioactive isotope, strontium-90, is formed as a *fission product in nuclear reactors. It has a *half-life of 28 years and consequently it is one of the most important and hazardous of the fission products from the point of view of long-term storage and disposal. Strontium-90 is particularly hazardous to humans because, like calcium, if ingested it is retained in the bone marrow.

Symbol: Sr; atomic number: 38; atomic weight: 87.6.
DB

structural formula. The formula of a chemical compound which purports to show in two dimensions, through the application of the rules of valency, the way in which the atoms in the molecule are linked together.
WG

subcritical assembly. *See* criticality.

subsidy. Payment by the government to a firm or individual in order to raise incomes, lower the price of the good or raise the use of a factor input. Agricultural commodities are often subsidized and in many countries temporary subsidies to labour have been given to firms in order to reduce unemployment. In some cases the payment of a subsidy can be somewhat indirect, as when the government finances research and development for subsequent commercial exploitation. This has happened in a number of countries with respect to nuclear power and is now happening with alternative energy sources such as solar power, wave power and wind power.

MC

substitute. Two commodities are substitutes when they have a cross price *elasticity of demand which is positive, i.e. with all other prices constant, a fall (rise) in the price of commodity one leads to a reduction (increase) in the quantity demanded of commodity two. For example, two fuels would generally be substitutes, though the effect may only show up in the long run due to the need to change fuel-using equipment. In the UK, gas and electricity have been substitutes in the domestic heating market. In some cases substitution between fuels may be impossible in response to changing *relative prices due to technical reasons. *See also* complementarity.

MC

substitution. In consumption or production, the partial replacement of one commodity or input by another in response to changing *relative prices. It is the essential process by which the *price mechanism works to allocate resources. For example, rising energy prices lead to the substitution for energy of other commodities in consumption (roof insulation in houses, bicycles for cars) and of other inputs in production (*capital and *labour). *See also* elasticity; isoquant; substitution effect.

MC

substitution effect. The effect on the consumption of a commodity by a household of a change in price of the commodity when the household's *real income is held constant. It is always negative in the sense of consumption increasing (decreasing) when the price falls (rises). The substitution and income effects are distinguished in the diagram using *indifference curves to represent the *utility function graphically for 'energy' and 'other' commodities. All income spent on energy initially buys an amount OB, while if all income is spent on 'other', it buys an amount OA; the line AB is the

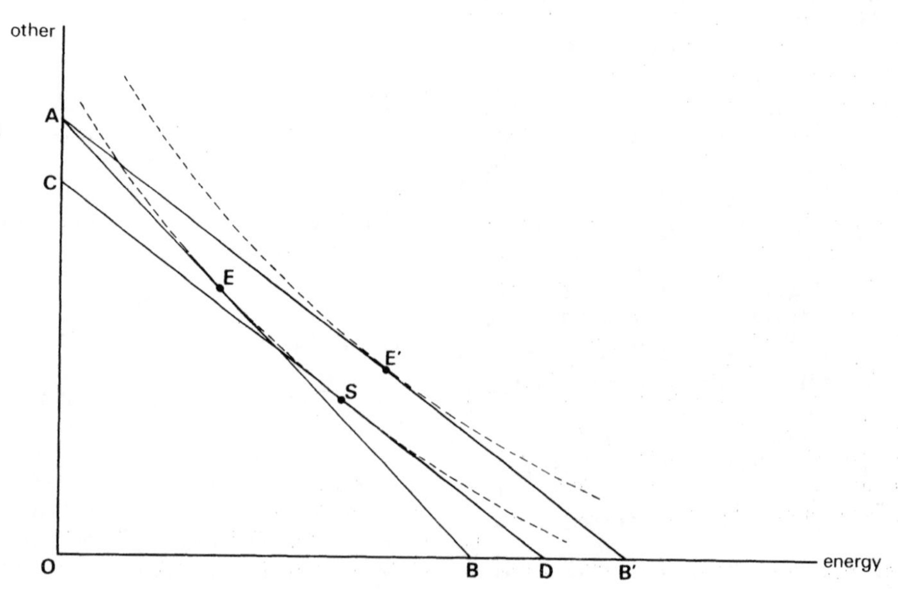

'budget constraint', the locus of energy/other combinations available to the household if it spends all its income. The slope of AB gives the relative prices of energy and other. Initially the household's preferred position is at E where it gets on the highest indifference curve possible, i.e. attains the highest possible utility level. A decrease in the price of energy shifts the budget constraint to AB′, with OB′ purchasable with all income spent on energy. The new preferred position is E′, where the level of utility is higher than at E; effectively the household's real income is increased by the price fall. To concentrate on the pure effect of the change in relative prices of energy and other, construct CD parallel to AB′ and tangential to I, at S. S is the position to which the household would have moved following the price change if it had been accompanied by an income change such as to hold the level of utility or real income constant at the original level. So the change in energy consumption from E to S is the substitution effect, the change due solely to the relative price change. It is, and must always be, negative, with (as here) a lower price leading to greater consumption. The *income effect is the difference in energy consumption between S and E′, being the change due solely to the real income effect arising from the price change.

MC

sugar-cane. A high-yielding tropical perennial crop (*Saccharum* species), a *4-carbon plant, which fixes about four times as much solar energy per unit area as do most temperate crops and twice as much as the most common tropical plants. Dry-matter yields of 90 t/ha.y have been obtained under ideal conditions in Hawaii, representing a mean annual *photosynthetic efficiency of 2.5%, with a peak of about 4%. It is therefore considered the best crop for the production of *ethanol fuel, particularly as its *bagasse content can be burned as fuel in the production process. The *fermentation of sugar-cane to ethanol is a clear *net energy gain and as such is used extensively in the large Brazilian National Alcohol Programme, aimed at substituting most of that country's imported oil with home-produced liquid fuel.

CL

sulphur [US: sulfur]. A yellow solid nonmetallic element occurring combined in both organic and inorganic compounds, and free as a native element. Metallic sulphides and sulphates occur as minerals in many coal seams. Native sulphur is found in large underground deposits, mainly in the USA. The element can exist in a number of forms (allotropes) which arise because the atoms tend to link together in different geometric arrangements of two, four, six and eight atoms. Thus at different temperatures sulphur can range from a crystalline powder, to a dark brown plastic solid, to a mobile yellow liquid.

Sulphur in coal is difficult to remove due to the fact that a proportion of it may be bound in the coal's organic structure, or because the sulphur-containing minerals may be very finely dispersed throughout the mass. Sulphur compounds in petroleum and natural gas can be removed in the petroleum refinery by *hydro-desulphurization processes. If not removed, then when the fuel is burned the sulphur appears in the flue gases as *sulphur oxides and gives rise to pollution.

Symbol: S; atomic number: 16; atomic weight: 32.06.

WG

sulphuric acid. A strong acid, H_2SO_4, in aqueous solution with strong oxidizing properties. Reaction with metals and alkalis forms salts called sulphates. It is formed in the atmosphere when combustion-formed oxides of sulphur react with water vapour and oxygen, and gives rise to the corrosion damage of structural materials.

WG

sulphur oxides. Two gaseous oxides, sulphur dioxide (SO_2) and sulphur trioxide (SO_3), formed when sulphur and sulphur-containing organic compounds are burned in air; sulphur dioxide predominates. Both oxides are considered to be air pollutants and to be responsible, together with smoke, for respiratory complaints of town dwellers. Sulphur dioxide is an irritant and kills a wide range of microorganisms. It is thus used as a fumigant. Sulphur trioxide reacts with water to form sulphuric acid, a strongly corrosive substance. The realization that these gases are injurious to health has led to legislation in some countries limiting the sulphur content of fuels and to the building of high chimneys for the better dispersion of pollutants.

WG

summer valley. The depression in demand for certain types of fuel during the summer.

MS

sunfuel. An informal term for a *biofuel or *bioenergy, i.e. for a fuel arising from some photosynthetic pathway via plants.

MS

sunshine. Energy from the radiation of the sun (*see* solar spectrum). Sunshine is incident in the direct beam, which casts sharp shadows, and also in diffuse radiation from solar radiation scattered in the earth's atmosphere and clouds and reflected from the local environment. The energy per day in bright sunshine might equal 30 MJ/m^2 arriving at a maximum intensity of 1 kW/m^2.

JWT

supercritical assembly. *See* criticality.

supply function. The relationship between the supply of some commodity and the price of the commodity. By supply is meant the quantity which sellers plan to sell at the ruling price. Also known as the supply curve when represented graphically, it slopes upwards from left to right when price is measured vertically and quantity horizontally. A supply function can refer to the plans of a single firm, or of all the firms which constitute an industry. The market or industry supply function is the summation of the supply functions of the individual firms.

The shape of a supply function reflects the behaviour of costs as output varies. The *elasticity of supply is a convenient representation of the properties of the supply function as they affect the determination of, and changes in, *market equilibrium. A large elasticity means that small price changes induce large supply responses. In the *short run many commodities have inelastic, i.e. a small elasticity, supply as compared with the *long-run situation, due to the fixity of the input levels of some *factors of production. For example, a sudden large increase in the demand for solar panels would lead, in the short run, to a large increase in their price and a small increase in the supply, due to the supply constraint given by the capacity of existing manufacturing plants. In the long run the price increase would be smaller and the supply increase larger as the capacity of the solar-panel industry was increased.

MC

surge tank. A large volume designed to accommodate pulses and erratic flows of water in pipes. Surge tanks prevent large sudden pressure increases that might otherwise damage the pipes.

JWT

sweating. The secretion of body fluid through the sweat glands of the skin.

Sweating occurs when the body temperature is unacceptably high; heat loss due to the evaporation of the sweat from the surface of the body thus supplements the normal conductive and radiative heat losses and helps reduce body temperature. In *air conditioning, the term is applied to the formation of *condensation on surfaces that are below the *dew-point temperature of the surrounding air.

TM

sweetening. A chemical process which removes the offensive smell of sulphur compounds, such as mercaptans in petroleum products, without necessarily lowering the sulphur content of the materials. Prior to their treatment the malodorous materials are said to be sour.

WG

symbiosis. A continued close association between two forms of life which is mutually beneficial. With reference to micro organisms, in an ectosymbiosis the microbe remains external to the cells and tissues of its host, as with bacteria that live on the surface of leaves and in the body cavities of animals. In an endosymbiosis the association is closer, with the microbe actually growing within the host cells and tissues, as with **Rhizobium* bacteria inhabiting the root nodules of *legumes where they are capable of *nitrogen fixation. The fixed nitrogen is made available to the plant, which in return provides photosynthate in the form of sugars as an energy and carbon source for the bacteria.

CL

synergism. An action where the combined effect of two active components in a mixture is greater than the sum of their individual effects. For example, a synergistic effect is demonstrated if the biological response to an exposure to two chemicals is greater than would be expected from the additive action of each individual chemical. In studies on air pollutants it has been found that the combination of sulphur dioxide and *benzopyrene results in respiratory tract tumours in rats, while no tumours were found following inhalation at similar concentrations of either agent alone.

PH

synroc. A process for converting high-level liquid *radioactive waste to a form of artificial rock called synroc (synthetic rock). The process involves the production of crystalline minerals which would incorporate the waste into a fine-grained rock similar to basalt. The minerals chosen are ones which are capable of dissolving different *radionuclides so that when the mix is melted at around 1300°C they become incorporated into the mineral crystals. *See also* radioactive waste management.

PH

synthesis gas. A mixture of carbon monoxide and hydrogen produced by the reaction of steam or oxygen with hydrocarbon materials such as coal, oil or natural gas. When passed over suitable catalysts, synthesis gas may be converted into a variety of chemicals. *See also* Fischer–Tropsch process.

WG

synthetic natural gas (SNG). A mixture of gases rich in methane produced when liquid petroleum feedstocks, or the product of the complete gasification of coal, is reacted with steam at elevated temperature and pressure over a nickel–alumina catalyst. Such a mixture may be used to meet current peak demands for natural gas, and eventually as other sources dry up provide all our future natural gas requirements.

WG

system boundary. The set of points at which the inputs to and outputs from systems are measured. For example, in order to supply fuel oil, crude oil must be extracted, conveyed to a refinery, fractionated and finally delivered. Each of these steps calls for energy expenditure. In a typical situation, for every unit of fuel oil delivered 1.15 units of crude oil will have to be extracted. Thus the amount of oil required to, say, furnish and deliver fuel oil will be different depending on which system boundary is used to measure the flows. *Energy statistics are computed at a range of system boundaries.

MS

systems analysis. (1) The analysis of the components of a system with a view to finding out how best to implement a desired course of action. It was first introduced by the Rand Corporation in the USA in order to establish least-cost ways of defeating an enemy. Definition of 'best' is left to the analyst or user. In this sense it is close to system engineering. It can lead to optimization by means of *linear programming.

(2) In a wider and more modern sense, the art and science of establishing the consequences of a chosen course of action, where no one criterion of 'best' can apply to all the many possible outcomes of the action. When the analysis has revealed the possible consequences of the action, it is something of a value judgement to decide which, among the various possible actions, should be chosen.

MS

T

tailings. *See* uranium tailings.

tantalum. A metallic element with a fairly high neutron capture *cross-section, used as the material for control rods in the Prototype Fast Reactor, Dounreay (*see* nuclear reactor).

Symbol: Ta; atomic number: 73; atomic weight: 180.9; density: 16,600 kg/m^3.

DB

tariff. A tax on imported goods. It can be either as a percentage of their value (an *ad valorem tariff) or on a unit basis (specific). Tariffs are imposed for a variety of reasons, including *protection, *balance of payments reasons and as a source of revenue.

MC

tar sands. Sedimentary deposits of coarse sand impregnated with up to 12% of a heavy bituminous oil, normally found near the earth's surface. The mode of formation is not clear, nor whether the oil has migrated into the sand or was laid down as organic matter with it. Deposits in various parts of the world are worked for the recovery of the oil. The sand may be dug out and heated with steam to cause the oil to distil off, but a more promising technique is the in situ heating of the deposit by pumping high-pressure steam through bore holes and collecting the oil from a conventional well. The product may be refined by conventional means to provide *fuel oil, be subjected to catalytic *cracking to provide motor spirit (gasoline), etc.

WG

tax reference price. The oil price which serves as the basis for calculating the tax liability of an oil company to the government of the nation in which the company operates an oil concession, per barrel of oil extracted. The tax reference price is not generally the same as the market price. *See also* posted price.

MC

technical progress. A concept in economics whereby technological developments are presumed to provide the means of producing more output per unit input. Usually, but not necessarily, technological change requires increased capital per unit of output and less labour, so that technical progress creates both investment and social consequences. For example technical progress in electricity generation is reflected in an improved conversion of *heat into *work. Through the *second law of thermodynamics one can demonstrate that there are limits to technical progress.

MS

technological forecasting. The science of predicting technological developments. There are two main forms: exploratory forecasting is based on the extrapolation of present trends, indicated by past and present data, into the near and far future; normative forecasting attempts to assess future goals, needs, desires, etc., and works backwards to the present, identifying what changes are required on the way. *See also* Delphi method.

PH

technology assessment. The study of the impact of a particular technology on the environment, the economy and the social structure and welfare of the community in which it is used. It can be of three kinds: reactive assessment is a reaction to problems already recognized and here the objective is to alter the technology if possible to prevent further damage; corrective assessment involves tracing problems to their causes and initiating research and development in an attempt to solve the problem before it becomes severe; anticipatory assessment is concerned with anticipating future problems which might be a threat from the proposed technology. *Technological forecasting obviously has a significant role in anticipatory technological assessment.

PH

temperature (Symbol: *T*). A measure of the hotness of a substance. A system at a higher temperature can always transmit a net flow of heat to a system at a lower temperature. For practical purposes this property defines the concept of temperature. On a molecular scale, the *kinetic energy of molecules is a measure of their temperature.

Various physical phenomena, such as the length of a mercury column or the volume of a gas, relate to temperature, thereby enabling *thermometers to be made. These thermometers require a scale defined by easily repeatable circumstances or defined by certain fundamental properties. For instance the *centigrade scale has historically been defined from the melting and boiling points of pure water at atmospheric pressure.

The *absolute scale of temperature defines temperature in terms of the energy possessed by matter; it thus has a thermodynamic basis. The unit of temperature on the absolute scale is the *kelvin (K), which is an SI base unit. All other scales are now defined by reference to this absolute scale. For practical purposes the Celsius and centigrade scales are equivalent, with the melting point of water at 0 °C and the boiling point at 100°C. An interval of 1°C is equivalent to 1 kelvin. In the *Fahrenheit scale, 32°F is equal to 0°C and 212°F to 100°C so that an interval of 9/5°F equals an interval of 1°C. The Rankine scale is the absolute temperature scale related to the Fahrenheit scale.

The perception of comfort is related to the temperature of the surrounding air and its thermal conductivity. For instance, high humidity at temperatures above skin temperature will produce a sensation similar to an increased temperature and the sun's rays falling on the skin or clothing may produce a surface temperature higher than that of the surrounding air. Environmental engineers attempt to quantify these effects by a variety of quasi-temperature scales, which include (a) air temperature: temperature as recorded by a standard mercury thermometer in a standard enclosure such as a Stevenson screen at a fixed height and shielded from direct radiation from the sun; (b) sol-air temperature: the outside air temperature which, in the absence of *solar radiation, would give the same temperature distribution and rate of heat transfer through the envelope of a building as exists with the actual outdoor temperature and the incident solar radiation; (c) effective temperature: the temperature of air in a closed *environment of specified humidity that would give the same sensation of comfort in a particular environment having, for example, a flux of solar energy, a wind or a different humidity.

JWT

temperature coefficient of reactivity. A characteristic of a nuclear reactor which quantifies the relationship between a change in *reactivity and temperature changes occurring in the main components of the reactor. Different components of the reactor, such as the *fuel, *moderator and *coolant, have different values of the temperature coefficient. A positive coefficient implies that as reactor power and temperature increase, so the reactivity increases, causing the power to increase still further. This is an unstable effect which if uncontrolled could lead to excessive power and temperature. A negative coefficient implies that as temperature increases, reactivity decreases, causing the power to drop. This is a stabilizing effect and is very desirable from the point of view of safe reactor operation.

DB

temperature gradient [thermal gradient]. The difference in temperature over unit thickness of a conducting material.

TM

temperature inversion. The situation existing when air temperature increases with altitude, unlike the usual relationship between air temperature and altitude. Temperature inversions can cause serious air pollution problems since pollutants are forced to accumulate in the lower atmosphere instead of dispersing, resulting in dramatic increases in *ground level concentrations. Such situations may remain unchanged for days until weather conditions alter and the inversion layer breaks up. The inversion layer is usually warm dry and cloudless so that it transmits the maximum amount of sunlight, which interacts with the pollutants to form *photochemical smog.

PH

tera-. A prefix denoting a multiple of one million million, i.e. 10^{12}.

teratechnology. A term used to describe a multidisciplinary approach to problem-solving. Postgraduate training in teratechnology involves the study of several different technologies which are generally applied to large-scale problems.

PH

teratogen. A substance or agent that causes the production of foetal abnormalities.

PH

terminal heat-recovery system. A system of *air conditioning a building in which a self-contained water-cooled electrically driven unit is sited in each room. In winter the unit works as a *heat pump; in summer, the changeover from heating to cooling is effected in the refrigerant circuit, the roles of the evaporator and condenser being reversed.

TM

terms of trade. The ratio of export prices (usually an *index number) to import prices (usually an index number). An improved (worsened) terms of trade means a country can now import more (less) for the same volume of exports. The terms of trade is to be distinguished from the *exchange rate which is the price of one currency in terms of another. Following the oil price rises of 1973–74, oil-exporting nations experienced improved terms of trade.

MC

test discount rate. A minimal discount rate, usually imposed by an official body, on a *cost–benefit analysis.

MS

tetraethyl lead (TEL). A toxic colourless liquid used as an *antiknock additive in gasoline. It makes possible *octane ratings of up to and over 100. Through its use it has been possible to develop more powerful internal combustion engines with high compression ratios. Unfortunately metallic lead and lead oxides are released in the exhaust gases. This may cause the fouling of exhaust valves and production of toxic fumes. As a result legislation has been introduced to reduce the lead concentration in petrol to a low value. Other methods of raising the octane rating are being sought.

WG

tetramethyl lead (TML). A liquid used as an *antiknock additive in gasoline in conjunction with *tetraethyl lead. It has a lower boiling point than the latter and thus aids in raising the octane rating of the more volatile components of a petrol, which predominate in the air-fuel mixture entering the cylinders of the engine when it is being started up from cold.

WG

theoretical air. The amount of air which contains the oxygen theoretically required to burn completely the combustible elements of a fuel.

WG

therm. A unit of energy equal to 100,000 *British thermal units.

JWT

thermal bridge. A route of a high thermal conductivity through a medium to provide thermal insulation. Commonly occurring thermal bridges are the metal ties between the external and internal skins of brickwork making up a building envelope.

TM

thermal capacity (Symbol: C). The property of a substance which allows it to store heat energy when its temperature is raised. Thermal capacity is a function of the specific heat capacity, density and dimensions. Its effect in, for instance, the wall of a building, is to effect a *decrement and *thermal lag in the flow of heat into and out of the building. *See also* heat storage.

TM

thermal column. An extension of the core of a thermal nuclear reactor, designed for research purposes, that has *moderator but no *fuel. Its purpose is to provide a source of *thermal neutrons for experiments.

DB

thermal comfort. The subjective experience of the state of thermal interaction between the human body and its immediate environment. Three modes of heat transfer between the body and its environment are relevant: evaporation (about 25%), radiation (about 45%) and convection (about 30%). The environmental properties which influence these modes are dry bulb temperature (affecting evaporation and convection), *relative humidity (affecting evaporation only), air velocity (affecting evaporation and convection) and mean radiant temperature (affecting radiation only).

TM

thermal conductivity (Symbol: k). A measure of the ability of a material to pass heat by conduction. If heat Q passes through unit area in unit time along a material, across faces at temperatures T_1 and T_2 and distance x apart, then

$$Q = k(T_1 - T_2)/x$$

JWT

thermal cracking. A petroleum-refinery

process in which a heavy residue or heavy distillate is heated to about 450°C at a pressure of between 2 and 5 atmospheres in order to cause large hydrocarbon molecules to break down (crack). A mixture of gas, gasoline, light distillate and coke is formed. The raw material and process conditions are selected to maximize the production of gasoline. The *feedstock may be heated at rest and the cracked vapours passed to a fractionating column until the cracking vessel is filled with *petroleum coke. Alternatively the feedstock is passed through a tubular furnace or brought in contact with a fluidized bed of coke particles at 500°C. They serve as seed for further coke production and are circulated to a furnace where a proportion is burned to provide reactions. *See also* catalytic cracking; visbreaking.

thermal cube. The dimensions – length, breadth and height – of a rectilinear building envelope which, in relation to the construction of each part of the envelope, minimizes the *steady-state conductive heat losses. With a uniform construction throughout, the resulting shape would be an actual cube; otherwise the total area of opposing faces would be inversely proportional to their aggregated *thermal conductivity.

TM

thermal cycling. Fluctuation in temperature, which if severe enough and repeated often enough can lead to damage or failure of the material concerned. For example, if metallic uranium suffers repeated temperature fluctuations above and below 660°C, at which temperature it undergoes a metallurgical phase change, there is considerable distortion and change of dimensions. In a nuclear reactor this effect would impose considerable stress on the *cladding of the uranium, with possible rupture. The effect is prevented by limiting the temperature of metallic uranium to less than 660 °C.

DB

thermal diffusivity (Symbol: K). A measure of the ability of a material to diffuse heat, defined as

$$K = k/(\rho c)$$

where k is the *thermal conductivity of the material, ρ the density and c the specific *heat capacity. Materials of high thermal diffusivity will transmit a temperature change rapidly, and vice versa.

JWT

thermal discharge. Warm or hot water discarded by a system or process, such as a steam-driven electrical generator. Hot water cannot hold as much oxygen as cold, and so when passed into a river may kill fish. However, properly mixed with cold water to a suitable temperature, such water can be used for rearing warm-water fish or for enhancing rates of fish growth.

MS

thermal gradient. *See* temperature gradient.

thermalization. The process of slowing down of *fission neutrons by repeated *scattering collisions until they reach thermal equilibrium with the moderator nuclei. *See also* *thermal neutrons.

DB

thermal lag. The temporal delay in the effect that a temperature change on one side of a barrier, such as the wall of a building, has on the temperature on the other side, due to the *thermal capacity of the barrier. Thermal lags for different densities and thicknesses of wall construction have been calculated by making the simplifying assumption that diurnal variations in climate are sinusoidal.

The lags, used in conjunction with *decrement factors, allow the dynamic thermal behaviour of a space within a building to be approximated.

TM

thermal load. The heat content of a waste stream from a manufacturing or other process that is discharged into a river or lake. The datum for assessment of heat content is the temperature of the river upstream of the discharge point. *See also* thermal pollution; waste (heat).

MS

thermal neutrons. *Neutrons which are in energy equilibrium with the medium in which they are being scattered. In a thermal reactor *fission neutrons undergo a succession of *scattering collisions in the *moderator until their energy is reduced to equilibrium with the energy of the moderator nuclei. At this point the neutrons are thermalized. In a moderator at 20 °C (293 K) the average speed of thermal neutrons is about 2500 m/s, corresponding to an energy of about 0.03 electronvolts.

DB

thermal pollution. The release of heat into the environment, normally as a result of human activities. The term is usually but not exclusively used to refer to the release of cooling water from power plants. For example, in a coal-fired power station about 60% of the energy content of the coal burnt is added to the environment as heat, some directly to the atmosphere via the stack, most to a body of water such as the sea or a lake or estuary via the cooling water. Some plants use *cooling towers to dissipate waste heat to the environment if sufficient water is not available locally to provide *once-through cooling.

The effects on the environment of thermal releases may be detrimental, beneficial or insignificant, depending on many factors, such as the way in which the cooling water is returned to its source or disposed of. Deleterious effects may result when the aquatic *ecosystem is disturbed. Such effects may include

(a) disruption of *food chains as a result of the loss of one or more key species, e.g. plankton;

(b) disruption of migration patterns;

(c) increased eutrophication rates;

(d) increased susceptibility of organisms to parasites, disease and chemical toxins;

(e) decreased oxygen concentration (solubility of oxygen in water decreases as temperature increases);

(f) death of certain species caused by corrosion inhibitors or biocides added to cooling water.

Waste heat may also have beneficial effects. For example, warmer water may be used for fish farming to cultivate shrimp, carp and other desirable food species; it can be used to irrigate food crops or to improve local recreational amenities; or it may lengthen commercial fishing seasons by attracting warm-water species to the heated areas.

PH

thermal reactor. *See* nuclear reactor.

thermal reforming. A petroleum-refinery process developed about 1930 to raise the *octane rating of *straight-run naphthas by converting *paraffins (alkanes) to *olefins. Only a small amount of aromatics is synthesized. The naphtha is passed through a tubular reactor at a temperature of about 600 °C and a pressure of 65 atmospheres, and the product mixture redistilled. It is found to be uneconomic to attempt to raise the octane rating of the product above 85–90. *See also* catalytic reforming; reforming

WG

thermal resistance. *See* resistance (thermal).

thermal response. *See* response factors.

thermal shield. A steel lining forming a shield between the *core and the concrete *biological shield of a nuclear reactor. Its purpose is to reduce the radiation level at the inside of the biological shield to a level which is low enough to prevent excessive *shield heating. DB

thermal wheel. A rotary air-to-air heat exchanger, used for *heat recovery from the air exhausted from a building. TM

thermie. A thermal unit frequently used in France and equal to 1000 kilocalories, i.e. 4.183 MJ. PH

thermionic emission. The liberation of electrons from a solid by virtue of its temperature. It is measured in amperes per square metre. TM

thermistor. A semiconductor device constructed of a material with an electrical resistance which is very sensitive to temperature. The device is used for the measurement of small temperature changes. TM

thermocouple. A device used to measure temperature difference or temperature from the electrical effects occurring as a result of the *Seebeck effect. Wires of different compositions, e.g. copper and iron, are joined at their ends to form two junctions, A and B. If A and B are at different temperatures, a potential difference develops across the junctions; the potential difference is a measure of the difference in temperature, which can be indicated on a calibrated scale. Thus if the temperature of B is known, the temperature of A may be deduced. Electronic circuits may be used to compensate for changes in the known or standard temperature of B so that a direct reading of temperature for a single external junction is displayed. JWT

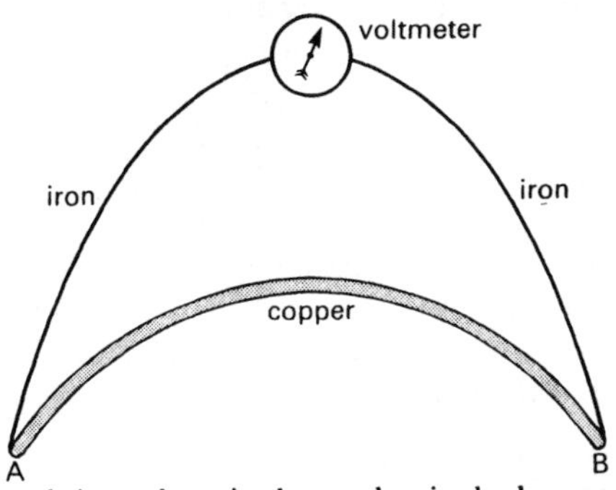

thermodynamic cycle. *See* heat engine.

thermodynamic limit. The minimum energy required to carry out a process involving chemical or physical change. For many processes this may be accurately calculated using thermodynamic tables of *free energies and *enthalpies. In thermodynamic theory the minimum energy requirement occurs only at a zero rate of conversion, thus no real-life process can operate at the thermodynamic minimum. In fact actual processes tend to operate, even at the highest level of efficiency, at well above the thermodynamic minimum, often as much as two times greater. The value of calculating the thermodynamic minimum is that it allows one to judge how far *technical progress can go.

thermodynamics. The study of energy processes, particularly but not necessarily involving heat. The analytical and theoretical methods of thermodynamics form the basis of all energy processes and machines. Strictly the term thermodynamics implies that the systems studied are in a state of movement or change; however the term is used for all energy processes, including those changing infinitely slowly. *See also* thermostatics. JWT

thermodynamic tables. Data listing the thermodynamic properties of substances, obtained as a result of accumulated scientific experimentation. Tables list *enthalpy, *free energy, *heat capacity, *heat of formation, *heat of combustion (where these are known). From such data it is possible to compute *heats of reaction, *free energy changes, *flame temperatures, and many other important heat and energy changes that occur whenever two or more substances interact.

MS

thermoelectricity. Electricity produced directly from the application of heat to a junction of two materials (*see* Seebeck effect). Power generation is possible from materials that are semiconductors having a high Seebeck coefficient. *See also* thermocouple.

JWT

thermometer. An instrument for measuring *temperature or temperature-related scales of *thermal comfort. Instruments for the measurement of temperature may be constructed on one of three principles: the expansion of a fluid (as in a mercury-in-glass thermometer), the change in electric resistance (as in a *thermistor) or the flow of electricity due to the *Seebeck effect (as in a *thermocouple). Mercury-in-glass thermometers exist in a number of variants appropriate to the measurement of thermal comfort: these include the dry-bulb thermometer and the wet-bulb thermometer (the comparison between which allows assessment of *relative humidity) and the globe thermometer, which allows the temperature effect of radiant heat to be measured.

TM

thermonuclear reaction. A reaction involving the *fusion of nuclei and requiring very high temperatures and densities. Such reactions provide the energies of stars. *See also* fusion reactor.

DB

thermophilic. Denoting or relating to heat-loving organisms which have optimum growth rates between 55–75°C and cannot grow below about 40°C and above around 80°C. Some indeed cannot grow at temperatures below 45°C or above 60°C. Thermophilic microbial *fermentations for energy production, as occur in some *biogas plants, operate at 55–60°C, thus speeding up reaction rates, lowering *retention times and reducing digester size and hence capital costs. However, there is a need to maintain this high temperature by expending some of the biogas output on digester heating, thus lowering the *net energy of the system.

CL

thermo-physical properties. Physical properties of substances that are altered by a change in temperature. Though there are very few properties that are not affected by temperature, in energy studies those of most interest are *heat capacity, *heat of formation, *heat of reaction and *heat of combustion.

MS

thermosiphon. A method of establishing circulation of a liquid by utilizing the difference in density between the hot and cold columns of the liquid.

TM

thermostat. An apparatus which maintains a system at a near constant temperature, which may be preselected.

TM

thermostatics. The study of heat processes under conditions of infinitely slow heat transfer. Many theoretical analyses assume thermostatic processes (e.g. the *Carnot cycle) and therefore relate poorly to practical working systems exchanging heat in meaningful times. Thus the practical study of real machines requires a dynamic analysis, not always given from a study of infinitely slow static processes.

JWT

third law of thermodynamics. An empirically based 'law' stating that the *entropy of a pure crystalline substance is zero at absolute zero temperature 0K. This provides a datum to which the entropy of each substrance can be evaluated.

MS

thorium. A metallic element of the *actinide series. Its only naturally occurring isotope, thorium-232, is radioactive, with a very long *half-life of 1.4×10^{10} years. It is the *fertile nuclide from which *fissile uranium-233 can be produced by the *breeding process. Thus thorium can be regarded as a fission-energy resource, provided *breeder reactors are developed to exploit the thorium–uranium-233 cycle. To date only a few reactors have been designed for this, but it is possible that in the future more reactors, and in particular the high-temperature gas-cooled reactor (fuelled initially with thorium and highly enriched uranium) will be developed to produce uranium-233. One advantage of the thorium–uranium-233 breeding cycle is that it is the only one capable of achieving a breeding ratio greater than unity in a thermal reactor. The prospect of a thermal breeder reactor is attractive. The naturally occurring thorium-232 is a fairly abundant element, being found as thorite (silicate), thorianite (oxide) and in the monazite sands.

3-carbon plants [C_3 plants]. Plants which are mostly temperate species and fix carbon dioxide through the *Calvin–Benson cycle, a reaction sequence involving mostly 3-carbon compound intermediates. Yields of 3-carbon plants are generally greatest at latitudes 40–50°, but are inferior to those of *4-carbon plants, which grow mainly in the tropics. CL

Three Mile Island. A nuclear power station near Harrisburg, Pennsylvania, USA, comprising two Babcock and Wilcox-designed pressurized water reactors. One of the two reactors suffered a serious accident in March 1979. The accident was caused by a feed pump trip and the malfunction of a pressure relief valve, and was compounded by errors and misjudgements on the part of the operators. It led to a partial meltdown of the fuel in the *core and the release of gaseous and volatile *fission products into the atmosphere. The consequence of this accident has been the destruction of the core of the reactor involved. This core is being dismantled, a task that will take years to complete. Despite the escape of radioactive gases and vapours into the atmosphere (in particular iodine-131), no identifiable casualties among the public have yet been attributed to this accident, although it has been estimated that the amount of radioactivity released into the atmosphere was enough to cause one cancer-related death in the following 30 years. DB

throttling. The process of reducing the pressure of a fluid during which its *enthalpy remains constant. TM

throwaway cycle. *See* open fuel cycle.

tidal flow power. Energy obtained in a certain period from tidal currents. JWT

tidal range. The difference in height between high and low tides. The range varies between a maximum (spring tide) and minimum (neap tide), with about two weeks between spring and neap tides. The period between a high tide and the following low tide is 12 hours 25 minutes. Local variations in tidal range can be very pronounced, especially as a result of estuaries and bays. Mid-ocean tidal ranges are less than 1 metre, but ranges may increase to about 8 metres at preferential sites, e.g. the Severn estuary, UK. *La Rance, France and the Bay of Fundy, Canada. JWT

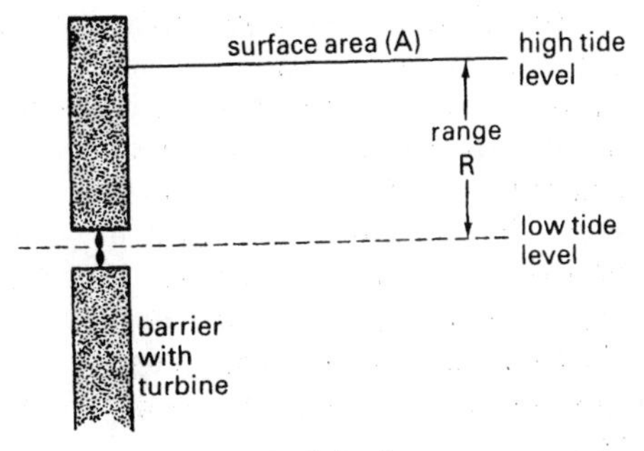

Tidal power from a single basin

tidal-range power. Energy extracted from the rise and fall of tides, i.e. as a result of *tidal range. The total peak power available from tidal generation at optimum sites throughout the world is estimated to be about 100,000 megawatts. This is about 10% of the corresponding figure for *hydropower. With widespread use of tidal power at sites of moderate potential, the total available could be greatly increased. Only one major installation has as yet been constructed (at *La Rance in Brittany, France) so the total potential is far from being realized. A simplified diagram is shown of a tidal basin behind a tidal *barrier. If all the water is captured at high tide and all released at low tide, the power available averaged over the tidal period t is

$$\rho A R^2 g/2$$

ρ is the density of the water, A the surface area of the basin, R the tidal range and g the acceleration of gravity. Thus if $R = 4$ m and $A = 10$ km^2, the average power is 17 MW.

The benefits and difficulties of tidal power can be deduced from this formula.

(a) The power is accurately predictable because R and τ (12 h 25 m) are known; R however varies by a factor of 3 each half month between the spring and neap tides.

(b) The availability of power does not correspond to the solar day (24 h) unless a multi-basin system is operated.

(c) The power varies as R^2 and so sites of large tidal range are preferred. These only occur when natural coastline contours, such as estuaries, allow the water movement to resonate with the period at increased amplitude; for example, at La Rance, Brittany, R varies between 6 m and 14 m in comparison with mid-ocean ranges of 0.5 m.

(d) Large basin area A is beneficial. This may cause significant ecological disruption as well as commercial advantages.

(e) The *turbines need to be suitable for large flow, low head operation.

(f) Tidal electricity can only be efficiently used if the power is fed to a large grid system that always accepts the power when it is available, or if the power can be used directly at the tidal frequencies (e.g. to produce hydrogen for energy storage), or lastly if a two- or three-basin system allows high- and low-head levels to be maintained so that water can pass between the basins at times determined by *consumer use rather than tidal periods. JWT

tidal wave. An oscillation of water from the surface to the bottom that travels as a wave. At sea this movement may be induced by earthquakes and propagates at very high speed in deep water with extremely large energy intensity. In deep water this wave (a tsunamis) has small unharmful amplitude, but when the wave reaches shore the energy is concentrated into a wave amplitude of immense size that can cause widespread damage. JWT

time preference. A positive rate of time preference is present when the current availability of something is valued more highly than its future availability. Typically, individuals and firms exhibit in their behaviour a positive rate of time preference. It is therefore necessary, in situations where there is no risk of default, to pay an *interest rate to induce a lender to forego money now for the same sum of money in the future. This in turn is the origin of the use of a *discount rate in *project appraisal. MC

tip-speed ratio. The ratio of the speed of the outer edge of a *wind turbine to that of the wind. The efficiency of the wind turbine depends on keeping this ratio constant. Multi-blade turbines have peak efficiency at a low tip-speed ratio of about 2, whereas two-bladed turbines optimize at values of about 10. JWT

tokomak. The toroidal *containment system developed in the USSR for *plasma studies in fusion energy research. The toroidal component of the magnetic field is maintained by windings around the *torus, and the poloidal component is maintained by the current in the plasma. *See also* fusion reactor. DB

toluene. An aromatic hydrocarbon, $C_6H_5CH_3$, in which one of the hydrogen atoms in the benzene ring has been replaced by a methyl group. Its properties are thus similar to those of *benzene although it is less volatile. A valuable solvent and chemical intermediate, it is the raw material for the manufacture of the explosive TNT. It may be separated from coal tar or may be produced by the *catalytic reforming of *straight-run petroleum fractions. It has a gross calorific value of 42.9 MJ/kg. WG

tonne (Symbol: t). A unit of mass equal to 1000 kg. It differs only slightly from the imperial ton (1016kg, 2240 lb) and is often called a metric ton.

ton of refrigeration (Symbol: TR). The amount of cooling produced by one US ton (2000 lb) of ice melting over a period of 24 hours. One TR is equivalent to 3.516 kW. TM

tonne of coal equivalent (TCE). The amount of heat equivalent to the *heat of combustion of one tonne of coal. Since there is no such thing as a standard coal, and since coal composition varies widely from one coalfield to another, the heat of combustion cannot be defined exactly. Generally this unit is chosen as a means of depicting energy use on a national or international scale, and the value attributed is listed within the accompanying text. Even so, it is rare to see a precise statement, e.g. whether the number so listed is a gross or net *calorific value. A typical figure is 29 GJ/tonne (gross calorific value). TCE is occasionally taken to imply an imperial ton (1016 kg). *See also* tonne of oil equivalent.

MS

tonne of oil equivalent (TOE). The amount of heat equivalent to the *heat of combustion of one tonne of oil. Since there is no such thing as a standard oil, and since oil composition varies from one oilfield to another, the heat of combustion cannot be defined exactly. Generally this unit is chosen as a means of depicting energy use on a national or international scale, and the value attributed to it is listed within the accompanying text. Even so, it is rare to see a precise statement, e.g. whether the number so listed is a gross or net *calorific value. A typical figure is 44 GJ/tonne of oil (gross calorific value). TOE is occasionally taken to imply one imperial ton (1016 kg). *See also* tonne of coal equivalent. MS

topped crude. A crude oil which has been heated to distil off the more volatile hydrocarbons. WG

torque. The rotational effect which results or can result from the application of a force. If the point of rotation is at A and the force acts at B a distance *d* away, such that the component of the force perpendicular to AB is *F*, then the torque is equal to

$$F \times d$$

The maximum torque on a particular rotating machine will depend on the frequency of rotation. For instance, multiblade *wind turbines have high torque at low frequency, but a machine with only a few blades reaches maximum torque at high frequency. In the latter case, the torque at low wind speeds might be insufficient to initiate rotation of the blades and an external starting mechanism may be necessary.

JWT

Torrey Canyon. The oil tanker which ran aground on the Seven Stones reef between the Scilly Isles and Lands End, England, on 18 March 1967, resulting in the spillage of more than 95,000 tonnes of crude oil. The oil caused extensive pollution to the Cornish and Breton coastlines. The tanker was American-owned, on charter to the British Petroleum Company, and was carrying a cargo of 119,000 tonnes of Kuwait crude oil to Milford Haven, South Wales.

After the ship broke up, the 20,000 tonnes of oil remaining in her tanks was burned after being ignited by precision bombing from the air. The oil slicks were sprayed with *dispersants over a period of 18 days to accelerate the dispersal of the oil. Floating booms were applied to harbours and estuaries in an attempt to contain the oil. Three methods were used to remove oil from polluted beaches and harbours: dispersing the oil by spraying with dispersants, burning the oil, either in situ or after collecting it in heaps, and mechanical removal of the oil or oil-contaminated sand.

The oil released by the *Torrey Canyon* caused widespread damage to intertidal marine plant and animal life, and was responsible for the death of large numbers (more than 25,000) of sea birds over a wide area. The damage to intertidal life was aggravated by the liberal use of dispersants which at that time contained inherently toxic hydrocarbon-based solvents. One hundred and forty miles of Cornish coastline were affected, including many holiday beaches and areas of scenic beauty. Commercial fisheries do not seem to have been significantly affected. PH

torus. The doughnut-shaped magnetic *containment that has been used in several of the most important *plasma confinement studies for fusion energy. The feature of the torus that makes it suitable for magnetic containment is that by virtue of its shape (a tube with no ends) the magnetic field produced by coils around it is continuous, and charged particles, i.e. the plasma, move within this field (*see* figure). *See also* fusion reactor. DB

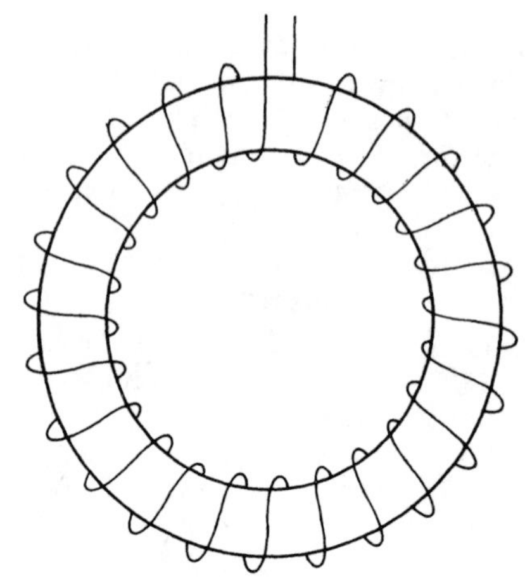

Torus with magnetic field coil

total energy system. A system by which all the energy requirements of a building or complex of buildings are met from the resource of a single fuel. Total energy systems may be configured in a large number of different ways and the optimum configuration will depend on the temporal variation in load demand. Typically, natural gas might be the energy resource brought onto the site. The demand for electricity might be met by generating it in a gas *turbine; exhaust gases from the turbine might be used to raise steam, which in turn is used for space heating, hot-water supply and, via an *absorption chiller, for cooling. TM

town gas. A fuel gas prepared from coal or oil by *carbonization or *thermal reforming. It contains a high proportion of hydrogen and has a correspondingly high *flame speed. *See also* coal gas. WG

tracking system. A controlled mechanical drive used to steer an *active solar energy device so that it continuously faces the sun. JWT

trade. Exchange between individuals, firms or nations either directly as in *barter or indirectly, as is now more usual, through the medium of *money. The benefits of trade arise from an extension of the range of commodities available for consumption, and from specialization by traders in the production of a given range of commodities according to the principle of *comparative advantage. International trade in oil, for example, is mainly the exploitation of the first source of possible advantage, whereas international trade in, say, motor cars is mainly the exploitation of the second source of advantage. MC

trade creation. A feature of *customs unions and free trade areas in which the reduction or elimination of tariffs between member states allows greater specialization and a larger demand and supply within the union or free trade area in comparison to the situation prior to the formation. MC

trade cycle. *See* business cycle.

'Tragedy of the Commons'. The title of a seminal article written by Garrett Hardin in 1968 (*Science*, 162) and reprinted in more than 20 anthologies. Hardin was Professor of Human Ecology at the University of California, Santa Barbara. In his article Hardin expounded the view that the *biosphere is like a common pasture where each individual, guided by a desire for personal gain, increases his herd until the pasture is ruined for all. From his analogy he concluded that freedom in a commons brings ruin to all, and that the freedom which must be constrained if ruin is to be avoided is the freedom to breed people. *See also* common property resource. PH

transaction costs. A cost incurred, other than price, in order to complete a transaction. It may include such items as transport costs, legal fees, telephone calls, trips to obtain information, the act of bargaining and the enforcement of property rights. It is sometimes argued that many *externality situations arise because their transactions costs are greater than the benefits which would accrue from market transactions concerning the situation. MC

transfer payment. A payment which is paid by a government body but is not matched by any current production of a good or service, i.e. it involves no exchange. Typical transfers are social security benefits and student grants. MC

transformation loss. The energy dissipated when energy of one quality is raised to a higher quality, for example, oil burnt to provide heat to produce electricity. MS

transformer. Electrical apparatus which enables the potential of a supply of *alternating current to be raised or lowered. It consists of coils of insulated wire wound on a core of laminated soft iron and works on the principle of electromagnetic *induction. TM

transient. The condition of any system which is not operating at steady state. For example, in power plant operation, a transient implies that the power, temperature, pressure or other operating parameters are varying. DB

transmission. (1) (mechanical) The system of mechanical linkages through which rotational movement is passed from the prime mover to other elements of the machine.

(2) (thermal) The *steady-state element of heat flow through the wall of an enclosure due to the difference in external and internal temperature. Transmission per unit area of wall per degree temperature difference is known as thermal *transmittance (U). Transmittance, which is the inverse of the net thermal *resistance of the wall materials, is measured in watts per square metre per kelvin. So-called U-values have been estimated for most construction materials.

(3) The distribution of electricity through the *grid. TM

transmission coefficient. *See* transmittance.

transmissivity. The property of a body to transmit radiant heat energy. Transmissivity is measured as that proportion of incident radiation which is not absorbed or reflected, but is transmitted. Its converse is *absorptivity.

TM

transmittance [transmission coefficient]. The fraction of radiation transmitted by a material over a specified range of wavelengths. Radiation is reflected, absorbed and transmitted. The transmittance of glass at solar wavelengths ranges from about 80% for window glass to 92% for pure 'iron oxide free' glass. Glass transmittance at infrared wavelengths from heated objects is very low, but polyethylene sheets transmit this infrared radiation with high transmittance. *See also* absorptance; reflectance; transmission (thermal).

JWT

transpiration. The loss of water by evaporation through plant leaf pores (stomata). The controlled reduction of transpiration in plants grown under greenhouse conditions allows the possibility of cultivating crops in regions of limited freshwater supply.

CL

transportation of nuclear materials. Such transportation occurs at several stages of the nuclear *fuel cycle. It involves the movement of materials in special containers by road, rail, air or water. The three most hazardous types of shipment are or will be irradiated fuel assemblies, high-level radioactive waste and plutonium from the reprocessing plant. For shipment, irradiated fuel assemblies are placed in massive air- or water-cooled casks, each weighing 25–100 tonnes, consisting of a series of shells of stainless steel and lead. These casks are designed to withstand severe transportation accidents as simulated by a series of damage tests which include:

(a) a free drop through 9.15 m onto a flat unyielding surface;
(b) a free drop through 1.02 m striking a vertical cylindrical steel bar 15 by 20 cm (puncture test);
(c) exposure to a fire of 800°C for at least 30 minutes;
(d) immersion in water to a depth of 0.915 m for at least 8 hours.

PH

transuranic elements [transuranium elements]. Elements with *atomic numbers greater than 92, and therefore members of the *actinide series. The transuranic elements do not exist naturally but are produced by nuclear reactions. The most common reaction is neutron capture (*see* absorption) in a nuclear reactor followed by radioactive decay by *beta particle emission, as in the process of *breeding. At present 11 transuranic elements have been produced, all radioactive. Some of them exist only in very small quantities and are of little importance in nuclear energy. The most important transuranic isotopes are listed in the table; of these nuclides, plutonium-239 and plutonium-241 are *fissile, and all of them present a long-term hazard in the disposal of *radioactive waste from *irradiated fuel.

Nuclide	*Atomic Number*	*Half-life*
neptunium-237	93	2.1×10^6 y
plutonium-238	94	87 y
plutonium-239	94	2.4×10^4 y
plutonium-240	94	6.6×10^3 y
plutonium-241	94	15 y
plutonium-242	94	3.9×10^5 y
americium-241	95	433 y
americium-243	95	7.4×10^3 y
curium-242	96	163 d
curium-244	96	18 y

DB

transuranium wastes (TRE wastes). A class of high-level *radioactive waste in which the alpha-emitting *transuranic elements, such as plutonium, americium and curium, predominate, with an activity greater than 10 microcuries per kg.

It is generated in nuclear fuel reprocessing plant. PH

trebles. Coal with an approximately uniform lump size, the pieces of which are about 75 mm in diameter. WG

tricarboxylic acid cycle (TCA cycle). *See* Krebs cycle.

trip. A switching arrangement in mechanical and electrical systems actuated automatically when certain parameters in the operation of the system (e.g., *load) fall outside predefined limits. TM

triple point. The temperature and pressure at which the solid, liquid and vapour of a substance are in equilibrium with one another. For example, the triple point of water is at 0.01°C and 0.00602 bar pressure. There can be more than one triple point for certain substances. PH

tritium. The radioactive *isotope of hydrogen, $^{3}_{1}H$, with a mass number of 3, i.e. its nucleus contains two neutrons and one proton. It has a *half-life of 12.3 years. It can be produced by several different processes, of which neutron capture in lithium-6 is the most important:

$$^{6}_{3}Li + ^{1}_{0}n \rightarrow ^{3}_{1}H + ^{4}_{2}He$$

Other processes are neutron capture in deuterium:

$$^{2}_{1}H + ^{1}_{0}n \rightarrow ^{3}_{1}H + \gamma$$

which has a very small *cross-section indeed, and the deuterium–deuterium *fusion reaction:

$$^{2}_{1}H + ^{2}_{1}H \rightarrow ^{3}_{1}H + ^{1}_{1}H$$

The most important use of tritium is in the fusion reaction

$$^{2}_{1}H + ^{3}_{1}H = ^{4}_{2}He + ^{1}_{0}n$$

The neutrons produced in this reaction can be absorbed in *lithium-6 to produce more tritium.

Tritium exists naturally in the environment as a result of the interaction between cosmic radiation and various atoms in the atmosphere. It is also present in the environment as a result of nuclear weapons testing and discharges from nuclear reactors. At present the levels of tritium in the atmosphere and tritiated water in the oceans are not high enough to cause concern, but this situation may change. DB

Trombe wall. A vertical sun-facing blackened building wall with an outer glass or transparent cover. Solar radiation is absorbed by the black surface, so heating the wall itself and the air between the cover and the surface. Heated air circulates upwards and enters rooms through ducts in the wall. The wall acts as both an *absorber and a storage system, being a *passive solar system. JWT

tsunamis. *See* tidal wave.

turbine. A device which converts the stored mechanical energy in a fluid into rotational mechanical energy. There are several major types of turbines, including steam turbines, gas turbines, water turbines, and wind turbines (or windmills). The broad principles of the steam and gas turbine are similar: the steam, raised in a boiler by the combustion of *fossil fuel, or the hot gases generated directly from the combustion of fossil fuel, expand through the blades on the turbine rotor causing them to move. The broad principles of the water turbine and the wind turbine are also similar, the water turbine converting the potential energy of the water into mechanical energy, the wind turbine converting the kinetic energy of the wind into mechanical energy. Turbines are widely used for the generation of electricity and for the supply of motive power. TM

turbulence. Erratic motions of a stream of liquid or gas. Turbines do not function successfully in turbulent motion of water or in turbulent gusting wind conditions.

JWT

two-stroke cycle. The cycle of events in a gasoline engine designed to have a power stroke for each revolution of the crankshaft. Intake of fuel vapour and air and its compression occur as the piston rises; power generation and exhaust take place as it descends. Although simpler, it is less efficient than the *four stroke cycle.

WG

U

U-gas. *See* gasification.

ultimate analysis. The determination of elementary composition, i.e. the chemical analysis of a material in terms of the percentages by weight of the elements present. For coals and petroleum fuels, the percentages of carbon, hydrogen, nitrogen, sulphur and oxygen are determined. The results of ultimate analysis make possible the calculations of the theoretical amount of air required to burn unit weight of the fuel. Ultimate analysis is also used in the classification of coals.

WG

uncertainty. A situation which involves no objective (calculable) probability but only a subjective one, i.e. the probability is based on belief. It is to be distinguished from *risk. The outcome of the next election or of OPEC remaining a *cartel are situations of uncertainty.

MC

underground gasification. A process for converting coal *in situ* underground into fuel gas and pumping it to the surface. Attempts have been made in various parts of the world to develop a commercial process, but with only a very limited success. The technique consists in drilling a number of bore holes down to the level of the coal seam, creating a fire at the base of one hole and attempting to extract combustible gases at the others from coal carbonized by the heat developed underground. However, it has been found difficult to link such holes successfully at their base. High-pressure air, explosives and electricity have all been tried and linkages over a distance of about 50 metres achieved. The success of the project depends greatly on the configuration of the seam and on finding an outlet for the fuel gas, which has a *calorific value lower than 6 MJ/m^3.

WG

unit. A popular term for one kilowatt-hour of electricity. *See also* units.

MS

United Kingdom Atomic Energy Authority (UKAEA). The national authority in the UK which is responsible for research and development of all non-military aspects of nuclear energy. The principal establishments of the Authority are the *Atomic Energy Research Establishment, Harwell, the *Atomic Energy Establishment, Winfrith, *Culham Laboratory, Oxfordshire, *Sellafield, Cumbria, *Dounreay, Caithness and *Risley, Cheshire. Other establishments include Springfields, Lancashire, involved in nuclear fuel element development, manufacture and testing, and Culcheth, Cheshire, the Safety and Reliability Directorate.

Main address: 11 Charles II Street, London SW1Y 4QP.

DB

United Nations Conference on Trade and Development (UNCTAD). An international organization formed under the auspices of the United Nations in 1964. Its aim is to consider the problems of trade specific to the developing countries. It has secured generalized trade preferences for developing countries and helped to stabilize commodity prices. *See also* commodity agreements.

Address: Palais des Nations, CH-1211, Geneva 10, Switzerland.

MC

United Nations Scientific Committee on the Effects of Atomic Radiation (UNSCEAR). A committee established in 1955 by the United Nations General Assembly as a result of the international concern about the effects of fallout from testing nuclear explosives. It was directed to assemble and study information on observed levels of radiation (both natural and man-made) in the environment, and on the effects of such radiation on man and his environment. It consists of representatives from 15 countries who meet annually in the headquarters of the *International Atomic Energy Agency in Vienna.

PH

unit of account. Something used as the standard of measurement, in terms of which records of economic activity are kept and prices expressed. In most economies today the unit of account is the local *currency.

MC

units. Standard quantities used to describe and define physical or economic phenomena in a precise way. Today there is a largely unified decimal system of physical units, the *SI units (Système Internationale d'Unités), accepted by all countries. Historical units, such as the calorie, horse-power and foot, linger on. Time, however, has retained its traditional non-metric break-down into hours, minutes and seconds. *See also* conversion tables p. viii.

MS

universal gas constant (Symbol: R). The constant occurring in the gas law equation for 1 mole of an ideal gas:

$$PV = RT$$

where P is the pressure, V the volume and T the temperature of the gas. R has the value 8.31 J/mol.K.

MS

unsaturated hydrocarbons. Hydrocarbons in which some of the carbon atoms in the molecular chain or ring are linked to their neighbouring carbon atoms by two or three valency bonds. Examples include the *alkenes.

WG

upwind. On the side of a device facing into the wind. The position onto which the wind is blowing.

JWT

uranium. A naturally occurring radioactive metallic element. It is essential for obtaining nuclear energy from *fission. Natural uranium contains three *isotopes: 99.285% uranium-238, 0.715% uranium-235 and a negligible amount of uranium-234. All are radioactive, with very long *half-lives, and give rise to the natural *radioactive decay chains. The half-lives of uranium-238 and uranium-235 are 4.5×10^9 years and 7.1×10^8 years respectively. Both are *fissionable, but only uranium-235 is *fissile with *neutrons of all energies; consequently it is the more important isotope as regards fission. Uranium-238 is important as the *fertile isotope from which fissile *plutonium-239 can be produced by *breeding. Another important man-made isotope is uranium-233 which is fissile and can be produced from *thorium-232 by breeding.

Uranium is quite widely dispersed in the earth's crust and oceans, although its concentration in ores is small, and in the ocean very small indeed – about 0.002 g/m^3. There are many uranium-bearing minerals, one of the most important being uraninite (an oxide). Uranium ores are treated to produce yellowcake, a mixture of oxides whose composition is usually expressed as U_3O_8. This is further processed to produce pure metallic uranium, uranium dioxide (UO_2) or uranium carbide (UC_2) for reactor fuel, or gaseous uranium hexafluoride (UF_6) for *enrichment purposes.

Metallic uranium has three distinct crystalline forms with transition temperatures of 660 °C and 770 °C. The transition from the alpha to the beta form at 660 °C is accompanied by swelling. This temperature is thus regarded as the maximum permissible for metallic uranium, which is used as the fuel in British Magnox reactors and French gas-cooled reactors (*see* nuclear reactor).

Uranium dioxide is extensively used as the fuel for most other types of reactor. It is a black powder with a melting point of 2750°C which is formed into pellets by being compacted at high pressure and temperature. The pellets of UO_2 are loaded into *Zircaloy or *stainless steel tubes to form fuel elements.

Enriched uranium contains more than 0.715% of uranium-235 and is produced from natural uranium by an enrichment process such as *gaseous diffusion or centrifuging. Many types of nuclear reactor require slightly enriched uranium with 2–3% of uranium-235 in order to achieve *criticality.

Depleted uranium contains less than 0.715% of uranium-235. It is produced as the tailings from enrichment plant and from the reprocessing of *irradiated fuel from natural-uranium reactors. In both cases the depleted uranium contains about 0.2–0.3% of uranium-235, so it can virtually be regarded as pure uranium-238. The most likely future use for depleted uranium is as the fertile material in fast breeder reactors (*see* nuclear reactor) and stockpiles now exist in several countries for this possible future use. It is also possible, and in certain circumstances might be economic, to pass depleted uranium from reprocessing plant to an enrichment plant to increase the content of uranium–235 to a high enough value to enable the uranium to be reused as the fuel for thermal reactors.

Symbol: U; atomic number: 92; atomic weight: 230.0. DB

uranium dioxide. *See* uranium.

uranium hexafluoride (hex). *See* uranium.

Uranium Institute. An international organization which serves the interests of uranium producers and consumers.
Address: New Zealand House, Haymarket, London SW1. MS

uranium milling. A process of uranium extraction by which uranium ore is crushed and ground to the consistency of fine sand, and the uranium extracted by a variety of complex chemical treatments. The result is a dry powder containing 70–90% by weight of uranium as a mixture of uranium oxides with a chemical formula equivalent to U_3O_8. When this is in the form of ammonium diuranate, it is referred to as *yellowcake; this forms the raw material for the fabrication of nuclear fuel. For each tonne of yellowcake extracted, there remains at least 100 tonnes of depleted ore, the *uranium tailings. *See also* uranium mining. PH

uranium mining. The process by which uranium ore is extracted from the earth's crust. Most mines are hardrock underground mines but some are open-cast. Most of the high-grade ores (up to 4% uranium) have already been worked out, and ores of 0.4% and less are now being worked. The radioactive gas *radon-222 is present in the atmosphere of underground uranium mines; in the early years of the mining industry, when the mines were poorly ventilated, many miners died from cancers of the respiratory system as a result of inhaling radon-222. *See also* uranium milling.

PH

uranium tailings. Crushed uranium ore from which most of the uranium has been extracted chemically. Vast piles of tailings have accumulated near uranium mills, e.g. in SW Colorado, USA. The tailings contain the radioactive gas *radon-222 and its radioactive products, which are readily released to the atmosphere since the tailings are very finely divided. In strong winds and dry conditions, tailings may be wind-blown and thus contaminate inhabited areas. Tailings exposed to weathering and leaching by rainwater contaminate waterways. In some towns (e.g. Grand Junction, Colorado) tailings were used for road material and as backfill for the construction of buildings, including schools and hospitals. This resulted in levels of radon within some buildings exceeding the dose limits set for uranium miners. Tighter legislation now requires that uranium tailings be properly managed; they should be stabilized with topsoil, concrete or asphalt, or stored in abandoned dry mines or other protected geological formations.

PH

US Bureau of Mines. A US Government research organization publishing on a wide variety of mining and related topics.

Address: 4800 Forbes Avenue, Pittsburgh, Pa, USA.

WG

user cost. In the context of the extraction of *natural resources, the *opportunity cost of extraction today rather than tomorrow. For example, if the owner of an oil well faces a situation in which the price of oil is rising, current extraction carries the cost that the price received per unit extracted is less than it will be in the future, future extraction possibilities being reduced by the amount of current extraction. The *optimal depletion programme for such an oil-well owner is the result of balancing such costs against the benefit of current over future revenue which arises from a positive rate of *time preference.

MC

utility function. A conceptual device in economics used for the representation of an individual's preferences over commodities. It ranks all possible combinations of commodity consumption according to the utility, or satisfaction, they yield the individual. For two commodities, the utility function can be represented graphically as a system of *indifference curves. The assumption that, in deciding on consumption patterns, individuals act so as to maximize utility gives rise to a number of testable hypotheses about consumption behaviour, and is the theoretical basis for *demand functions and for the analysis of policy measures to improve economic *welfare. *See also* income effect; substitution effect.

MC

utilization factor. The ratio of the useful number of *lumens to the total lumens emitted by a *lamp.

TM

***U*-value.** A coefficient of heat transfer. The term is used in architectural jargon. *See* transmission (thermal).

TM

V

vacuum distillation. A process by which a liquid is distilled in equipment in which the pressure is considerably less than atmospheric. This enables boiling to take place at a much lower temperature and thus the possibility of thermal degradation is reduced. In the distillation of crude petroleum the residue from the primary atmospheric distillation unit may be redistilled under vacuum for the recovery of lubricating oils.

WG

value. A term which has gone through a variety of uses in economics, many now redundant. It now refers to the total sum of price multiplied by quantity. Thus the value of output is the sum, over the products produced, of price times quantity produced; the value of input is the sum, over inputs, of price times quantity input. For a firm then, output value is sales receipts and input value is expenditure on inputs. In economics, measurement in value terms is necessary for meaningful aggregation and performance indication. For example, the information that for constant input levels the output of a coal industry changed between two years from 100 tonnes of grade 1 coal and 50 tonnes of grade 2 coal to 50 tonnes of grade 1 and 100 tonnes of grade 2 is not any indication of whether performance improved or not. If output in value terms increased, because grade 2 coal sells for more than grade 1, then performance improved.

MC

value added. The *value of a firm's output less the value of the *intermediate products bought in from other firms. It is necessarily equal to the payments by the firm to the *factors of production it employed in producing the output. *National product is measured on a value-added basis so as to avoid double counting with respect to the amount of *final products produced in the economy. It is because national product is so measured that it is numerically identical to *national income. *See also* national income accounts.

MC

value-added tax (VAT). In principle, a tax levied on the *value added on all the firms in the economy, at a uniform rate. It is seen as a way of raising revenue which does not discriminate or favour particular commodities as do most other types of *indirect taxation. In practice the tax is not levied at a uniform rate on all firms. In the UK, for example, value added in production for *export is exempt from VAT, and some commodities (including coal, gas and electricity) are zero rated so that not only are they not subject to VAT but their producers can reclaim any tax they have paid on purchased inputs.

MC

value judgement. A judgement which attributes a value, such as good, evil or desirable, to a thing, action or policy.

PH

vaporizing oil. A fuel for the spark-initiated internal combustion engine with a boiling range similar to that of *kerosene. It contains a high proportion of aromatic hydrocarbons and is unsuitable as a lamp oil. The low volatility makes it of little value as a fuel for the normal high-speed gasoline engine. It has thus been used as a fuel in specially designed slow-speed or constant-speed engines, which find application in farm tractors and fishing boats. The engine is started on petrol to obtain sufficient exhaust heat to vaporize the less volatile fuel. When the system has reached the desired temperature the operator switches over to the supply of vaporizing oil. With the development of small diesel engines the demand for vaporizing oil has declined. Vaporizing oil has a gross calorific value of 46 MJ/kg.

WG

vapour compression chiller. A type of refrigeration plant in which the refrigerant, a specially selected volatile liquid known as Refrigerant 12, provides a cooling effect during evaporation. The refrigerant vapour is removed from the evaporator, compressed by a fan or pump (known as the compressor), condensed and returned through an expansion valve to the evaporator. The energy input to the vapour compression chiller is in the form of mechanical power, as opposed to the low-grade heat input to the *absorption chiller.

TM

vapour pressure. The numerical expression of the tendency of a liquid to vaporize. Liquid as it is heated has an increasing tendency to vaporize, until at its boiling point, its vapour pressure equals the ambient (usually atmospheric) pressure.

MS

variable air volume system. A system of *air conditioning a building in which dehumidified cooled air is ducted around the building and is delivered to each room through a variable geometry diffuser or slot. Variation in the diffuser or slot geometry controls the volume of air delivered to each room.

TM

visbreaking. Short for viscosity breaking. A mild *thermal cracking process employed in petroleum refineries to reduce the *viscosity of a heavy petroleum residue with the minimum production of gasoline and gas. The feedstock is passed through a tubular furnace at 480°C and 7 atmospheres, and the product mixture is distilled to separate gas, gasoline and light distillate from a fuel oil with a viscosity considerably lower than that of the heavy residue.

WG

viscosity. The resistance to flow of a liquid, which may be said to have a high or low viscosity. It is measured in terms of the coefficient of dynamic viscosity, which is the ratio of applied shear stress to the rate of shear.

WG

viscosity index. A measure of the variation of the viscosity of a lubricating oil with temperature. The viscosity of the oil under test is measured at two temperatures and compared with the values given to two reference oils with good (i.e. viscosity index = 100) and poor (viscosity index = 0) viscosity/temperature characteristics.

WG

visible trade. Trade in goods rather than services, the latter being *invisible trade. The visible balance, also called the balance of trade, is the difference between receipts and payments from goods. *See also* balance of payments.

MC

vitrain. One of the recognizable constituents of coal. It is a glossy black brittle material which breaks with a concoidal fracture, i.e. the break has a curved surface showing concentric rings. Under the microscope it shows a faint cellular structure and is almost free of other plant remains. It is thought to have been formed from the bark of the trees growing in the Carboniferous forests. In the lump of coal it is seen as thin bright black bands. *See also* clarain.

WG

vitrification. *See* glassification.

void coefficient of reactivity. The measure of the change in reactivity of a *nuclear reactor due to the formation of voids within the core of the reactor. Such voids are formed by the boiling of the coolant, and the effect is only encountered in liquid-cooled reactors, such as pressurized water reactors, boiling water reactors, CANDUs and sodium-cooled fast reactors. The formation of voids is caused by power and temperature increases in the reactor, and if it results in an increase in reactivity, causing a further power rise, an unstable condition exists that is potentially dangerous (positive void coefficient). If the formation of voids causes a decrease of reactivity, resulting in a decrease of power, a stable condition exists that is inherently safer (negative void coefficient). It is an important design feature of pressurized and boiling water reactors that they have negative void coefficients. A positive void coefficient in the RBMK reactor at *Chernobyl, USSR led to the serious accident in that power station in 1986.

DB

volatile matter. Matter which changes readily from a solid or liquid form to a vapour. In the coal industry it is quantified as the percentage loss in weight of a sample of coal when heated under standard conditions in a covered crucible to 925°C. It consists of water vapour and volatile products of carbonization which escape from the crucible. The 'true' volatile matter is obtained by subtracting the moisture content of the coal from the result.

WG

volatility. A measure of the ease with which a liquid evaporates or vaporizes. A low boiling material, i.e. one with a high *vapour pressure, is said to have a high volatility. A substance with a high boiling point is said to have a low volatility.

WG

volt (Symbol: V). The SI unit of *potential difference. One volt is the difference of electrical potential between two points of a conductor carrying a constant current of one *ampere when the power dissipation between the points is one *watt.

TM

voltage (Symbol: *V*). The *potential difference between two points in a circuit. It is measured in volts.

TM

W

wake. The movement of fluid (liquid or gas) on the downstream side of a device. Thus air may be disturbed and turbulent downwind of an *aerogenerator for distances many times greater than the tower height.

JWT

waste. Something discarded from a process. Wastes fall into four categories – biological, chemical, nuclear and heat – within a broader all-embracing term, pollution, though not all wastes may be considered polluting. Some wastes have value, or come to have value as economic circumstances change.

Biological wastes consist of organic carbon containing substances resulting from biochemical or *fermentation processes. Such waste can be polluting by virtue of creating *biological oxygen demand in river water, and so killing aquatic life. Occasionally treatment can not only reduce pollution but can yield economically useful products.

Chemical wastes are usually low-concentration but often toxic byproducts of chemical processes.

Nuclear wastes cannot be disposed of freely in the environment, except at very low concentrations. *See also* Dumping at Sea Act; radioactive waste; radioactive waste management.

Waste heat is usually discarded in the form of a fluid, at such a low temperature as to be of no economic value to industry. Electrical generating stations are the principal generators of waste heat, a byproduct of their main concern to produce electricity. Waste heat is an inevitable consequence of the fact that heat sources are being turned into work (electricity) with an inevitable rejection of low-grade heat (*see* Carnot cycle). As energy becomes more expensive, it becomes economic to introduce new systems for utilizing waste heat, e.g. better *heat exchangers, *heat pumps, or creation of warm lagoons to breed fish. *See also* combined heat and power; thermal pollution.

MS

waste gas. The mixed gaseous products of combustion issuing from a furnace, chimney, or flue.

WG

water gas. A fuel gas which is essentially a mixture of *carbon monoxide and *hydrogen and is generated by blowing steam through a bed of red-hot *coke. The gas has a relatively low calorific value (11 MJ/m^3) and thus requires to be enriched before being fed into a town-gas supply system based on a coal gas. Generally a petroleum fraction, *gas oil, is injected into the hot freshly made water gas and undergoes *thermal cracking to provide low molecular weight hydrocarbons; these raise the overall calorific value of the mixture to about

18 MJ/m^3. This enriched gas is known as carburetted water gas.

Water gas burns with a blue flame due to the presence of carbon monoxide and gives a high*flame temperature. It was thus formerly used in welding operations. Its production is only economic if very cheap coke is available as a by-product of coal gas manufacture.

WG

water hyacinth. A prolific tropical water-weed, *Eichhornia crasspipes*, which exerts high levels of *nutrient uptake and is capable of yielding over 150 t/ha.y on a dry weight basis. Previously considered a serious pollutant of waterways, its potential as an energy source is now being exploited. On harvesting, the biomass can be treated in several ways. It can undergo aerobic *composting, *anaerobic digestion or can be used as a cattle feed supplement. One dry weight tonne of water hyacinth can be digested to 400 m^3 of biogas, equivalent to a gross energy output of 8 GJ.

CL

water power. Energy harnessed from the fall of water. *See* hydroelectric generation.

JWT

water turbine. *See* turbine.

water wheel. A large wheel-like *hydropower *turbine which is powered by a stream of water moving buckets or paddles arranged around the wheel's circumference. Water may reach the wheel across its top (overshot), half way up (breastshot) or by flowing underneath the wheel (undershot). Most water wheels have a horizontal axis, but vertical-axis machines with a horizontal wheel can be constructed.

JWT

watt (Symbol: W). The SI unit of *power. One watt is equal to one *joule per second (J/s).

TM

wave energy. *See* wave power.

wave power. Energy present in the height (*potential energy) and movement (*kinetic energy) of water waves. Deep-water long-wavelength ocean rollers have energy fluxes that may reach 100 kilowatts per metre of wave front. For instance, off the west coast of Scotland the average energy in such waves is 70 kW/m. The energy in waves may be considered as energy obtained from the wind, but stored and smoothed to form a more constant and concentrated energy supply than the wind itself.

Wave-energy extraction devices may reach 80% efficiency over the wide spectrum of wave conditions met in the real sea. The attraction of wave-power systems (none of which now operate on more than reduced scale models or low-power applications) is that extended devices of 100 km length could supply about 5000 MW of electricity for national *grids. Such investment is attractive for marine nations with ship-building industry and with favourable wave regions.

In a wave of period T and amplitude A, with density of water ρ and acceleration of gravity g, the power in the wave is

$$P = \rho g^2 A^2 T/(8\pi)$$

Common ocean waves with a period of 6 to 10 seconds, 100 m wavelength and amplitude about 1.5 m therefore have large energy fluxes. However in storm conditions the effective amplitude (called the significant wave height) may rise to 4 m or more, so increasing the energy flux about 10 times. Such conditions produce extreme forces on mechanical structures so that considerable and costly overdesign is required.

A great variety of devices have been

constructed or proposed to absorb this mechanical energy and transform it for electricity production; however the great majority of power developments are still at the development stage. Research and development is occurring mainly in Japan and the UK, where there is considerable debate about the type and scale of design. Proposals centre on a range of devices, including

(a) structures that transform the wave surface movement into a flow of air, subsequently powering an air *turbine (e.g. the Lanchester clam);

(b) structures that float in or on the surface and whose movement is coupled hydraulically or by gyroscopic motion to generators (e.g. Salter's duck);

(c) structures held beneath the surface and activated by subsurface movement and pressure changes (e.g. the Bristol cylinder);

(d) catchment systems where waves breaking over barriers maintain a height differential for a low-head hydroelectric system (e.g. the Mauritius artificial lagoon proposal).

JWT

wax. A light-coloured petroleum fraction which is solid at ambient temperature and contains high molecular weight hydrocarbons.

WG

wealth. The value at a point in time of the stock of *assets owned by an economic agent, whether individual, firm or nation.

MC

weapon-grade nuclear materials. Nuclear materials from which it is possible to make nuclear weapons. The three most important *nuclides which could be used for nuclear weapons are the *fissile materials used as fuel in nuclear reactors, i.e. uranium-233, uranium-235 and plutonium-239. To produce an explosive, it is necessary to assemble a quantity of fissile material which exceeds a minimum amount, known as the critical mass: this is the smallest mass in which a *chain reaction can be sustained. The critical mass for a given nuclide depends on its chemical purity, physical surroundings and the method of assembly of the weapon.

With 100% pure fissile material surrounded by a reflector of natural uranium 15 cm thick, the critical masses are 5.8 kg for U-233, 15 kg for U-235, 4.4 kg for Pu-239. Under less ideal conditions, the critical mass would be considerably larger. For example, for U-235 at a concentration of 40%, the critical mass is 75 kg while at 20% it goes up to 250 kg. Therefore only uranium with a U-235 concentration greater than 20% is considered weapon-grade. In contrast, a reduction in the concentration of Pu-239 to 50% results in only a doubling of the critical mass.

The presence of other plutonium nuclides has a strong influence on the explosive yield of a Pu-239 weapon. A significant quantity of Pu-240, for example, is highly detrimental. In a nuclear reactor there is a gradual accumulation of Pu-240 which may reach 25% in three years. For this reason the nuclear industry maintained for many years that reactor-grade plutonium is unsuitable for weapons. However it is now recognized that with sophisticated high-speed implosion technology for assembly of the weapon a high-explosive yield can be achieved.

PH

weathering. The deterioration that takes place in a material when subjected to atmospheric conditions over a long period. In a low-rank coal which is porous, the effect is to cause the pieces to crumble to a powder, together with a loss of calorific value due to oxidation.

WG

Weibull distribution function. A

mathematical equation used to describe the variation in speed of the wind. The function F(v) gives the probability that the wind speed will exceed a specified value v:

$$F(v) = \text{exponential}\,[-(v/c)^k]$$

where c and k are constants which are used to fit the equation to particular wind patterns.

JWT

welfare. A term often used rather loosely in economics to indicate an interest in the economic wellbeing of society, rather than merely an analysis of how the economic system works. For example, an analysis of the welfare implications of an oil price rise concerns the way such a rise finally impinges on the individual members of society, as opposed to an analysis of the way firms respond in terms of changes in input mixes, output levels, selling prices for output. A project which passed the *compensation test would be said to improve welfare as would an intervention in the market system in response to *market failure. *See also* second best problem.

MC

West Valley Reprocessing Plant. The world's first commercial nuclear fuel reprocessing plant, operated by Nuclear Fuel Services Inc. at West Valley in western New York State. The plant operated from 1966 until 1972 when it closed for extensive modifications in order to meet more stringent health and safety regulations of the *Nuclear Regulatory Commission. The parent company, Getty Oil, later decided not to reopen the plant. In 1980 federal legislation was passed to clean up the site at a cost of $300 million.

PH

wet gas. (1) A mixture of permanent gases containing water vapour which condenses on cooling.

(2) A mixture of low molecular weight hydrocarbon gases, e.g. natural gas, from which liquid hydrocarbons condense when the gas is subjected to moderate pressure.

WG

white damp. The name given to carbon monoxide when found in the air of the tunnels, etc., of a coal mine. It may be formed as the result of a fire or an explosion.

WG

white spirit. A petroleum fraction which boils in the range 150–200°C. It is used as a solvent and thinner for paint.

WG

whole-body scanner. A scintillation monitor designed to measure the quantity of *radioactivity present in a human body.

PH

wide-cut petrol [US: wide-cut gasoline]. A petroleum fraction with a wider boiling range (20–280°C) than permitted for normal motor spirit (30–200°C). It serves as a fuel for the gas turbines of jet aircraft.

WG

Wigner energy. Energy stored in graphite due to the dimensional changes in its crystalline structure (Wigner growth) which result from neutron irradiation. The stored Wigner energy can be released if the graphite is maintained at the appropriate temperature, this process being known as annealing. Being an *exothermic reaction, graphite annealing must be carried out under carefully controlled conditions, otherwise overheating will result.

DB

wildcat. An exploratory drilling for oil without certainty of success.

MS

willingness to pay. What an individual is prepared to give up in order to achieve or obtain something, usually expressed in terms of a sum of money. Markets are the means by which willingness to pay is made manifest, so that the justification for having markets allocate resources via the *price mechanism is that the outcome reflects people's willingness to pay and that resources are allocated in line with people's preferences. Where markets cannot work, as with the provision of *public goods or in the presence of *externalities, economists generally believe that, where possible, allocation should be made according to the willingness to pay principle. The difficulty is then, in *cost benefit analysis for example, in measuring willingness to pay. Consider the decision to build a dam for hydroelectricity generation in an area of outstanding natural beauty: the cost of the harm involved should in principle be entered into the cost benefit analysis as the total willingness to pay to avoid the harm.

MS

wind energy. *See* wind power.

wind energy conversion system (WECS). A term commonly used for a wind generator, *aerogenerator, windpump or windmill, especially when integrated into a commercial energy system as single units or arrays of machines.

JWT

winding. *See* armature.

windmill. A mechanical mill, powered by the wind, for the grinding of grain. Windmill is also a general term for wind-powered machines, including *aerogenerators.

JWT

wind power. Energy extracted from wind passing through a *turbine. Traditional windmill design has been radically improved by modern technology so that wind turbines have become power sources for electricity generations as *aerogenerators, or sources for mechanical power in pumps and hydraulic equipment. Modern wind-power machines are built on aerodynamic principles developed from airplane and helicopter design. They use modern materials including those of composit structure and are controlled by electronics with sophisticated switchgear. Despite these improvements, however, the science and technology of wind machines is still at an elementary stage. Nevertheless the potential for wind power at economic costs is so great that research and development in both the industrialized and developing countries is progressing rapidly.

Machines producing electricity – aerogenerators – vary in size from large machines with a maximum capacity of about 3–4 megawatts (swept blade diameter about 100 m), to small machines with a capacity of about 30 kilowatts (swept diameter about 10 m). Turbines may rotate on a horizontal axis in the traditional manner, or may rotate on a vertical axis as with the *Darrieus wind turbine.

The power in a wind of speed v, density ρ, across an area A is

$$\rho A v^3/2.$$

Note that the power varies as v^3 (*see* cube factor), so a doubling in wind speed produces an eight-fold increase in power. The proportion of this power extracted by the turbine is the *power coefficient, C_p. Since the wind must remain with enough energy to leave the turbine, C_p has a theoretical maximum value of about 60% (*see* Betz theory). In practice the maximum value of C_p for an efficient wind turbine is about 40%, and moreover this will only be obtained when the ratio of the speed of the blade tips to the speed of the wind is an optimum (*see* tip-speed ratio). This optimum is usually designed from analysis of the local wind characteristics. High power is obtained at high wind speeds, which unfortunately occur less frequently than low speeds. Thus a wind turbine is usually designed for optimum performance at about twice the average wind speed. All these wind speeds refer to speeds at the turbine height, which may have to be derived from meteorological data at other heights. For some particular functions, e.g. water pumping, it may be more important to increase the frequency of supply than maximize the available power.

Wind turbines tend to be used only at sites where the average wind speed is greater than about 4 m/s and the most economic use is obtained when the wind speed is considerably greater than this (*see* Beaufort scale). Sites are considered excellent for wind power if the wind is steady, without frequent gusts or variation, and the average speed is above 7 m/s. Such sites are commonly on island and coastal regions with dominant prevailing winds, such as the N and NE boundaries of the North Atlantic Ocean. Wind speed increases with height so that attractive sites may be found in mountainous regions, where unfortunately gusting and turbulence may also increase.

In gale and hurricane conditions damage to the turbines must be avoided. Various methods are used to reduce the power and forces at the blades. The turbines may be turned away from facing the wind on a horizontal-axis machine, the blades may be designed to decrease their rate of rotation at very high wind speed (perhaps in a manner similar to the stalling condition of an airplane wing) or a brake may be applied. Some vertical-axis machines are arranged so that the normally vertical blades move towards a horizontal position in high wind speed; only a small swept area is then presented to the wind. This happens in the Musgrove design. The danger of damage to a wind turbine by gale-force winds and continuous flexing leading to metal fatigue is a major factor in design and operation.

Aerogenerators producing electricity may operate as independent units in an autonomous mode, often called stand alone operation, or may be connected to a local electrical *grid network. The design of the machine will be different for each type of operation, particularly with regard to the generator and control mechanisms. Grid connection allows a cheaper form of generation and control, usually with an induction generator. Moreover excess electricity may be sold to the utility or company operating the grid. Autonomous operation may be at fixed rotational frequency for immediate fixed-frequency electricity generation, in which case control is difficult and expensive. Variable-frequency generation allows an optimum tip speed ratio and hence power coefficient to be maintained. Control of the aerogenerator may also be accomplished by automatically connecting and disconnecting various electrical *loads so that the best total load for a particular wind speed is obtained. Such load control with variable-frequency turbine operation allows the highest overall efficiency of wind-power extraction to be maintained, and is now possible with relatively cheap electronic methods.

National wind-power programmes were first developed in the USA and Denmark. In the USA a carefully structured research programme is aimed at the successful development of large 3–4 megawatt aerogenerators. In Denmark emphasis has also been placed on a strong commercial development of small 20–100 kW wind generators for use at farms and rural housing. The prime objective is thereby to reduce dependence on imported oil for heating and electricity generation. Most industrialized countries now have similar research and development programmes aimed at a 10% or greater penetration of conventional energy or electricity production.

JWT

Windscale. *See* Sellafield.

wind speed duration curve. A graph of the probability of specified wind speeds occurring at a site during a specified period. *See also* Weibull distribution function.

JWT

wind turbine. A rotating machine for generating power from the wind. It is usually composed of blades fixed to a horizontal or vertical axis.

JWT

Winkler process. *See* gasification

Wissenschaftlich: Technische Gesellschaft für Energiewirtschaft der Kammer der Technik. A publishing society concerned with solid fuels, electricity, gas and general energy utilization.

Address: Clara-Zetkin Str. 115-117, 108 Berlin, German Democratic Republic.

Wobbe index. The thermal output of a gas burner is determined by the diameter of the orifice from which the gas is issuing and by the gas pressure, calorific value and relative density. For a given gas appliance, diameter and pressure are fixed and the thermal output may be shown to be proportional to the expression

$$\frac{\text{calorific value}}{\sqrt{(\text{relative density})}}$$

This is known as the Wobbe index and its value for given fuel gases may be used to calculate changes in orifice diameter and pressure necessary to keep thermal output constant when the available fuel gas is changed, e.g. from coal gas to natural gas.

WG

wood. *See* firewood.

work (Symbol: *W*). A measure of effort. Mechanical work is computed from the product of *force on a body times the distance that body is moved. Other forms of work are the chemical reduction of an ore to a metal, the action of an electrical force, work against gravity and work of compression. Work is measured in *joules, the resulting energy or *power output in a particular period being measured in *watts. Humans produce useful work at powers of about 20 watts for long extended periods and about 1 kilowatt for brief intensive periods (*see* metabolic energy). Thus humans perform far less intensively than most powered machines.

MS

workshop on alternative energy strategies (WAES). An attempt on the part of Professor Carroll Wilson of Massachussets Institute of Technology to bring together an authoritative group of people from many countries to make an assessment of the world energy supply and demand situation, and what could be done about it. Their report is widely quoted as an important comment on the world energy situation. It is published as a book: *Energy: the Global Prospect, 1985–2000* (Nimrod Press, Boston, USA).

MS

world bank. *See* International Bank for Reconstruction and Development.

X

xeno currency market. *See* eurocurrency.

xenon. An inert gas which has many *isotopes, both naturally occurring and radioactive. The most important in nuclear reactors is xenon-135, which is produced as a *fission product. It has a very high neutron capture *cross-section and acts as a *reactor poison.

Symbol: Xe; atomic number: 54; atomic weight: 131.3.

DB

X-rays. High-energy *electromagnetic radiation, but less energetic than *gamma radiation, which is emitted from atoms as the result of changes in electron energy levels.

DB

xylene. An aromatic hydrocarbon, $C_6H_4(CH_3)_2$, in which two of the hydrogen atoms in the *benzene ring have been replaced by methyl groups. It may be obtained by the distillation of coal or by the *catalytic reforming of *straight-run petroleum fractions. It occurs in three isomeric forms (orth-, meta-, and para-) which have similar properties. All are flammable. It has a gross calorific value of 43.4 MJ/kg.

WG

Y

yaw angle. The angle between the optimum direction for a vertical-axis *wind turbine and the actual direction of the wind. Since wind direction fluctuates continuously and rapidly, such a turbine has to correct its position to reduce the yaw angle.

JWT

yellowcake. A mixture of oxides of *uranium whose composition is expressed as U_3O_8. It is the first stage of refining uranium ore (*see* uranium milling) and after extraction is shipped to factories at which *fuel element manufacture or *enrichment is carried out.

DB

yield. (1) The useful product of a natural, agricultural or industrial process. It may, for example, be applied to the harvest of a crop.

(2) In finance, the income received from a *security, expressed as a percentage of the current market value. In the case of an irredeemable government bond with a coupon payment of X per period and with a current market price of p, then the yield is equal to X/p.

JWT, MC

Z

zenith. The direction perpendicular to a surface. The zenith to a horizontal plane is vertical. *See also* zenith angle.

JWT

zenith angle. The angle between the sun's direction onto a horizontal plane and the vertical, i.e. angle *z* in the figure.

JWT

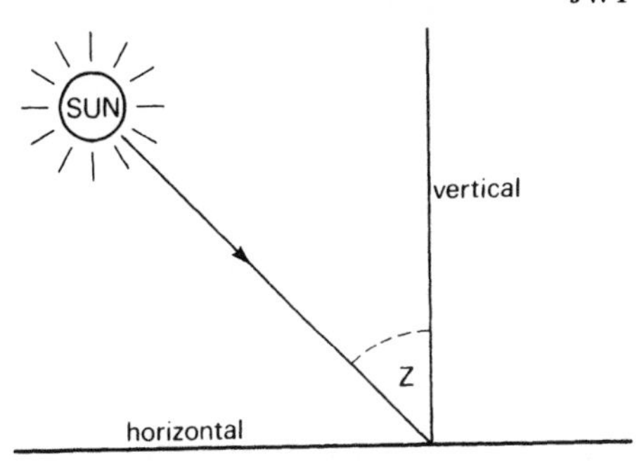

zero crossing period. The time between successive movements of a mixed sea-wave surface crossing the average level. With a single-frequency wave, this period is the normal period of the wave. However with a mixture of waves of different frequency and amplitude, the zero crossing period is obtained from an average of events (*see* diagram).

JWT

zero sum game. A situation where when one participant or group of participants gain, some other groups lose an equal amount. For example, there may be a fixed supply of some resource. Hence if one group overconsumes, another will be deprived. There are situations where this does not hold, and *synergy results.

MS

zeroth law of thermodynamics. The law stating that systems in thermal equilibrium with each other have the same temperature. Systems not in thermal equilibrium with each other have different temperatures. This law, therefore, gives an operational definition of temperature that does not depend on the sensations of hotness or coldness. The law was so called since it was enunciated after the *first and *second laws of thermodynamics had become well established.

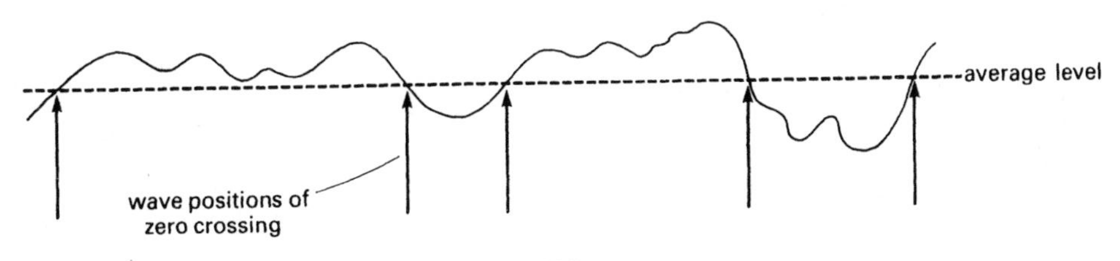

zinc-65. A radioactive *isotope of the metal zinc, usually present in the liquid radioactive effluent from power stations and reprocessing plants. It emits both *beta particles and *gamma radiation with a *half-life of 244 days. In certain situations it may constitute the critical radionuclide which determines the radiation exposure of a *critical group in the human population. For example, zinc-65 released by the nuclear station at Bradwell, UK is taken up by local oysters with a *concentration factor of around 10^5; thus local consumers of oysters form the critical group.

PH

zircaloy. An alloy of zirconium (98%) with small amounts of tin, iron, chromium and nickel to give better corrosion resistance and high-temperature strength. It is used extensively as *cladding material for the fuel in pressurized and boiling water reactors and in CANDU reactors, where it is also used for the pressure tubes. *See also* nuclear reactor.

DB

zirconium. A metallic element whose low neutron capture *cross-section, excellent corrosion resistance, reasonably good strength and good fabrication characteristics make it a very suitable *cladding and structural material for nuclear reactors. The corrosion resistance and strength can be improved by alloying (*see* Zircaloy).

Symbol: Zr; atomic number: 40; atomic weight: 91.2; density: 6500 kg/m³.

DB